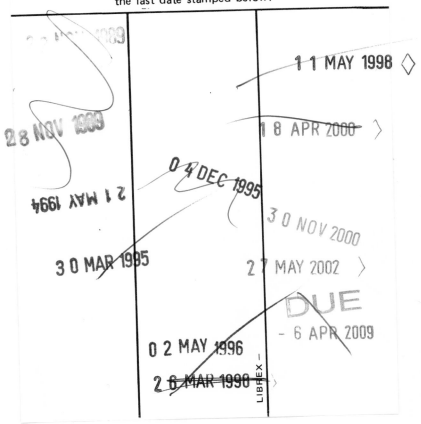

Air Pollution Control and Design Handbook

POLLUTION ENGINEERING AND TECHNOLOGY

A Series of Reference Books and Textbooks

EDITORS

RICHARD A. YOUNG

Editor, Pollution Engineering
Technical Publishing Company
Barrington, Illinois

PAUL N. CHEREMISINOFF

Associate Professor
of Environmental Engineering
New Jersey Institute of Technology
Newark, New Jersey

1. Energy from Solid Wastes, *Paul N. Cheremisinoff and Angelo C. Morresi*
2. Air Pollution Control and Design Handbook (in two parts), *edited by Paul N. Cheremisinoff and Richard A. Young*
3. Wastewater Renovation and Reuse, *edited by Frank M. D'Itri*
4. Water and Wastewater Treatment: Calculations for Chemical and Physical Processes, *Michael J. Humenick, Jr.*
5. Biofouling Control Procedures, *edited by Loren D. Jensen*

Additional Volumes in Preparation

Air Pollution Control and Design Handbook

(IN TWO PARTS)

PART 2

Edited by

PAUL N. CHEREMISINOFF

Associate Professor
of Environmental Engineering
New Jersey Institute of Technology
Newark, New Jersey

and

RICHARD A. YOUNG

Editor, Pollution Engineering
Technical Publishing Company
Barrington, Illinois

MARCEL DEKKER, INC. New York and Basel

Library of Congress Cataloging in Publication Data (Revised)

Main entry under title:

Air pollution control and design handbook.

(Pollution engineering and technology ; 2)
Includes index.
TD885.A37 628.5'3 76-588
ISBN *0-8247-6448-X*

MARCEL DEKKER, INC.

270 Madison Avenue, New York, New York 10016

Current printing (last digit):
10 9 8 7 6 5 4 3 2 1

PRINTED IN THE UNITED STATES OF AMERICA

PREFACE

Air pollution is a result of technological progress, and of human activities, intensified by the demands of twentieth-century life-styles, and given political, legal, and economic impetus by legislation and its enforcement in the last decade. Air pollution has been acknowledged as a disquieting phenomenon, and a widespread commitment has been made to control the problem. To be successful, such a commitment requires application of sound engineering practice, economics, and a spark of creativity. To succeed in controlling air pollution regardless of its source, the engineer needs to be able to manipulate the appropriate technologies.

This book addresses one of the major problems facing industry. Analysis and recommendations are presented in various problem areas by respective experts and stress practical aspects of pollution control engineering as related to air quality. The editors and contributors identify the major elements of the industrial air pollution situation. Adverse environmental impact will, of course, be dependent on the nature of a particular business, its raw materials, products, and processes.

From the large spectrum of technological possibilities, development of techniques to control air pollution from stationary sources has received great impetus and made rapid strides since enactment of federal and state air pollution laws. The threat of fine and plant shutdown in addition to social responsibilities have made manufacturing enterprises not only more cognizant of problems in this area but have goaded them to action.

Recent international events have included sharp increases in the prices of essential fuels, creating the widely perceived "energy crisis" which in turn has additionally exacerbated problems in air pollution control. Cheaper, more abundant fuels such as coal do not burn clean, so use of such fuels presents air pollution problems.

High priority and urgency have been assigned to vigorous programs in air pollution control.

This handbook is intended as a further step in closing the gap between knowledge and practice. Technological progress requires continual reexamination of priorities and experience. It is intended for use by experts and novices; engineers, managers, and students who are interested in learning the language, and as a reference for those who are faced with air pollution problems from stationary sources. The editors extend heartfelt thanks to their many friends in industry who cooperated and helped in the publication of this book.

Paul N. Cheremisinoff
Richard A. Young

CONTRIBUTORS TO PART 2

MAHESH V. BHATIA, Hooker Chemicals & Plastics Corp., Grand Island, New York

MURRAY BORENSTEIN, Environmental Elements Company, Division of Koppers Inc., Ramsey, New Jersey

JACK D. BRADY, FMC Corporation, Atlanta, Georgia

PAUL N. CHEREMISINOFF, New Jersey Institute of Technology, Newark, New Jersey

ROBERT T. FELLMAN, Ebasco Services, Inc., New York, New York

V. FREGA, The Ceilcote Company, Berea, Ohio

W. J. GILBERT, Croll-Reynolds Company, Westfield, New Jersey

THOMAS G. GLEASON, Swemco, Inc., New York, New York

JEFFREY W. GOLLER, Nassau Recycle Corporation, Staten Island, New York

THOMAS M. HELLMAN, Allied Chemicals Corporation, Morristown, New Jersey

WARREN KLUGMAN, The Ceilcote Company, Berea, Ohio

L. KARL LEGATSKI, FMC Corporation, Atlanta, Georgia

J. W. MacDONALD, Air Pollution Control Division, The Ceilcote Company, Berea, Ohio

ROBERT W. McILVAINE, The McIlvaine Company, Northbrook, Illinois

YEI-MIN PENG, MicroPul Division, United States Filters Corporation, Summit, New Jersey

NORMAN D. PHILLIPS, Fuller Company, Catasauqua, Pennsylvania

DAVID RIMBERG, MicroPul Division, United States Filters Corporation, Summit, New Jersey

KEN SCHIFFTNER, Environmental Elements Company, Division of Koppers Inc., Ramsey, New Jersey

S. V. SHEPPARD, The Ceilcote Company, Berea, Ohio

F. G. SHINSKEY, The Foxboro Company, Foxboro, Massachusetts

J. W. SIMONINI, Air Pollution Control Division, Rexnord, Inc., Louisville, Kentucky

SRINI VASAN, Peabody Process Systems, Inc., Stamford, Connecticut

HOWARD P. WILLETT, Peabody Process Systems, Inc., Stamford, Connecticut

CONTENTS OF PART 2

CONTENTS OF PART 1

1. SYSTEMS FOR CONTINUOUS STACK EMISSION MONITORING
   P. C. Wolf, Instrument Division, Ethyl Corporation, Princeton, New Jersey

2. CONTINUOUS PARTICULATE MASS MEASUREMENTS IN STACK EMISSIONS
   P. C. Wolf, Instrument Division, Ethyl Corporation, Princeton, New Jersey

3. ISOKINETIC STACK TESTING
   J. W. Leatherdale, Consulting Engineer, North Plainfield, New Jersey

4. INVESTIGATIONS AND EVIDENCE OF AIR POLLUTION
   A. H. Principe and D. G. Stroz, Northern Illinois Police Crime Laboratory, Highland Park, Illinois

5. FEDERAL AIR DATA BASE SYSTEMS—AEROMETRIC AND EMISSIONS REPORTING SYSTEMS
   J. C. Bosch, Jr., and G. Nehls, U.S. Environmental Protection Agency, Research Triangle Park, North Carolina

6. MODELING ATMOSPHERIC DISPERSION OF POLLUTANTS
   J. T. Yeh, Pittsburgh Energy Research Center, Energy Research and Development Administration, Pittsburgh, Pennsylvania

7. DESIGN OF CHIMNEYS AND STACKS FOR POLLUTION CONTROL
   R. T. Fellman, Ebasco Services, Inc., New York, New York

   P. N. Cheremisinoff, New Jersey Institute of Technology, Newark, New Jersey

8. PARTICLE DYNAMICS
   A. C. Morresi, Hoffman-La Roche, Inc., Nutley, New Jersey

9. GRAVITY SETTLING CHAMBERS
   N. P. Cheremisinoff, Clarkson College of Technology, Potsdam, New York

   V. Van Brunt, University of South Carolina, Columbia, South Carolina

10. CYCLONES
   M. V. Bhatia, Engelhard Minerals & Chemicals Corporation, East Newark, New Jersey

   P. N. Cheremisinoff, New Jersey Institute of Technology, Newark, New Jersey

11. FABRIC FILTER DUST COLLECTORS
M. G. Kennedy, American Air Filter Company, Louisville, Kentucky

P. N. Cheremisinoff, New Jersey Institute of Technology, Newark, New Jersey

12. EUROPEAN BAGHOUSE DESIGN
L. Bergmann, Duren, West Germany

13. ELECTROSTATIC PRECIPITATORS
R. W. Ziminski, Consulting Engineer, Somerville, New Jersey

14. ELECTROSTATICS—OVERVIEW
M. J. Freeman, New York, New York

P. N. Cheremisinoff, New Jersey Institute of Technology, Newark, New Jersey

15. CARBON ADSORPTION APPLICATIONS
T. E. Cannon, VIC Manufacturing Company, Minneapolis, Minnesota

16. DUST RETARDANTS
W. Canessa, Golden Bear Division, Witco Chemical Corporation, Bakersfield, California

17. THERMAL INCINERATION
R. D. Ross, Read-Ferry Company, Haddonfield, New Jersey

18. THERMAL INCINERATION ECONOMICS
W. Elnicki, AER Corporation, Ramsey, New Jersey

19. HEAT RECOVERY
J. H. Mueller, Regenerative Environmental Equipment Company (REECO), Morris Plains, New Jersey

20. HEAT RECOVERY COST JUSTIFICATION AND SYSTEMS
K. I. Erlandsson, Kelley Company, Inc., Milwaukee, Wisconsin

21. CATALYTIC INCINERATION
T. H. Snape, Matthey Bishop, Inc., Malvern, Pennsylvania

22. WASTE TREATMENT BY FLUIDIZED BED THERMAL OXIDATION
J. W. Knox and C. M. Wheeler, Copeland Systems, Inc., Oak Brook, Illinois

23. NO_x CONTROL IN CENTRAL STATION STEAM BOILERS
W. J. Aghassi, Leeds & Northrup, New York, New York

P. N. Cheremisinoff, New Jersey Institute of Technology, Newark, New Jersey

Air Pollution Control and Design Handbook

Chapter 24

PACKED TOWER AND ABSORPTION DESIGN

Mahesh V. Bhatia

Hooker Chemicals & Plastics Corp.
Grand Island, New York

Scrubbers used in air pollution control are available in a wide range of types and sizes. They are used mainly to eliminate one or more objectionable gaseous or particulate components from a gas stream. Their principle of design is based on mass transfer (diffusion), inertial impaction, or electrostatic attraction.

In mass transfer, gaseous components are dissolved in liquid. When gas and liquid are brought into intimate contact, the concentration gradient is established between two phases, and diffusion takes place. Objectionable components, higher in concentration in the gaseous phase, are transferred to the liquid phase, having a lower concentration. During this diffusion process, solute is transferred in liquid with or without chemical reaction.

THEORY

When fluid passes over a solid surface, its velocity at the surface of the solid is zero. The velocity of the fluid thus changes from that in the bulk stream to the solid surface across which it is flowing. The velocity rises sharply in a zone between the interface and a very small distance perpendicular to the solid interface. This small zone is called the laminar region. The fluid in the bulk stream can be in the turbulent region. The zone between the laminar and turbulent regions is known as the transition or buffer region.

Intimate contact between gas and liquid is established in the laminar region over solids known as packing. The laminar region consists of stagnant gas and liquid films. As diffusing fluid passes from the main stream, it has to pass through the main stream, buffer zone, and laminar regions.

Diffusion through the laminar film is on a molecular scale and is known as molecular diffusion. Molecules in gases move in random directions, and as they do so, they collide against each other. The resultant distance is, therefore, very small; hence, molecular diffusion is a slow process. If, on the other hand, the temperature of the gas is higher, molecules travel at a higher velocity and can cover larger distances, thereby increasing the rate of diffusion. At lower pressure, there is a greater distance between molecules; this can also increase the rate of diffusion.

The main concentration gradient is established in the laminar region, so the mechanism of gas absorption by diffusion is a molecular diffusion.

RATE OF DIFFUSION

At steady state, gas-gas molecular diffusion of component A from position 1 to position 2 through stagnant nondiffusing component B is given by

$$\frac{D_{AB}P_t}{RTZ_g P_{BM}} (P_{A_1} - P_{A_2}) \tag{1}$$

where

D_{AB} = diffusivity (cm^2/sec or ft^2/hr)

R = gas constant (82.06 cm^3 atm/g mole) or 0.729 ft^3 atm/lb mole)

P_t = total pressure (atm)

T = temperature (K or R)
P_{BM} = log mean pressure of nondiffusing component B (atm)
Z_g = distance in direction of diffusion (cm or ft)
P_A, P_B = partial pressure of A or B (atm)
Subscripts 1, 2 indicate position

Estimation of D_{AB} for Gases

The Hirschfelder-Bird-Spotz [1] relation to estimate D_{AB} is

$$D_{AB} = \frac{0.0009292\ T^{3/2}[(1/M_A) + (1/M_B)]^{1/2}}{P(\gamma_{AB})^2[f(kT/\epsilon_{AB})]} \qquad (2)$$

In this formula, D_{AB} is in square centimeters per second and T is in degrees Kelvin; γ_{AB} is the molecular separation at collision in angstroms [= $(\gamma_A + \gamma_B)/2$]; M_A and M_B are the molecular weights of components A and B, respectively; ϵ_{AB} is the energy of molecular interaction in ergs [= $(\epsilon_A\epsilon_B)^{1/2}$]; k is Boltzmann's constant (1.38×10^{-16} ergs/K); and $f(kT/\epsilon_{AB})$ is the collision function given in Fig. 24-1. Values of ϵ/k (K) and γ (Å) for common components are given in Table 24-1.

For components for which values of ϵ/k and γ are not given, the following empirical relations are recommended:

$$\frac{\epsilon}{k} = 0.77T_c \qquad (3)$$

$$\frac{\epsilon}{k} = 1.15T_b \qquad (4)$$

$$\frac{\epsilon}{k} = 1.92T_m \qquad (5)$$

where T_c, T_b, and T_m are the critical temperature, boiling point temperature, and melting temperature, respectively, in degrees Kelvin; and

$$\gamma = 1.18V_o^{1/3} \qquad (6)$$

where V_o is the molal volume given in cubic centimeters per gram mole of liquid at the normal boiling point.

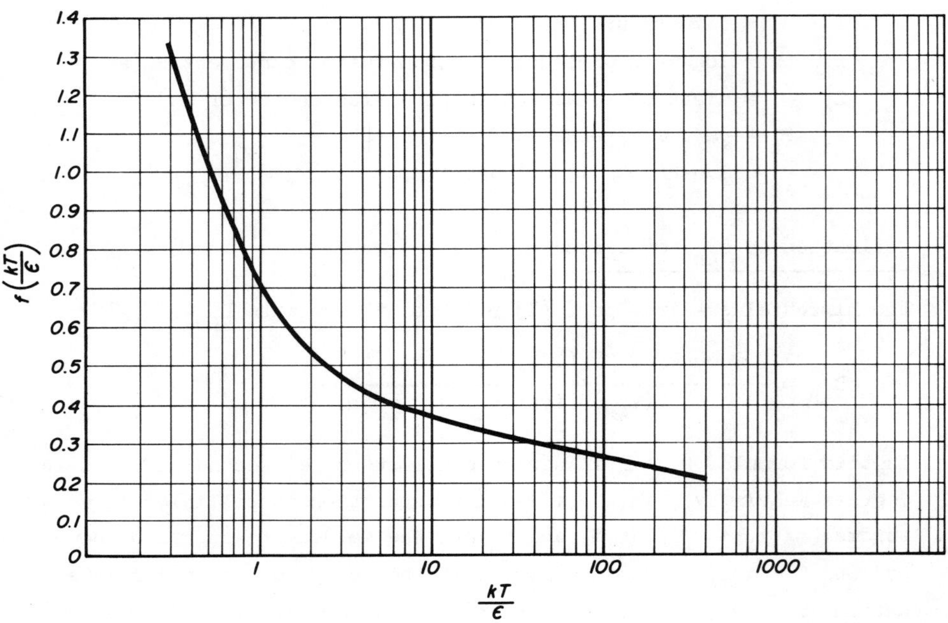

FIG. 24-1. Collision function $f(kT/\epsilon_{AB})$.

For larger molecules, molal volume can be determined as a summation of atomic volumes. Values of atomic volumes can be obtained from the literature, such as Perry's Chemical Engineer's Handbook [4].

Diffusivity Coefficient in Liquids

By the Wilke correlation [2], diffusivity of components in liquids is given by

$$D_{AB} = \frac{T}{\mu_B F} \tag{7}$$

where

T = temperature (°K)

μ_B = viscosity of solvent (cps)

TABLE 24-1

Force Constants and Collision Diameters [2]

Gas	ε/k (K) (from viscosity)	γ_0 (angstroms) (from viscosity)
Air	97.0	3.617
Ammonia	315	2.624
Argon	124.0	3.418
Benzene	440	5.270
CO_2	190	3.996
CO	110.3	3.590
CCl_4	327	5.881
Diphenyl	600	6.223
Ethane	230	4.418
Ethanol	391	4.455
Ethyl ether	350	5.424
Ethylene	205	4.232
Fluorocarbon F-12	288	5.110
Helium	6.03	2.70
n-Heptadecane	800	7.923
Hydrogen	33.3	2.968
HCl	360	3.305
Iodine	550	4.982
Methane	136.5	3.882
Neon	35.7	2.80
Nitrobenzene	539	4.931
NO	119	3.47
Nitrogen	91.5	3.681
N_2O	220	3.879
n-Octadecane	820	7.963
n-Octane	320	7.451
Oxygen	113.2	3.433
Propane	254	5.061
SO_2	252	4.290
Water	363	2.655

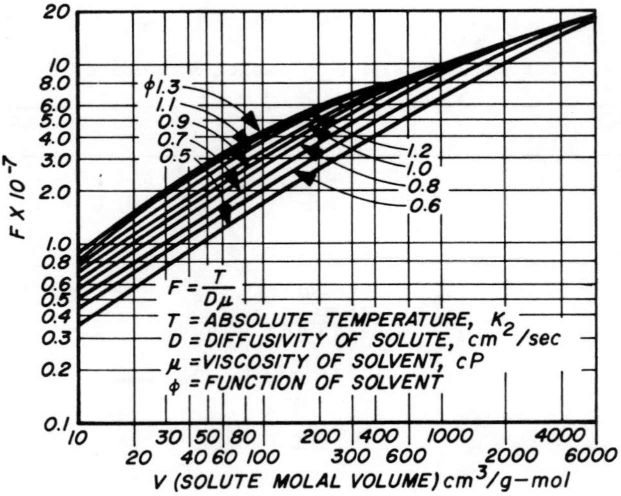

FIG. 24-2. (From Ref. 3, with permission of McGraw-Hill Book Company.)

D_{AB} = diffusivity of component A through solvent B

F = a function of the molecular volume of the solute

φ, as shown in Fig. 24-2, is a solvent characteristic, used as parameter in Fig. 24-2, and is equal to 1.0, 0.82, and 0.70 for water, methanol, and benzene, respectively. For other solvents where data are unavailable, φ may be assumed to be 0.9 [3].

Example 1

Estimate the diffusion coefficient between SO_2 and air at 15 psia and 25°C (gas absorption of SO_2 by water is controlled by resistances both in gas film and liquid film).

$$T = 25 + 273 = 298°K$$

$$P = \frac{15}{14.7} = 1.020 \text{ atm}$$

$$\frac{\epsilon_{SO_2\text{-air}}}{kT} = (0.326 \times 0.846)^{1/2} = 0.525$$

From Fig. 24-1,

	Air	SO_2	
ϵ/k	97.0	252	(Table 24-1)
ϵ/kT	0.326	0.846	
r (Å)	3.617	4.290	(Table 24-1)
M	29	64	

For $\frac{\epsilon}{kT} = 0.525$, $f(\frac{\epsilon}{kT}) = 1.0$ (See Fig. 24-1.)

$$r_{SO_2\text{-air}} = \frac{3.617 + 4.290}{2} = 3.953$$

Using Eq. (2),

$$D_{SO_2\text{-air}} = \frac{0.0009292 \times 298^{3/2}[(1/29) + (1/64)]^{1/2}}{1.02(3.953)^2(1.0)}$$

$$= 0.067 \text{ cm}^2/\text{sec}$$

Example 2

Estimate diffusivity of chlorine in water at 25°C and 1 atm pressure.

$T = 273 + 25 = 298$ K

$\mu_{H_2O} = 0.8937$ cps

φ = molal volume of solute, Cl_2 is 48.2 (from Table 24-2), solute characteristic for H_2O = 1.0

Using Fig. 24-2,

$$F = 2.1 \times 10^7$$

Using Eq. (7),

$$D = \frac{T}{\mu F} = 1.58 \times 10^{-5} \text{ cm}^2/\text{sec}$$

TABLE 24-2

Atomic and Molecular Volumes [3]

Atomic volume		Molecular volume	
Carbon	14.8	H_2	14.3
Hydrogen	3.7	O_2	25.6
Chlorine	24.6	N_2	31.2
Bromine	27.0	Air	29.9

TABLE 24-2 (Continued)

Atomic volume		Molecular volume	
Iodine	37.0	CO	30.7
Sulfur	25.6	CO_2	34.0
Nitrogen	15.6	SO_2	44.8
Nitrogen in primary amines	10.5	NO	23.6
Nitrogen in secondary amines	12.0	N_2O	36.4
Oxygen	7.4	NH_3	25.8
Oxygen in methyl esters	9.1	H_2O	18.9
Oxygen in higher esters	11.0	H_2S	32.9
Oxygen in acids	12.0	COS	51.5
Oxygen in methyl ethers	9.9	Cl_2	48.4
Oxygen in higher ethers	11.0	Br_2	53.2
Benzene ring: subtract	15	I_2	71.5
Naphthalene ring: subtract	30		

CONCEPT OF MASS TRANSFER COEFFICIENT

The process of gas absorption can be further explained by two-film theory: two-phase flow essentially exists in gas absorption, and gas and liquid are considered to be diffusing through films. The gas concentration is changed from P_A in bulk to P_{Ai} at the interface between gas and liquid, and the liquid concentration is changed from C_A in bulk to C_{Ai} at the interface. These are essentially the driving forces, and are represented by

$$N_A = k_g(P_A - P_{Ai}) \tag{8}$$

$$= k_L(C_{Ai} - C_A) \tag{9}$$

where N_A is the rate of diffusion of component A (moles/hr ft^2), and k_g and k_L are known as the gas and liquid mass transfer coefficients, respectively. Comparing these with Eq. (1), we get

$$k_g = \frac{D_{AB_g} P}{RTZ_g P_{BM}} \tag{10}$$

$$k_L = \frac{D_{AB_L}}{Z_L} \tag{11}$$

Although one can estimate diffusivities fairly well, it is difficult to estimate Z_g, Z_L, and the interfacial concentrations C_{Ai} and P_{Ai}. The thickness of the film coefficient depends on the turbulence outside the film, which can be expressed as a function of the Reynolds number. The problem of determining interfacial concentration can be eliminated by using the overall mass transfer coefficient as explained below.

The rate of absorption depends on the concentration gradient. Another way of expressing this concentration gradient is in terms of the difference between bulk and equilibrium concentrations.

The equilibrium concentration is determined from the rate at which a gas component is transferred to the liquid phase and also from the rate at which it is transferred back to the gas phase. Thus, no matter how long the gas and liquid are in contact, there will be no change in concentration of the diffusing component in the gas or liquid phases. Equilibrium may be defined by

$$P_A^* = HC_A \tag{12}$$

where H is the Henry's law constant, P_A^* is the partial pressure of the solute in the gas at equilibrium conditions, and C_A is the concentration of the solute in the liquid.

When the equilibrium concentration is used in the driving-force concentration gradient, the local mass transfer coefficient is replaced by an overall mass transfer coefficient:

$$\begin{aligned} N_A &= k_L(C_{Ai} - C_A) = K_L(C_A^* - C_A) \\ &= k_g(P_A - P_{Ai}) = K_G(P_A - P_A^*) \end{aligned} \tag{13}$$

where K_G and K_L are overall mass transfer coefficients for the gas and liquid films, respectively. It is difficult to determine the interfacial concentrations C_{Ai} and P_{Ai}, but one can estimate the equilibrium concentrations C_A^* and P_A^* from the equilibrium relationship. The relationships between local and overall mass transfer coefficients are given by

TABLE 24-3

Diffusivities of Pairs of Gases and Vapors (1 atm) [4]

Substance	Temp. (°C)	Air	A	H_2	O_2
Acetic acid	0	0.1064		0.416	
Acetone	0	0.109		0.361	
n-Amyl alcohol	0	0.0589		0.235	
sec-Amyl alcohol	30	0.072			
Amyl butyrate	0	0.040			
Amyl formate	0	0.0543			
i-Amyl formate	0	0.058			
Amyl isobutyrate	0	0.0419		0.171	
Amyl propionate	0	0.046		0.1914	
Aniline	0	0.0610			
	30	0.075			
Anthracene	0	0.0421			
Argon	20				
Benzene	0	0.077		0.306	0.0797
Benzidine	0	0.0298			
Benzyl chloride	0	0.066			
n-Butyl acetate	0	0.058			
i-Butyl acetate	0	0.0612		0.2364	
n-Butyl alcohol	0	0.0703		0.2716	
	30	0.088			
i-Butyl alcohol	0	0.0727		0.2771	
Butyl amine	0	0.0821			
i-Butyl amine	0	0.0853			
i-Butyl butyrate	0	0.0468		0.185	
i-Butyl formate	0	0.0705			
i-Butyl isobutyrate	0	0.0457		0.191	
i-Butyl propionate	0	0.0529		0.203	
i-Butyl valerate	0	0.0424		0.173	
Butyric acid	0	0.067		0.264	
i-Butyric acid	0	0.0679		0.271	
Cadmium	0				
Caproic acid	0	0.050			
i-Caproic acid	0	0.0513			
Carbon dioxide	0	0.138		0.550	0.139
	20				
	25				
	500‡				0.9
Carbon disulfide	0	0.0892		0.369	
Carbon monoxide	0			0.651	0.185
	450‡				1.0
Carbon tetrachloride	0			0.293	0.0636
Chlorobenzene	30	0.075			
Chloroform	0	0.091			
Chloropicrin	25	0.088			

N_2	CO_2	N_2O	CH_4	C_2H_6	C_2H_4	$n\text{-}C_4H_{10}$	$i\text{-}C_4H_{10}$
	0.0716						
	0.0422						
	0.0347						
0.194							
	0.0528						
	0.0425						
	0.0476						
	0.0483						
	0.0327						
	0.0364						
	0.0366						
	0.0308						
	0.0476						
	0.0471						
0.17							
		0.096	0.153				
0.163							
		0.0996*	0.00215†				
	0.063						
	0.137				0.116		

TABLE 24-3 (Continued)

Substance	Temp. (°C)	Air	A	H_2	O_2
m-Chlorotoluene	0	0.054			
o-Chlorotoluene	0	0.059			
p-Chlorotoluene	0	0.051			
Cyanogen chloride	0	0.111			
Cyclohexane	15		0.0719	0.319	0.0744
	45	0.086			
n-Decane	90			0.306	
Diethylamine	0	0.0884			
2,3-Dimethyl butane	15		0.0657	0.301	0.0753
Diphenyl	0	0.0610			
n-Dodecane	126			0.308	
Ethane	0			0.459	
Ethanol	0			0.377	
Ether (diethyl)	0	0.0778		0.298	
Ethyl acetate	0	0.0715		0.273	
	30	0.089			
Ethyl alcohol	0	0.102		0.375	
Ethyl benzene	0	0.0658			
Ethyl n-butyrate	0	0.0579		0.224	
Ethyl i-butyrate	0	0.0591		0.229	
Ethylene	0			0.486	
Ethyl formate	0	0.0840		0.337	
Ethyl propionate	0	0.068		0.236	
Ethyl valerate	0	0.0512		0.205	
Eugenol	0	0.0377			
Formic acid	0	0.1308		0.510	
Helium	0		0.641		
	20				
n-Heptane	38				
n-Hexane	15		0.0663	0.290	0.0753
Hexyl alcohol	0	0.0499		0.200	
Hydrogen	0	0.611			0.697
	25				
	500				4.2
Hydrogen cyanide	0	0.173			
Hydrogen peroxide	60	0.188			
Iodine	0	0.07			
Mercury	0	0.112		0.53	
Mesitylene	0	0.056			
Methane	500				1.1
Methyl acetate	0	0.084		0.333	
Methyl alcohol	0	0.132		0.506	
Methyl butyrate	0	0.0633		0.242	
Methyl i-butyrate	0	0.0639		0.257	
Methyl cyclopentane	15		0.0731	0.318	0.0742
Methyl formate	0	0.0872			

N_2	CO_2	N_2O	CH_4	C_2H_6	C_2H_4	$n\text{-}C_4H_{10}$	$i\text{-}C_4H_{10}$
0.0760							
0.0841							
0.0751							
0.0813							
	0.0686						
	0.0546						
	0.0487						
	0.0685						
	0.0407						
	0.0413						
	0.0573						
	0.0450						
	0.0367						
	0.0874						
0.705							
			0.066‡				
0.0757							
	0.0351						
0.674	0.550	0.535	0.625	0.459	0.486	0.272	0.277
	0.646			0.537	0.726		
0.070							
0.13							
	0.0567						
	0.0879						
	0.0446						
	0.0451						
0.0758							

TABLE 24-3 (Continued)

Substance	Temp. (°C)	Air	A	H_2	O_2
Methyl propionate	0	0.0735		0.295	
Methyl valerate	0	0.0569			
Naphthalene	0	0.0513			
Nitrogen	0				0.181
	25				
Nitrous oxide	0			0.535	
m-Octane	0	0.0505			
	30		0.0642	0.271	0.0705
Oxygen	0	0.178		0.697	
Phosgene	0	0.095			
Propionic acid	0	0.0829		0.330	
Propyl acetate	0	0.067			
n-Propyl alcohol	0	0.085		0.315	
i-Propyl alcohol	0	0.0818			
	30	0.101			
n-Propyl benzene	0	0.0481			
i-Propyl benzene	0	0.0489			
n-Propyl bromide	0	0.085			
i-Propyl bromide	0	0.0902			
Propyl butyrate	0	0.0530		0.206	
Propyl formate	0	0.0712		0.281	
n-Propyl iodide	0	0.079			
i-Propyl iodide	0	0.0802			
n-Propyl isobutyrate	0	0.0549		0.212	
i-Propyl isobutyrate	0	0.059			
Propyl propionate	0	0.057		0.212	
Propyl valerate	0	0.0466		0.189	
Safrol	0	0.0434			
i-Safrol	0	0.0455			
Sulfur hexafluoride	25			0.418	
Toluene	0	0.076	0.071		
	30	0.088			
Trimethyl carbinol	0	0.087			
2,2,4-Trimethyl pentane	30		0.0618	0.288	0.0688
2,2,3-Trimethyl heptane	90			0.270	
n-Valeric acid	0	0.050			
i-Valeric acid	0	0.0544		0.212	
Water	0	0.220		0.75	
	450				1.3

Source: From Ref. 4, with permission of McGraw-Hill Book Company.

N_2	CO_2	N_2O	CH_4	C_2H_6	C_2H_4	$n\text{-}C_4H_{10}$	$i\text{-}C_4H_{10}$
	0.0528						
	0.165			0.148	0.163	0.0960	0.0908
	0.096						
0.0710							
0.181	0.139						
	0.0588						
	0.0577						
	0.0364						
	0.0490						
	0.0388						
	0.0395						
	0.0341						
0.0705							
0.0684							
	0.0376						
	0.138						

TABLE 24-4

Diffusivities in Liquids (25°C) [4]

(Dilute solutions and 1 atm unless otherwise noted; use $D_{L}\mu/T$ = constant to estimate effect of temperature; reference gives effect of concentration)

Solute	Solvent	$D_L \times 10^5$ (cm^2/sec)	Estimated possible error (± %)
Acetal*	Ethanol	1.25	5
Acetamide*	Ethanol	0.68	5
Acetamide*	Water	1.19	3
Acetic acid	Acetone	3.31	
Acetic acid	Benzene	2.11	
Acetic acid	Carbon tetrachloride	1.49	
Acetic acid	Ethylene glycol	0.13	
Acetic acid	Toluene	2.26	
Acetic acid*	Water	1.24	3
Acetonitrile	Water	1.66	5
Acetylene	Water	1.78, 2.11	
Allyl alcohol*	Ethanol	1.06	5
Allyl alcohol	Water	1.19	6
Ammonia*	Water	1.7, 2.0, 2.3	
i-Amyl alcohol*	Ethanol	0.87	5
i-Amyl alcohol	Water	1.0	8
Benzene	Carbon tetrachloride	1.53	
Benzene (50 mole %)	n-Decane	1.72	
Benzene (50 mole %)	2,4-Dimethyl pentane	2.49	
Benzene (50 mole %)	n-Dodecane	1.40	
Benzene (50 mole %)	n-Heptane	2.47	
Benzene (50 mole %)	n-Hexadecane	0.96	

Benzene (50 mole %)	n-Octadecane	0.86	
Benzoic acid	Acetone	2.62	
Benzoic acid	Benzene	1.38	
Benzoic acid	Carbon tetrachloride	0.91	
Benzoic acid	Ethylene glycol	0.043	
Benzoic acid	Toluene	1.49	
Bromine	Benzene	2.7	
Bromine	Carbon disulfide	4.1	
Bromine	Water	1.3	
Bromobenzene	Benzene	2.30	
Bromoform*	Acetone	2.90	
Bromoform	i-Amyl alcohol	0.53	
Bromoform	Ethanol	1.08	5
Bromoform*	Ethyl ether	3.62	
Bromoform	Methanol	2.20	
Bromoform	n-Propanol	0.94	
n-Butanol	Water	0.96	5
Caffeine	Water	0.63	6
Carbon dioxide	Ethanol	4.0	6
Carbon dioxide	Water	1.96	1
Carbon disulfide (50 mole %, 200 atm)	n-Butanol	3.57	
Carbon disulfide (50 mole %, 200 atm)	i-Butanol	2.42	
Carbon disulfide (50 mole %, 218 atm)	Chlorobenzene	3.00	
Carbon disulfide (50 mole %, 200 atm)	2,4-Dimethyl pentane	3.63	
Carbon disulfide (50 mole %, 100 atm)	n-Heptane	3.0	
Carbon disulfide (50 mole %, 50 atm)	Methyl cyclohexane	3.5	
Carbon disulfide (50 mole %, 200 atm)	n-Octane	3.10	
Carbon disulfide (50 mole %)	Toluene	2.06	
Carbon tetrachloride	Benzene	2.04	3
Carbon tetrachloride*	Cyclohexane	1.49	2
Carbon tetrachloride	Decalin	0.776	2
Carbon tetrachloride	Dioxane	1.02	2
Carbon tetrachloride*	Ethanol	1.50	2
Carbon tetrachloride	n-Heptane	3.17	2

TABLE 24-4 (Continued)

Solute	Solvent	$D_L \times 10^5$ (cm^2/sec)	Estimated possible error (± %)
Carbon tetrachloride	Kerosene	0.961	2
Carbon tetrachloride	Methanol	2.30	2
Carbon tetrachloride	i-Octane	2.57	2
Carbon tetrachloride	Tetralin	0.735	2
Chloral*	Ethanol	0.68	5
Chloral hydrate	Water	0.77	7
Chlorine	Water	1.44	4
Chlorobenzene	Benzene	2.66	
Chloroform	Benzene	2.50	6
Chloroform	Ethanol	1.38	3
Cinnamic acid	Acetone	2.41	
Cinnamic acid	Benzene	1.12	
Cinnamic acid	Carbon tetrachloride	0.76	
Cinnamic acid	Toluene	2.41	
1,1′-Dichloropropanol	Water	1.0	6
Dicyanodiamide*	Water	1.18	4
Diethyl ether	Benzene	2.73	
Diethyl ether	Water	0.85	
2,4-Dimethyl pentane (50 mole %)	n-Dodecane	1.44	
2,4-Dimethyl pentane (50 mole %)	n-Hexadecane	0.88	
Ethanol*	Water	1.28	4
Ethyl acetate	Ethyl benzoate	0.94	
Ethylene dichloride	Benzene	2.8	
Formic acid	Acetone	3.77	
Formic acid	Benzene	2.28	
Formic acid	Carbon tetrachloride	1.89	
Formic acid	Ethylene glycol	0.094	

Formic acid	Toluene	2.65	
Formic acid	Water	1.37	10
Glucose	Water	0.69	6
Glycerol	i-Amyl alcohol	0.12	
Glycerol	Ethanol	0.56	
Glycerol*	Water	0.94	6
n-Heptane (50 mole %)	n-Dodecane	1.58	
n-Heptane (50 mole %)	n-Hexadecane	1.00	
n-Heptane (50 mole %)	n-Octadecane	0.92	
n-Heptane (50 mole %)	n-Tetradecane	1.29	
Hexamethylene tetramine	Water	0.67	
Hydrogen chloride*	Water	3.10	3
Hydrogen	Water	5.85 (4.4?)	
Hydrogen sulfide	Water	1.61	
Hydroquinone*	Ethanol	0.53	5
Hydroquinone*	Water	0.88, 1.12	
Iodine	Acetic acid	1.13	
Iodine	Anisole	1.25	
Iodine	Benzene	1.98	
Iodine	Bromobenzene	1.25	10
Iodine	Carbon disulfide	3.2	
Iodine	Carbon tetrachloride	1.45	8
Iodine	Chloroform	2.30	3
Iodine	Cyclohexane	1.80	
Iodine	Dioxane	1.07	
Iodine*	Ethanol	1.30	
Iodine	Ethyl acetate	2.2	
Iodine	Ethyl ether	3.61	
Iodine	Ethylene bromide	0.93	
Iodine	n-Heptane	3.4, 2.5	
Iodine	n-Hexane	4.15	
Iodine	Mesitylene	1.49	
Iodine	Methanol	1.74	
Iodine	Methyl cyclohexane	2.1	

TABLE 24-4 (Continued)

Solute	Solvent	$D_L \times 10^5$ (cm^2/sec)	Estimated possible error (± %)
Iodine	n-Octane	2.76	
Iodine	Tetrabromoethane	2.0	
Iodine	n-Tetradecane	0.96	
Iodine	Toluene	2.1	
Iodine	m-Xylene	1.82	
Iodobenzene	Ethanol	1.09	3
Lactose*	Water	0.49	5
Maltose*	Water	0.48	5
Mannitol*	Water	0.65	5
Methanol	Water	1.6	
Micotine*	Water	0.60	8
Nitric acid*	Water	2.98	2
Nitrobenzene	Carbon tetrachloride	1.00	
Nitrogen	Water	1.9	
Nitrous oxide	Water	1.8	
Oxalic acid*	Water	1.61	2
Oxygen	Glycerol*-water (106 poise)	0.24	
Oxygen	Sucrose*-water (125 poise)	0.25	
Oxygen	Water	2.5	20
Pentaerythritol*	Water	0.77	4
Phenol	i-Amyl alcohol	0.2	
Phenol	Benzene	1.68	
Phenol	Carbon disulfide	3.7	
Phenol	Chloroform	2.0	
Phenol	Ethanol	0.89	
Phenol	Ethyl ether	3.9	
n-Propanol	Water	1.1	

Pyridine*	Ethanol	1.24	3
Pyridine	Water	0.76	7
Pyrogallol	Water	0.74	7
Raffinose*	Water	0.41	4
Resorcinol*	Ethanol	0.46	5
Resorcinol*	Water	0.87	4
Saccharose*	Water	0.49	4
Stearic acid*	Ethanol	0.65	5
Succinic acid*	Water	0.94	
Sucrose	Water	0.56	6
Sulfur dioxide	Water	1.7	
Sulfuric acid*	Water	1.97	3
Tartaric acid*	Water	0.80	10
1,1,2,2-Tetrabromoethane	1,1,2,2-Tetrachloroethane	0.61	4
Toluene	n-Decane	2.09	
Toluene	n-Dodecane	1.38	
Toluene	n-Heptane	3.72	
Toluene	n-Hexane	4.21	
Toluene	n-Tetradecane	1.02	
Urea	Ethanol	0.73	
Urea	Water	1.37	2
Urethane	Water	1.06	
Water	Glycerol	0.021	

Source: From Ref. 4, with permission of McGraw-Hill Book Company.

$$\frac{1}{K_G} = \frac{1}{k_g} + \frac{H'}{k_\ell} \tag{14}$$

$$\frac{1}{K_L} = \frac{1}{k_\ell} + \frac{1}{H'k_g} \tag{15}$$

where H' is Henry's law constant adjusted by dimensions.

If a gas is very soluble in liquid, a small change in the gas concentration can be effected in a large change in the liquid concentration. This means that H is small. So, from Eq. (14), $k_g \simeq K_G$; in such cases, absorption is said to be gas film resistance controlled. On the other hand, when the gas is not soluble in liquids, $k_\ell \simeq K_L$; this is called liquid film resistance controlled absorption. In most air pollution control scrubbers, the solvent used has a high solubility and operation is gas film resistance controlled. Table 24-5 indicates the types of resistances for some of the common systems [4].

TABLE 24-5

Source of Major Resistance to Mass Transfer for Some Common Systems

Liquid phase	Both phases	Gas phase
Oxygen-water	Sulfur dioxide-water	Ammonia-water
Hydrogen-water	Nitrogen dioxide-sulfuric acid	Ammonia-acid
Carbon dioxide-water		Sulfur dioxide-alkali
Chlorine-water	Acetone-water	Hydrochloric acid-water
Carbon dioxide-sodium hydroxide	Hydrogen sulfide-alkali	Water-acid
Carbon dioxide-amines		Water-calcium chloride, brine
		Evaporation
		Condensation

DESIGN — CAPACITY AND EFFICIENCY

Operating Line

In the design of a scrubber, it is necessary to construct an operating line on the equilibrium diagram. It is recommended that engineers work with the concentration in terms of moles of solute per mole of pure solvent or moles of solute per mole of inert gas. This makes the material balance calculations much easier.

Let

x = mole fraction of solute in liquid
y = mole fraction of solute in gas
X = mole fraction of solute in pure liquid
Y = mole fraction of solute in inert gas
L_s = flow rate of pure liquid (lb moles/hr ft^2)
G_s = flow rate of inert gas (lb moles/hr ft^2)

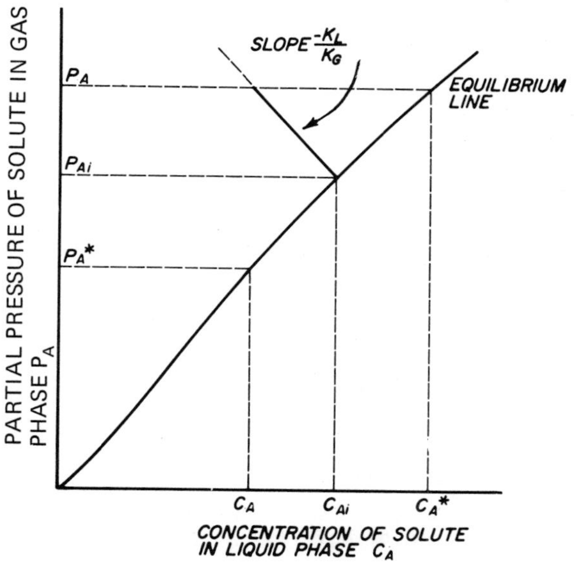

FIG. 24-3.

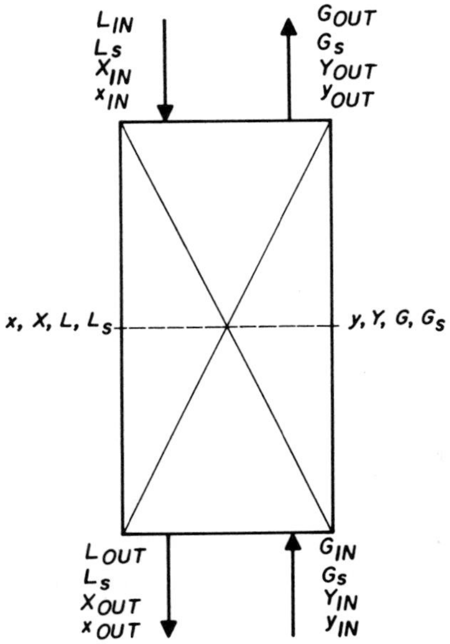

FIG. 24-4. Flow schematic across packed tower.

Hence,

$$X = \frac{x}{1 - x} \tag{16}$$

$$Y = \frac{y}{1 - y} \tag{17}$$

Consider a countercurrent scrubber with flow and concentration conditions given in Fig. 24-4. Taking the material balance across the scrubber, we have

$$G_s(Y_{in} - Y_{out}) = L_s(X_{out} - X_{in}) \tag{18}$$

$$Y_{in} - Y_{out} = \frac{L_s}{G_s}(X_{out} - X_{in}) \tag{19}$$

When the line represented by Eq. (19) is drawn on an equilibrium diagram, it is called an operating line (Fig. 24-5).

When transfer of solute is from gas to liquid, the operating line is above the equilibrium line and the operation is called ab-

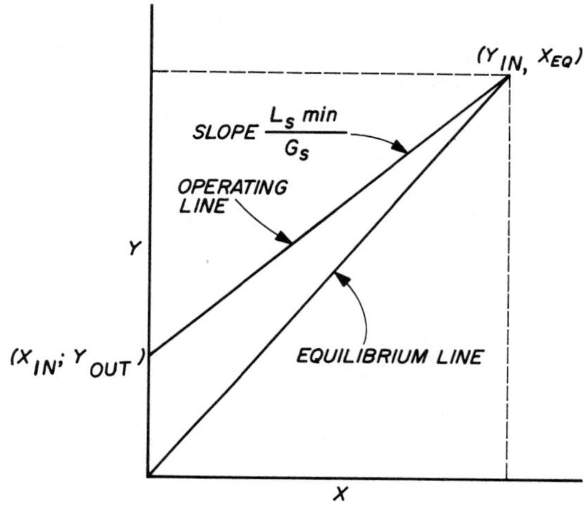

FIG. 24-5. Minimum liquid:gas ratio.

sorption. On the other hand, when transfer is from liquid to gas, the operating line is below the equilibrium line and the operation is known as stripping.

Liquid Requirement

Construction of an operating line as discussed is required to determine the liquid requirement in scrubbing. While designing the scrubber, concentration of solute both in gas to be treated and solvent to be used are known. Furthermore, minimum acceptable concentrations of objectionable materials in the gas and gas flow rates are also known. For minimum liquid rate, concentrations Y_{in} and X_{out} will be at equilibrium. Hence, that condition is represented by a point on the equilibrium curve (Y_{in}, X_{eq}). The slope of the operating line between (Y_{in}, X_{eq}) and (Y_{out}, X_{in}) represents the condition at minimum liquid flow rate. The slope is $(L_s/G_s)_{min}$. Although the gas flow rate is expressed in terms of pound-moles per hour per square foot, the dimensions of area will be cancelled in the ratio L_s/G_s. Knowing G_s, $L_{s(min)}$ can be found.

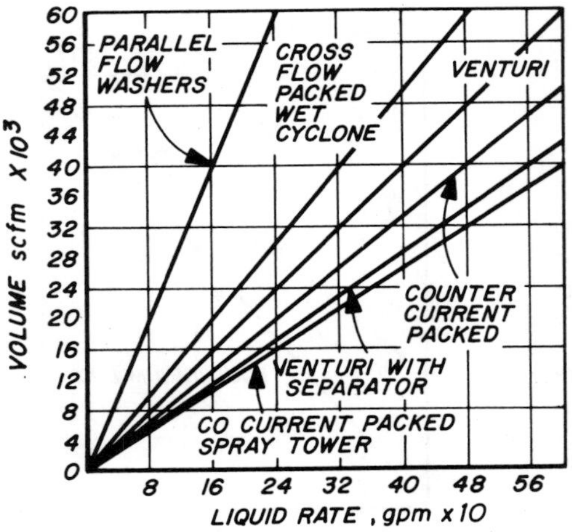

FIG. 24-6. Note: Actual make-up water rate or water consumed will depend on the physical condition of the inlet gas stream, the concentration of contaminants, requirements of the waste-treatment system, and the nature of the chemical reactions in general. A rule of thumb of 10% ± 5% of the recycle rate can be used as fresh-water requirement. (Courtesy of American Institute of Chemical Engineers.)

If liquid is not available in sufficient quantity, it can be recirculated. Refer to Fig. 24-6 [5].

Example 3

Develop an equilibrium diagram for an SO_2-H_2O system from the following data: Sulfur dioxide is to be scrubbed by water at 30°C, 1 atm pressure; gas flow rate is 3,000 scfm; and SO_2 concentration in air is 1.4% by weight. It is desired to reduce the sulfur dioxide concentration by 95 percent.

1. Equilibrium Data:

SO_2 percent by weight in water	SO_2 pp mmHg
7.5	688
5.0	452
2.5	216

SO_2 percent by weight in water	SO_2 pp mmHg
1.5	125
1.0	79
0.7	52
0.5	36
0.3	19.7
0.1	4.7
0.05	1.7
0.02	0.6

2. Convert the above data into terms of X and Y.

x	$X = \frac{x}{1 - x}$	$y = \frac{PP}{P_t}$	$Y = \frac{PP}{P_t - PP}$
0.075	0.075	0.9052	9.555
0.050	0.050	0.5947	1.4675
0.025	0.025	0.2842	0.3970
0.015	0.015	0.1644	0.1968
0.010	0.010	0.1039	0.1160
0.007	0.007	0.06842	0.07344
0.005	0.005	0.0473	0.04965
0.003	0.00300	0.02592	0.02661
0.001	0.00100	0.00618	0.00622
0.0005	0.00500	0.002236	0.002241
0.0002	0.00020	0.0007894	0.000790
0	0	0	0

Inlet concentration is given as 1.25 percent by weight, if total gas weight is 100 lb. SO_2 concentration weight is 1.4 percent. Weight of air in 100 lb of gas mixture = 98.6 lb.

3. Concentration at inlet is

$$Y_{in} = \frac{1.40}{64} \times \frac{29}{98.60} = 0.0064 \ \frac{\text{lb moles of } SO_2}{\text{lb moles of air}}$$

at Y = 0.0064, equilibrium concentration;

at X = 0.00101 moles of SO_2/moles of H_2O.

4. Ninety-five percent of the SO_2 is to be removed. Therefore, the gas concentration leaving the scrubber is

$$0.05 \times 0.0064 \frac{\text{lb moles of } SO_2}{\text{lb moles of air}} = 0.00032 \frac{\text{lb moles of } SO_2}{\text{lb moles of air}}$$

The slope of the operating line for minimum liquid flow rate is

$$\frac{0.0064 - 0.00032}{0.00101 - 0} = \frac{\text{lb moles of } SO_2}{\text{lb moles of air}} \times \frac{\text{lb moles of } H_2O}{\text{lb moles of } SO_2}$$

$$= 6.019 = \frac{\text{lb moles of } H_2O}{\text{lb moles of } SO_2}$$

The total gas flow rate is 3,000 scfm; that is, total gas flow rate is

$$\frac{3{,}000}{378} = 7.936 \frac{\text{lb moles}}{\text{min}}$$

Therefore, air flow rate is

$$(1 - 0.0064) \times 7.936 = 7.8852 \frac{\text{lb moles}}{\text{min}}$$

Minimum liquid flow rate is

$$7.8852 \times 6.019 = 47.46 \text{ lb moles } H_2O/\text{min or } 854 \text{ lb } H_2O/\text{min}$$

The above calculations were based on theoretical minimum water requirements. Practical conditions, however, may require different minimum values for scrubbing; that is, practical considerations are based on efficiency and economics.

Efficiency of scrubbing depends on the degree to which the packing is wetted; at higher flow rates, more packing surface is wetted. This gives a higher value of active area, and hence a better rate of mass transfer. Also, higher liquid flow rates mean that more liquid must be stripped off the solute, which means that regeneration of scrubbing liquid is more expensive. On the other hand, lower liquid flow rates mean larger scrubbing towers and higher initial capital costs.

Generally, liquid rates are 25 to 100 percent greater than the theoretically calculated liquid rate. Manufacturers of packing suggest that liquid flow rates be 2 to 5 gpm/ft^2 depending on the type

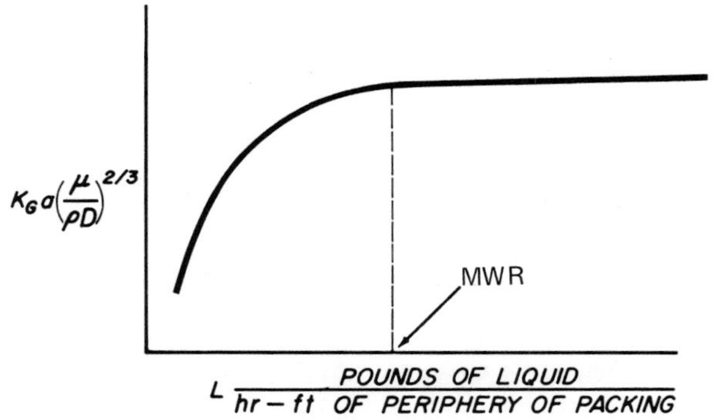

FIG. 24-7. Minimum wetting rate.

and size of the packing. This value is based on the fact that above a certain liquid flow rate known as minimum wetting rate (MWR) or minimum effective wetting rate, the increase in the mass transfer rate expressed by $K_g a(Sc)_g^{2/3}$ is not substantial. $(Sc)_g$ is the Schmidt dimensional number $\mu/\rho D_{AB}$ where D_{AB} is gas diffusivity. This is illustrated in Fig. 24-7, which is based on work by Pratt [6].

Before considering how the capacity of the scrubber is determined, it is essential to go through the process of selecting the packing to be used for making intimate contact with the fluids.

PACKINGS

Packed towers are commonly used in gas absorption. The packings are used to establish an intimate contact between vapors and liquid used to wash the objectionable components.

About 60 years ago, packings used were broken stone, coke, and Raschig rings. Twenty years thereafter, other types of packings became established. Now there are a formidable number of different types of packings used in mass transfer operations. General overall dimensions of packings range from 1/4 to 4 in. and they are made of a variety of materials including stoneware, porcelain, plastics, and metals.

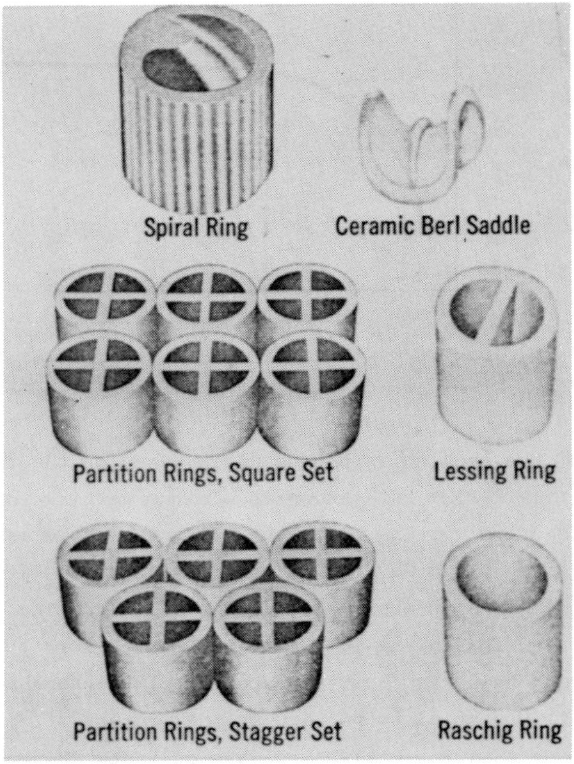

FIG. 24-8. Tower packings. (Courtesy of Maurice A. Knight Co.)

Generally speaking, packings can be classified into two types:

1. Random packings--for this type of packing the specific surface area (ft^2/ft^3 of tower volume) is higher, but pressure drop through them is greater.
2. Regular packings--these are systematically stacked, like grids, give larger openings for gas, and hence cause less pressure drop. They give less surface area, but provide better liquid distribution. They are usually more costly to install.

FIG. 24-9. Metal pall rings. (Courtesy of Koch Engineering.)

FIG. 24-10. Plastic pall rings. (Courtesy of Koch Engineering.)

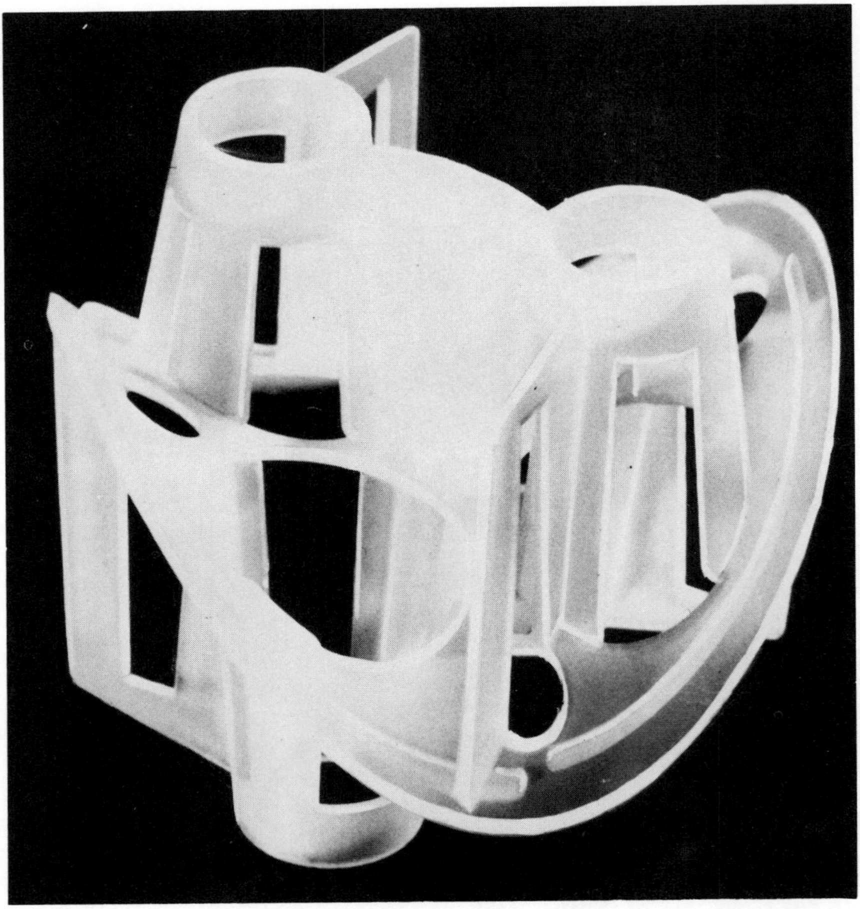

FIG. 24-11. Heilex-300 tower packing. (Courtesy of Heil Process Co.)

Factors Governing the Selection of Packing

The following factors govern packing selection for packed tower absorption design:

1. Cost: Cost per cubic foot must be determined; generally, plastic packings are cheaper than metal packings.

2. Low pressure drop: Packing should not create excessive pressure drop across the packed bed. Pressure drop is caused by resistance to gas flow. Resistance to flow is indicated by volume

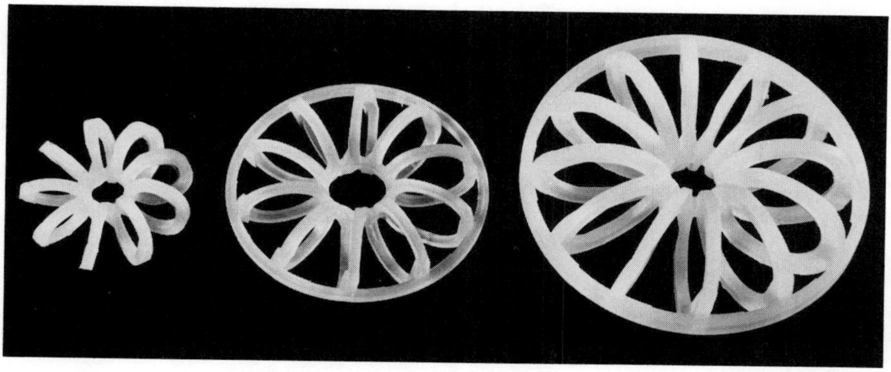

FIG. 24-12. Tellerette tower packing. Nominal sizes 1, 2, and 3 in. (Courtesy of Ceilcote Co., Inc.)

void fraction when towers are packed. One may notice that the void fraction of all of the spheres in a large-diameter tower is the same; this, however, does not mean that pressure drop is the same across small as well as large spheres. For smaller spheres, channel diameters are smaller compared to those created by larger diameter spheres, so the pressure drop is lower for larger packings.

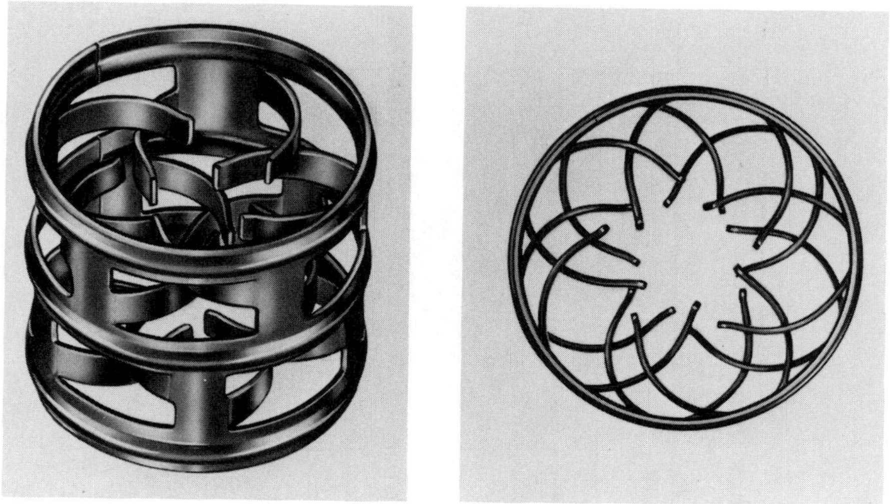

FIG. 24-13. Hy-pak packing. (Courtesy of Norton Co.)

TABLE 24-6

Corrosion Chart

(This chart is for use as a guide only. Figures in parentheses indicate temperature at which data were collected.)

	304 SS	316 SS	Polypropylene	High-density polyethylene	PVC	Polyester resins
Acetaldehyde	S (140)	S (140)	S (140)	U	U	ND
Acetic acid, 5%	L	L	S (140)	S (140)	S (140)	S (220) (bisphenol polyester)
Acetic acid, 80-100%	U	L	S (140)	S (140)	S (70)	S (100) (chlorinated ester)
Acetone	S	S	U	U	U	U
Ammonium carbonate	S	S	S (70)	S (140)	S	ND
Ammonium hydroxide	S	S	S (70)	S (140)	U	U
Ammonium chloride	L	S	S (140)	S (140)	S (140)	S (210) (bisphenol polyesters)
Aniline	S	S	S (180)	S (70)	U	U
Aqua regia	U	U	U	L (70)	U	U
Benzaldehyde	S (212)	S (212)	S (70)	S (120)	U	U
Benzene	S	S	S (70)	U	U	U
Boric acid	U	U	S (180)	S (120)	S (140)	S (170) (bisphenol and isophthalic esters)

Butane	S	S	S (70)	U	S (140)	U
Butanol	S	S	S (180)	S (140)	S (140)	S (100) (vinyl ester)
Calcium chloride	S (330)	S (330)	S (180)	S (120)	S (140)	S (170) (bisphenol and isophthalic polyesters)
Calcium hydroxide	S (212)	S (212)	S (180)	S (120)	S (140)	S (170) (bisphenol and isophthalic polyesters)
Carbon dioxide	S (235)	S (235)	S (180)	S (120)	S (140)	S (200) (vinyl ester and bisphenol polyesters)
Carbon disulfide	S (93)	S (93)	U	U	U	S (80) (bisphenol and chlorinated polyesters)
Carbon monoxide	S (392)	S (500)	S (150)	S (120)	S (140)	
Carbon tetrachloride	S (77)	S (77)	U	U	L (70)	S (120) (bisphenol and chlorinated polyesters)
Carbonic acid	S (212)	S (212)	S (120)	S (120)	S (140)	S (200) (bisphenol polyester)
Chlorine gas (dry)	U	U	U	S (120)	U	S (200) (bisphenol, chlorinated, and vinyl polyesters)
Chlorine gas (wet)	U	U	U	S (120)	U	S (200) (bisphenol, chlorinated, and vinyl polyesters)
Chlorine dioxide	U	U	ND	ND	ND	S (200) (bisphenol polyester)

TABLE 24-6 (Continued)

	304 SS	316 SS	Polypropylene	High-density polyethylene	PVC	Polyester resins
Chloroform	U	U	U	U	U	U
Chromic acid	S (120)	U	S (150)	S (120)	S (70)	S (150) (chlorinated polyester)
Copper chloride	U	U	S (120)	S (120)	S (140)	S (210) (bisphenol polyester)
Copper sulfate	S (200)	S (200)	S (150)	S (120)	S (140)	S (210) (bisphenol and chlorinated polyesters)
Ethanolamine (mono- and di-)	S (120)	S (120)	S	ND	U	S (80) (chlorinated polyester)
Diethylene glycol	S	S	S (120)	S (120)	S (140)	S (200) (bisphenol polyester)
Ethyl alcohol	S	S	S (120)	U	S(140)	S (140) (bisphenol polyester)
Ethylene chloride	ND	ND	S (70)	U	U	U
Fatty acids	U	U	S (150)	U	S (140)	S (200) (chlorinated and bisphenol poly-esters)
Ferric chloride	U	U	S (180)	S (120)	S (140)	S (200) (bisphenol and vinyl poly-esters
Ferrous sulfate	S (130)	S (130)	S (180)	S (70)	S (140)	S (200) (bisphenol & vinyl polyesters)

Formaldehyde	S	S	S (70)	S (140)	S (140)	S (170) (bisphenol polyester)
Furfural	S	S	U	U	U	U
Glycerine	S	S	S (140)	S (140)	ND	S (200) (bisphenol & vinyl polyesters)
Heptane	S (212)	S (212)	U	U	S (140)	S (125) (chlorinated & vinyl polyesters)
Hexane	S (250)	S (250)	U	U	S (70)	S (140) (bisphenol and isophthalic polyesters)
Hydrochloric acid, 10%	U	U	S (140)	S (140)	S (140)	S (200) (bisphenol polyester)
20%	U	U	S (140)	S (140)	S (140)	S (200) (bisphenol & vinyl polyesters)
30%	U	U	S (140)	S (140)	S (140)	S (180) (chlorinated polyester)
(conc.)	U	U	S (70)	S (140)	ND	ND
Hydrocyanic acid	S (77)	S (77)	S (140)	S (120)	S (140)	ND
Hydrofluoric acid	U	U	S (150)	S (120)	S (70)	ND
Hydrogen	S (950)	S (950)	S (120)	S (140)	S (140)	S (200) (bisphenol and chlorinated polyesters)
Hydrogen chloride (dry gas)	U	U	S (140)	S (140)	S (140)	S (210) (bisphenol and isophthalic polyesters)
Hydrogen chloride (wet gas)	U	U	S (140)	S (140)	S (140)	S (200) (bisphenol polyester)

TABLE 24-6 (Continued)

	304 SS	316 SS	Polypropylene	High-density polyethylene	PVC	Polyester resins
Hydrogen fluoride gas	U	U	U	U	U	S (80) (bisphenol and isophthalic polyesters)
Hydrogen peroxide	U	L (212)	S (120)	S (120)	S (70)	S (100) (chlorinated polyester)
Hydrogen sulfide	S (170)	S (170)	S (150)	S (120)	S (140)	S (170) (bisphenol and isophthalic polyesters)
Hypochlorous acid	U	U	S (140)	S (140)	S (140)	S (150) (bisphenol & vinyl polyesters)
Kerosene	S (400)	S (400)	U	S (70)	S (140)	S (150) (bisphenol, vinyl, isophthalic, and chlorinated polyesters)
Kraft liquor	ND	ND	S (120)	S (120)	S (140)	ND
Lactic acid	U	U	S (120)	S (120)	S (140)	S (200) (bisphenol and chlorinated polyesters)
Magnesium carbonate	U	U	S (120)	S (120)	S (140)	S (200)(bisphenol and chlorinated polyesters)
Magnesium chloride	U	U	S (120)	S (120)	S (140)	S (200) (bisphenol & vinyl polyesters)

Magnesium hydroxide	U	U	S (180)	S (120)	S (140)	S (200) (bisphenol polyester)
Maleic acid	U	S (125)	S (140)	S (140)	S (140)	S (200) (vinyl polyester)
Mercuric chloride	U	U	ND	ND	ND	S (200) (bisphenol and chlorinated polyesters)
Methanol	S (180)	S (180)	S (140)	S (140)	ND	S (80) (bisphenol and chlorinated polyesters)
Methyl aldehyde	S (200)	S (200)	ND	ND	ND	ND
Methyl chloride	S (190)	S (190)	U	U	U	S (40) (chlorinated polyester)
Methylene chloride	S (90)	S (90)	U	U	U	U
Methylethyl ketone	S (170)	S (170)	U	U	U	U
Naphtha	S (100)	S (100)	S (70)	U	S (140)	S (200) (bisphenol, vinyl, & chlorinated polyesters)
Naphthalene	S (440)	S (400)	S (70)	S (120)	U	S (200) (bisphenol & vinyl polyesters)
Nitric acid, 5%	S (140)	S (140)	S (180)	S (120)	S (140)	S (160) (bisphenol and chlorinated polyesters)
10%	S (120)	S (120)	S (180)	S (120)	S (140)	S (200) (chlorinated polyesters)
20%	S (200)	S (98)	S (180)	ND	S (140)	S (80) (bisphenol polyester)

TABLE 24-6 (Continued)

	304 SS	316 SS	Polypropylene	High-density polyethylene	PVC	Polyester resins
Nitric acid, 40%	U	S (98)	L (180)	U	S (140)	S (80) (bisphenol polyester)
90%	U	S (80)	U	U	U	U
Nitrous acid, 10%	S (95)	S (95)	U	ND	ND	S (150) (bisphenol polyester)
Oleic acid	U	U	ND	ND	ND	S (210) (vinyl polyester)
Oxalic acid	U	U	S (120)	S (140)	S (140)	S (200) (vinyl polyester
Perchloric acid	U	U	S (120)	S (140)	S (70)	S (80) (chlorinated polyester)
Phenol	S (300)	S (300)	S (180)	U	S (70)	U
Phosphoric acid, 10%	S (175)	S (175)	S (180)	S (120)	S (140)	S (200) (bisphenol, isophthalic, and chlorinated polyesters)
80%	U	S (220)	S (180)	S (120)	S (140)	S (200) (bisphenol, isophthalic, and chlorinated polyesters)
Phthalic acid	U	S (300)	ND	ND	ND	S (200) (vinyl polyester)

Phthalic anhydride	S (430)	S (430)	ND	ND	ND	S (200) (bisphenol polyester)
Picric acid	U	U	S (150)	S (120)	U	S (100) (bisphenol, vinyl, and chlorinated polyesters)
Potassium aluminum sulfate	ND	ND	ND	ND	ND	S (200) (bisphenol & vinyl polyesters)
Potassium bicarbonate, 30%	S (212)	S (212)	S (140)	S (120)	S (140)	S (150) (bisphenol & vinyl polyesters)
Potassium bromide	S (65)	S (65)	S (180)	S (70)	S (140)	S (170) (isophthalic and bisphenol polyesters)
Potassium carbonate	S (240)	S (240)	S (180)	S (70)	S (140)	S (150) (bisphenol & vinyl polyesters)
Potassium chloride	S (120)	S (120)	S (180)	S (70)	S (140)	S (170) (bisphenol and isophthalic polyesters)
Potassium hydroxide	U	U	S (180)	S (70)	S (140)	S (150) (vinyl polyester)
Potassium permanganate	U	U	S (150)	S (120)	S (140)	S (150) (vinyl and chlorinated polyesters)
Propylene glycol	S (260)	S (260)	ND	S (140)	ND	S (200) (bisphenol polyester)
Sea water	S (80)	S (80)	S (140)	S (140)	S (140)	S (90) (chlorinated polyester)
Silver nitrate	U	S (80)	S (70)	S (140)	ND	S (200) (chlorinated polyester)

TABLE 24-6 (Continued)

	304 SS	316 SS	Polypropylene	High-density polyethylene	PVC	Polyester resins
Sodium carbonate	S (150)	S (150)	S (180)	S (140)	S (140)	S (125) (bisphenol, chlorinated, and vinyl polyesters)
Sodium bicarbonate	S (85)	S (85)	S (180)	S (140)	S (140)	S (140) (bisphenol and chlorinated polyesters)
Sodium chloride	S (100)	S (100)	S (180)	S (70)	S (140)	S (170) (isophthalic and bisphenol polyesters)
Sodium dichromate	ND	ND	S (150)	S (120)	S (140)	S (170) (isophthalic and bisphenol polyesters)
Sodium hydroxide	S (275)	S (275)	S (180)	S (120)	S (140)	S (140) (chlorinated polyester)
Sodium hydrochlorite	U	U	S (120)	S (120)	S (140)	S (180) (chlorinated polyester)
Sodium sulfate	S (170)	S (170)	S (180)	S (120)	S (140)	S (170) (isophthalic and bisphenol polyesters)
Sodium sulfide, 10%	U	S (100)	S (180)	S (120)	S (140)	S (200) (bisphenol polyester
Sulfur dioxide	S (70)	S (70)	S (70)	S (70)	S (140)	S (200) (bisphenol polyester)

Sulfur trioxide, 10%	S (212)	S (212)	U	U	S (140)	S (220) (bisphenol polyester
Sulfuric acid, 10%	U	U	S (180)	S (120)	S (140)	S (200) (bisphenol, chlorinated, and vinyl polyesters)
50%	U	U	S (140)	S (70)	S (140)	S (200) (bisphenol, chlorinated, and vinyl polyesters)
90%	U	U	S (70)	S (70)	S (70)	U
Sulfurous acid	S (189)	S (189)	S (140)	S (140)	S (140)	ND
Trichloroethylene	S (115)	S (115)	S (70)	U		S (90) (bisphenol, isophthalic, and chlorinated polyesters)
Toluene	S (150)	S (150)	S (70)	U		S (75) (chlorinated, isophthalic, and vinyl polyesters)
Turpentine	S (200)	S (200)	S (70)	S (70)		ND
Urea	S (90)	S (90)	S (70)	S (140)		S (120) (isophthalic and bisphenol polyesters)
Water	S (200)	S (200)	S (180)	S (140)		S (80) (all polyesters)
Xylene	U	S (212)	U	U		S (80) (bisphenol, chlorinated, isophthalic, and vinyl polyesters)

TABLE 24-6 (Continued)

	304 SS	316 SS	Polypropylene	High-density polyethylene	PVC	Polyester resins
Zinc chloride	L (160)	L (160)	S (70)	S (140)		S (170) (bisphenol and isophthalic polyesters)
Zinc sulfate	S (212)	S (212)	S (70)	S (140)		S (170) (bisphenol and isophthalic polyesters)

S = satisfactory.
U = unsatisfactory.
L = limited.
ND = no data.

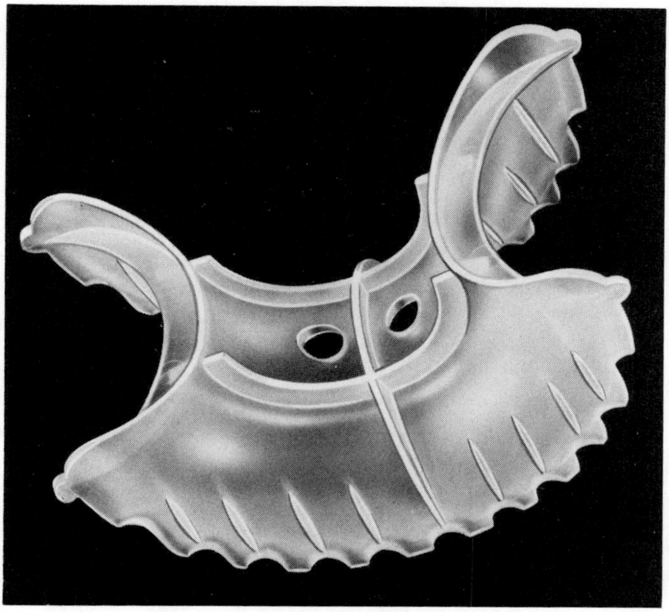

FIG. 24-14. Plastic Super Intalox saddle. (Courtesy of Norton Co.)

Pressure drop is also a function of liquid rates. As the liquid rate is increased, liquid film over packings becomes thicker and gives less area for gas flow.

3. Corrosion resistance: Materials of construction should be such that neither the gas nor the liquid will have any corrosive effect on packing materials. They should be inert; refer to Table 24-6 for chemical resistance to materials. Ceramic packings and plastics are commonly used.

4. Large specific area: Packings should have large specific area, that is, large area per cubic foot. Larger specific area means larger area available for contact between gas and liquid, and therefore more efficient absorption operation.

5. Shape and fluid distribution characteristics: Shape of the packings should be such that they do not create a channeling effect and do not nest into each other; utilization of the surface area should be maximum. Packings must let the gas flow through the voids

without loading or flooding. For efficient absorption, pressure drop through a packed bed should be due to skin friction rather than contraction or expansion of the gas stream. Shapes should create turbulent contact between the gas and the liquid and should minimize stagnant areas and fouling.

6. Temperature resistance: When scrubbing is to be done at high temperatures, thermal resistance of the packing material must be considered.

7. Structural strength or embrittlement resistance: Packings must be strong enough to withstand normal loads during installation, service, and physical handling, and also thermal fluctuations. Packings in the lower section of a tower must be capable of withstanding the loads of the packings in the upper section.

8. Weight: The weight of the packing should be light enough so as not to substantially add to the materials required for supporting grids, shell and foundation, and tower supports.

9. Design flexibility: Efficiency of scrubbing changes as liquid and gas rates are changed. A scrubber must be designed to account for these variations so that they do not affect the efficiency substantially.

LOADING AND FLOODING

The diameter of packed towers is restricted by pressure drop and liquid entrainment by the gas. These conditions are termed loading and flooding.

When gas passes through packings without liquid flowing over them, the graph of log ΔP versus log G is a straight line with a slope of 2. Here G is mass flow rate in pounds per square foot per hour. As the packing is irrigated, the pressure drop increases due to there being less area available for flow, but the slope of log ΔP versus log G remains almost the same. As the liquid flow rate is further increased, there is a break in the slope, which increases. This point is called the loading point. The disadvantage of loading

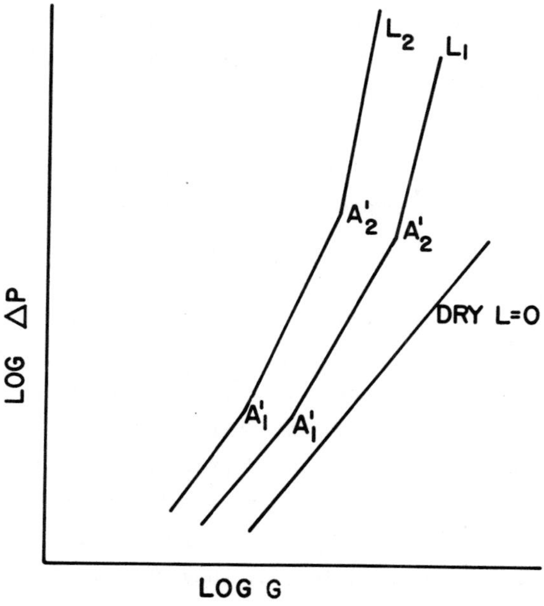

FIG. 24-15. Loading and flooding conditions in packed tower. A_1' is loading point. A_2' is flooding point.

is the increased pressure drop. Further increases in gas velocity, however, cause a second break in the slope of the curve. The slope again increases--this time substantially. This point is called "flooding." At this condition, liquid fills the voids in the packings and becomes a continuous phase with gas bubbling through. Besides increasing pressure drop, high entrainment of liquid in the gas occurs.

In Fig. 24-15, loading is indicated by point A_1 and flooding is indicated by point A_2.

PACKING FACTOR CONCEPT

It has been indicated that capacity of the bed is limited by flooding and a further correlation was given to estimate flooding velocity. The term F in the ordinate (Fig. 24-16) is called packing factor.

Sherwood first gave the correlation for a flooding condition in 1938 [7]. In that correlation, a_p/ϵ^3 was taken as the packing factor, where a_p is the specific area of packing in square feet per cubic foot and ϵ is the void fraction. But the active area of packings changes substantially with variations in liquid loading and changes in packing characteristics such as surface shielding, wettability, support plate blockage, and wall effects. Therefore, a_p/ϵ^3 is really not a correct measure of packing factor. Instead, packing factor should be calculated so that values obtained fit the observed flooding curve. This was done by Lobo in 1945 [8]; Leva, thereafter, added curves for fixed pressure drops [9]. Just as Lobo improved the packing factor worked out by Sherwood, Eckert found in 1961 that packing factors obtained for fixed pressure drop curves change from those obtained for pressure drop at the flooding point. Such variations are shown in Table 24-7 [10]. If such data are available for packing the engineer has selected, they should be used for design purposes. Packing factors are also given in Table 24-8.

It may be observed that packing factors, in general, are less than the actual values of a_p/ϵ^3. However, one may wonder why Raschig rings of 1/4-in. and 3/8-in. size give much larger values of packing factor compared to actual values of a_p/ϵ^3. It should be remembered that these values are obtained only after extrapolation. No other explanation can be given for such a large positive variation between two values. When packing factors are not available, values of a_p/ϵ^3 may be taken without much error when the diameter is one-fourth power dependent.

Capacity of a scrubber as represented by its diameter can be determined based on its limitations due to flooding conditions for gas flow rates.

Figure 24-16 indicates the generalized correlation for estimating gas velocity at flooding and at different pressure drops. Values of packing factor F should be obtained from Table 24-8. There are no correlations to enable one to estimate "loading" condition theoretically. However, it is safe to assume that conditions below 40 to 70 percent of flooding are those below loading as well.

TABLE 24-7

Packing Factors

Sherwood abscissa (ΔP = in.H_2O/ft depth)	→ 0.5			0.5-3.75			3.75 → ∞		
	0.5	1.0	1.5	0.5	1.0	1.5	0.5	1.0	1.5
Intalox saddles, ceramic									
1 in.	–	97	98	–	91	93	–	84	75
1-1/2 in.	50	49	52	53	49	48	40	39	35
2 in.	38	39	40	39	34	33	31	31	30
Raschig rings, ceramic									
1 in., 1/8-in. wall	165	155	156	175	153	144	135	120	110
1-1/2 in., 3/16-in. wall	92	88	90	90	82	80	75	65	60
2 in., 1/4-in. wall	73	73	72	69	64	62	52	50	47
Pall rings, steel									
1 in., 0.024 wall	52	48	47	54	52	53	45	42	42
1-1/2 in., 0.030 wall	30	28	29	39	36	34	34	31	27
2 in., 0.036 wall	25	28	23	26	24	22	21	20	20
Raschig rings, steel									
1 in., 1/16-in. wall	158	144	142	139	122	101	89	77	67
1-1/2 in., 1/16-in. wall	94	85	80	75	65	65	40	43	42
2 in., 1/16-in. wall	72	71	66	60	54	50	38	38	37

TABLE 24-8

Packing Selection Data
(Data to be used for guide purpose only. Manufacturers should be contacted for exact data.)

Packing	Size (in.)	Approx. pcs/ft³	Weight (lb/ft³)	Surface area (ft²/ft³)	Percent void fraction	Packing factor	Price ($/ft³)
Heilex (polypropylene)	2	212	4	30	93	21	8.33
	3	52	3	23	95	14	5.54
Hypak (metal)	1	850	19[a]	--	96	43	32.50[b]
	2	107	14[a]	--	97	18	31.70[b]
	3	31	13[a]	--	97	15	28.00[b]
Pall rings (metal)	5/8 x 0.018 thick	5,950	38[a]	104	93	70	81.25[b]
	1 x 0.024 thick	1,400	30[a]	63	94	48	
	1-1/2 x 0.030 thick	375	24[a]	39	95	28	33.23[b]
	2 x 0.036 thick	170	22[a]	31	96	20	
	3-1/2 x 0.048 thick	33	17[a]	20	97	16	
Pall rings (polypropylene)	5/8	6,050	7	104	87	97	
	1	1,440	5.5	63	90	52	
	1-1/2	390	4.5	39	91	40	8.25
	2	180	4.5	31	92	25	
	3-1/2	33	4	28	92	16	3.75
Partition rings, stoneware, square set	3 x 3	64	63	37	55	--	150.00
	4 x 3	36	63	29	55	--	240.00
	4 x 4	27	54	28	61	--	300.00
	6 x 4	12	48	19	66	--	
	6 x 6	8	48	18	66	--	900.00
Lessing rings, stoneware or porcelain (dumped)	1	1,300	50	69	66	--	7.50[c], 10.00[d]
	1-1/4	650	56	53	62	--	6.90[c], 9.20[d]

	1-1/2	350	58	40	60	--	6.80[c], 9.10[d]
	2	150	49	32	67	--	6.70[c], 9.00[d]
Berl saddles	1/4	113,000	55	274	63	900	
(porcelain and ceramic)	1/2	17,600	54	155	64	240	24.00[c], 32.00[d]
	3/4	5,000	48	98	68	170	15.00[c], 20.00[d]
	1	2,300	48	79	68	110	9.50[c], 12.00[d]
	1-1/2	660	45	46	70	65	8.70[c], 11.00[d]
	2	250	38	32	75	45	7.00[c], 9.30[d]
Berl saddles (steel)	1	2,500	87	85	83	--	
	2	825	60	58	88	--	
Intalox saddle (ceramic)	1/4	117,500	54	300	75	725	
	3/8	49,800	50	240	76	330	
	1/2	20,700	45	190	78	200	28.35
	3/4	6,500	44	102	77	145	
	1	2,385	44	78	77	98	10.85
	1-1/2	709	42	59	80	42	
	2	265	42	36	79	40	
	3	53	37	28	80	22	7.25
Intalox saddle (polypropylene)	1	1,620	6	63	91	30	
	2	190	3-3/4	33	93	20	7.65
	3	42	3-1/4	27	94	15	
Tellerettes							
(1-13/16 in. OD x 3/4 in. H)	1	1,000	7-1/2	55	87	40	12.50
(2-3/4 in. OD x 1 in. H)	2	275	3.9	38	93	20	7.30
(3-3/4 in. OD x 1-1/2 in. H)	3	102	5	30	92	15	5.25
Raschig rings	1/4	88,000	58	220	61	1,600	
(porcelain and ceramic)	3/8	26,000	60	148	60	1,000	
	1/2	10,700	52	114	65	580	14.00[c], 19.00[d]
	5/8	5,800	52	80	65	380	12.00[c], 16.00[d]
	3/4	3,000	44	72	70	255	10.00[c], 13.30[d]
	1	1,330	44	58	70	155	7.00[c], 9.30[d]

TABLE 24-8 (Continued)

Packing	Size (in.)	Approx. pcs/ft³	Weight (lb/ft³)	Surface area (ft²/ft³)	Percent void fraction	Packing factor	Price ($/ft³)
[Raschig rings (porcelain and ceramic)]	1-1/4	670	40	44	73	125	6.40[c], 8.50[d]
	1-1/2	380	42	36	72	95	6.30[c], 8.40[d]
	2	162	38	28	75	65	6.20[c], 8.30[d]
	3	46	34	19	77	37	6.50[c], 9.50[d]
	4	25	42	18	72	--	
Raschig rings (steel)	1/4 x 1/32 wall	88,000	150	236	69	700	
	3/8 x 1/32	26,000	102	158	81	390	
	1/2 x 1/32	11,800	77	128	84	300	105.50
	1/2 x 1/16	11,000	132	118	73	410	
	5/8 x 1/32	7,300	66	112	86	170	
	5/8 x 1/16	7,000	120	106	75	290	
	3/4 x 1/32	3,410	55	83	88	155	64.00
	3/4 x 1/16	3,190	100	72	78	220	
	1 x 1/32	1,440	40	63	92	115	75.10[b]
	1 x 1/16	1,345	73	57	85	137	
	1-1/4 x 1/16	725	62	49	87	110	
	1-1/2 x 1/16	420	50	41	90	83	
	2 x 1/16	180	38	31	92	57	39.10
Single spiral rings, stoneware (square set)	3 x 3	64	58	36	59	--	10.88
	4 x 4	27	52	28	64	--	8.10
	6 x 6	8	48	19	66	--	8.80

[a]Weights indicated are for carbon steel packing. Multipliers: copper 1.20, aluminum 0.37, nickel 1.15, stainless steel 1.05.

[b]Prices for 304 S.S. material packings.
[c]Prices for stoneware material packings.
[d]Prices for porcelain packings.

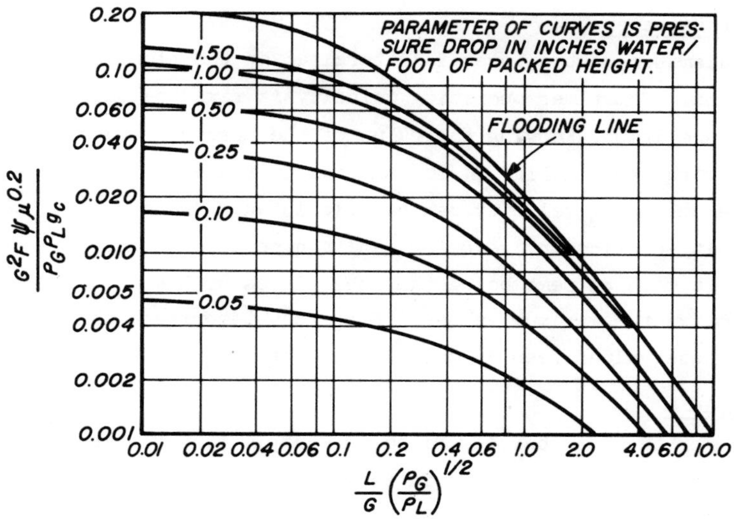

FIG. 24-16. Correlation between gas velocity and pressure drop. L = liquid rate (lb/sec ft^2), G = gas rate (lb/sec ft^2), ρ_L = liquid density (lb/ft^3), ρ_G = gas density (lb/ft^3), F = packing factor, μ = viscosity of liquid (cP), ψ = ratio of density of water to density of liquid, g_c = gravitational constant = 32.2.

Mathematically, gas velocity at flooding is given by [11]

$$\left[\left(\frac{28.6V_F}{(\rho_L/\rho_G)^{1/2}}\right)\left(F\mu_L{}^{0.2}\right)^{1/2}\right]^{1/3} + \left[\frac{Q_F}{7.481}\left(F\mu_L{}^{0.2}\right)^{1/2}\right]^{1/2} = 18.91 \tag{20}$$

where

V_F = gas rate at flooding (ft^3/min ft^2)

Q_F = liquid rate at flooding (ft^3/min ft^2)

F = packing factor

μ_L = liquid viscosity (cps)

ρ_L = liquid density (lb/ft^3)

ρ_G = gas density (lb/ft^3)

Tower Diameter

The diameter of the tower is determined by the flooding-loading conditions. Sometimes operating velocity is fixed by the allowable pressure drop. Knowing the value of the abscissa and the pressure drop parameter in Fig. 24-16, the value of the ordinate is fixed and hence operating velocity can be calculated. The ratio of operating velocity to flooding velocity can then be determined.

It must be remembered that the values of L and G used in Fig. 24-16 are the maximum values. For a countercurrent absorber they are those values seen at the bottom section of the absorber.

Example 4

For the scrubber described in Example 3, determine the diameter of the scrubber. Liquid flow rate is twice the minimum rate. Two-inch plastic Pall ring packings are to be used.

Liquid entering is 2 x 854 = 1,708 lb of H_2O/min

$$SO_2 \text{ absorbed} = 0.95 \times 0.0064 \frac{\text{lb moles of } SO_2}{\text{lb moles of air}}$$

$$= 0.00608 \frac{\text{lb moles of } SO_2}{\text{lb moles of air}}$$

$$\text{Gas flow rate} = \frac{3{,}000}{396} = 7.575 \frac{\text{lb moles of gas}}{\text{min}}$$

For 1.4 percent SO_2 by weight, the gas molecular weight is

$$\frac{100}{(1.4/64) + (98.6/29)} = \frac{100}{0.02187 + 3.4} = \frac{100}{3.42187} = 29.22$$

Therefore,

$$\text{gas density} = \frac{29.22}{396} = 0.0738 \frac{\text{lb}}{\text{ft}^3}$$

$$G = 0.0738 \times 3{,}000 = 221.4 \text{ lb/min}$$

$$G_s = \text{air flow rate (lb/sec)}$$

$$G_s = \frac{3.4}{3.42187} \times 7.575$$

$$= 7.527 \frac{\text{lb moles of air}}{\text{min}}$$

or 218.28 lb of air/min

Therefore,

$$SO_2 \text{ absorbed} = 7.527 \times 0.00608$$

$$= 0.04576 \frac{\text{moles of } SO_2}{\text{min}}$$

$$= 2.928 \frac{\text{lb of } SO_2}{\text{min}}$$

Therefore,

$$\text{water leaving scrubber} = 1{,}708 + 2.928 = 1{,}710.928 \text{ lb of liquid/min}$$

$$\simeq 1{,}711$$

$$\frac{L}{G}\left(\frac{\rho_g}{\rho_L}\right)^{1/2}$$

$$= \frac{1{,}711}{221.4}\left(\frac{0.0738}{62.4}\right)^{1/2}$$

$$= 0.2657$$

For the plastic Pall ring from Table 24-8, F = 25. For abscissa = 0.2657, ordinate = 0.065. Therefore,

$$\frac{G^2_{\text{flooding}}\psi F \mu^{0.2}}{\rho_g \rho_L g_c} = 0.065$$

For water, ψ = 1.0, μ = 0.8 cps.

$$G^2_{\text{flooding}} = \frac{0.065 \times 62.4 \times 0.0738 \times 32.2}{1 \times (0.8)^{0.2} \times 25}$$

$$= 0.403$$

$$G_{\text{flooding}} = 0.634 \frac{\text{lb}}{\text{sec ft}^2}$$

$$G_{\text{operating}} = 50\% \text{ that at flooding}$$

Therefore,

$$G_{\text{operating}} = 0.317 \frac{\text{lb}}{\text{sec ft}^2}$$

$$G_{\text{entering}} = 221.4 \frac{\text{lb}}{\text{min}}$$

$$= 3.69 \frac{\text{lb}}{\text{sec}}$$

Therefore the area of cross section = 3.69/0.317 = 11.64 ft^2, and the diameter is

$$\frac{\pi}{4}D^2 = 11.64$$

$$D^2 = 14.82 \text{ ft}^2$$

$$D = 3.84 \text{ ft}$$

Pressure Drop

Pressure drop across the packed bed is very low. Figure 24-16 can be used to estimate the pressure drop. Pressure drop at flooding condition is between 1/2 and 4 in. per foot of packed length, depending on type and size of packing used [12]. The average value of the pressure drop is approximately 1 in.wc/ft of packed length. Packing manufacturers generally are able to give pressure drops for different gas and liquid flow rates for air-water systems.

Example 5

Estimate the pressure drop for the conditions mentioned in Example 4.

$$\text{Ordinate} = \frac{G^2 F \psi \mu_L^{0.2}}{\rho_g \rho_L g_c}$$

$$G = \frac{3.69}{11.64} = 0.317 \frac{\text{lb}}{\text{sec ft}^2}$$

$$G^2 = 0.100$$

Therefore,

$$\text{Ordinate} = \frac{0.1 \times 25 \times 1 \times 0.8^{0.2}}{0.0738 \times 62.4 \times 32.2}$$

$$= 0.01612$$

and

$$\text{Abscissa} = 0.2657$$

ΔP is slightly less than 1/4 in.wc/ft.

Equivalent Diameter

Packings are sold in different sizes. The sizes indicated by manufacturers are actually nominal sizes and do not necessarily represent any particular dimension. It would therefore not be proper to compare performance of different packings based on nominal sizes. In calculations of HTU or pressure drop, an important design parameter is the Reynolds number, defined by DeG/μ where De is the equivalent diameter in feet and G is the gas flow rate (lb/hr ft^2). The equivalent diameter for packings can be defined by

$$De = \frac{6(1 - \epsilon)}{a_p} \tag{21}$$

When a single sphere or cylindrical pellets are considered, the surface area:volume ratio is given by 6/D. When gas-phase catalytic reactors are considered, a_p is easy to calculate, it being the total area of packings. However, in the case of a scrubber, a_p varies depending on the surface wetted. It may be remembered that this was also the consideration in determining packing factor. Hence, it is reasonable to assume that a_p is a product of the packing factor and ϵ^3. Since packing manufacturers give data on packing factor as well as ϵ, specific surface area (a_p) can be calculated and the equivalent diameter of the packing can be determined.

Height of Packed Bed

The capacity of the packed tower is given by flooding and pressure drop considerations. Efficiency and rate studies lead to the determination of packed height.

If solvent and solute are allowed to remain in contact for a long period of time, they will attain an equilibrium condition. At equilibrium, the concentration at the interface and the concentration in bulk are equal. This means that the driving force causing absorption is zero. For 100 percent efficiency, tower packing height must be infinite. Packed towers, therefore, cannot be designed for 100 percent efficiency.

The mass transfer rate at which absorption takes place depends on the amount of packing surface exposed. This is a function of packing size, shape, liquid rate, and distribution over the packing surface. The height of a packed tower at given gas and liquid rates indicates both rate and efficiency of absorption.

From a material balance over a small section of a countercurrent scrubber,

$$d(Gmy) = K_G a(y - y^*)dZ \tag{22}$$

Gm is gas flow in lb moles/hr ft^2. From this equation, it follows that

$$\frac{Gmdy}{1 - y} = \frac{[K_G a(1 - y)_m][y - y^*]dZ}{(1 - y)_m} \tag{22a}$$

In air pollution studies, most cases involve mass transfer factor based on gas film resistance, and the concentration of objectionable component is very small.

$$K_G = k_G \text{ and } 1 - y \simeq 1$$

$$No_G = N_G = \int_{y_2}^{y_1} \frac{dy}{y - y^*} \tag{23}$$

For such dilute concentrations, Henry's law follows:

$$y^* = H'x \tag{12a}$$

This reduces the equation to

$$No_G = \frac{y_1 - y_2}{(y - y^*)_{1m}} \tag{24}$$

where $(y - y^*)_{1m}$ is log mean difference

$$= \frac{(y_1 - y^*) - (y_2 - y^*)}{\ln(y_1 - y^*)_1/(y_2 - y^*)_2} \tag{25}$$

The overall height of the transfer unit is related to individual film resistance by

$$H_{O_G} = H_G + \frac{HGm}{Lm} H_L \tag{26}$$

$$H_{O_L} = H_L + \frac{Lm}{HGm} H_G \tag{27}$$

(Be sure to convert units of H to those of the slope of the equilibrium line.) A summary of the formulas used to determine height is given in Table 24-9 [13]. Height is calculated as the product of the number of transfer units and the height of the transfer units.

If available, the value of HTU should be obtained from scrubbing at process conditions. If specific data are not available, they can be reasonably predicted by comparing data from a similar system. Adjustments will be necessary due to the effects of variations in the Schmidt and Reynolds numbers.

Most of the packing manufacturers give data on HTU for typical controlled gas and liquid film resistance, and for both films, resistance-controlled absorptions at different flow rates. When data for a specific system are not available, one may use the manufacturers' figures for other systems to predict the required data for a specific case.

Air-water or ammonia-water systems are mainly gas film resistance controlled. Packing manufacturers have data available for these systems. If liquid rates are above MWR, a change in liquids will not significantly affect $k_g a$ (Fig. 24-7). Therefore, the change in HTU will be due mainly to gas flow rate and the properties of the gas.

$$(\mathrm{HTU})_g \propto \left(\frac{DeG}{\mu_g}\right)^{1/2} (Sc)_g^{(0.5-0.66)} \tag{32}$$

where De is the equivalent diameter of the packing.

If liquid rates are below the preloading condition,

$$(\mathrm{HTU})_g \propto \left(\frac{DeG}{\mu_g}\right)^{(1.4-1.3)} L^{-(1.4-1.3)} (Sc)_g^{1/2} \tag{33}$$

The effect of variations in temperature on $(\mathrm{HTU})_g$ is also not significant. In general, for liquid-film resistance-controlled absorption systems,

$$(\mathrm{HTU})_L \propto \left(\frac{DeL}{\mu_L}\right)^{1/4-1/2} (Sc)_\ell^{1/2} \tag{34}$$

TABLE 24-9

NTU	HTU	Height Z = HTU x NTU	
$N_G = \int_{y_2}^{y_1} \frac{(1 - y)_{ln}}{(1 - y)(y - y_i)} dy$	$H_G = \frac{Gm}{k_g aP(1 - y)_{ln}}$	$Z = \frac{Gm}{k_g aP(1 - y)_{ln}} \int_{y_2}^{y_1} \frac{(1 - y)_{ln} dy}{(1 - y)(y - y_i)}$	(28)
$N_L = \int_{x_2}^{x_1} \frac{(1 - x)_{ln}}{(1 - x)(x_i - x)} dx$	$H_L = \frac{Lm}{k_\ell aC(1 - x)_{ln}}$	$Z = \frac{Lm}{k_\ell aC(1 - x)_{ln}} \int_{x_2}^{x_1} \frac{(1 - x)_{ln} dx}{(1 - x)(x_i - x)}$	(29)
$N_{O_G} = \int_{y_2}^{y_1} \frac{dy}{y - y^*} + \frac{1}{2} \ln \frac{1 - y_2}{1 - y_1}$	$H_{O_G} = \frac{Gm}{K_G aP(1 - y)_{ln}}$	$Z = \frac{Gm}{K_G aP(1 - y)_{ln}} \int_{y_2}^{y_1} \frac{(1 - y)_{ln} dy}{(1 - y)(y - y^*)}$	(30)
$N_{O_L} = \int_{x_2}^{x_1} \frac{dx}{x^* - x} + \frac{1}{2} \ln \frac{1 - x_2}{1 - x_1}$	$H_{O_L} = \frac{Lm}{K_L aC(1 - x)_{ln}}$	$Z = \frac{Lm}{K_L aC(1 - x)_{ln}} \int_{x_2}^{x_1} \frac{(1 - x)_{ln} dx}{(1 - x)(x^2 - x)}$	(31)

$(1 - y)_{ln}$ is log mean of $(1 - y)$ and $(1 - y^*)$.
$(1 - x)_{ln}$ is log mean of $(1 - x)$ and $(1 - x^*)$.

TABLE 24-10

Summary of Qualitative Effects of Operating Variables on Film Capacity [14]

	k_La	H_L	k_Ga	H_G
Gas mass velocity	No effect if below loading point. Increases above loading point.	No effect if below loading point. Decreases above loading point.	Increases very markedly.	Increases less markedly.
Liquid mass velocity	Increases very markedly.	Increases less sharply.	Little or no effect.	Little or no effect.
Packing size	Usually increases slightly as packing size decreases.	Usually decreases slightly as packing size increases.	Increases as packing size decreases.	Decreases as packing size decreases.
Packing arrangement	Usually no effect.	Usually no effect.	Usually no effect.	Usually no effect.
Packed height	No effect.	No effect.	Increases slightly.	Increases very little.
Liquor distribution	Sensitive with some packings.	Sensitive with some packings.	Less sensitive than k_La.	Less sensitive than H.
Temperature	Increases very markedly.	Decreases quite markedly.	Little or no effect.	Little or no effect.

In contrast to gas-film resistance-controlled absorptions, the effect of temperature variation on HTU_ℓ is substantial.

$$HTU_\ell \propto \exp(-0.013\ \Delta T) \tag{35}$$

where ΔT in degrees Fahrenheit is the variation from the reference temperature. References for data on different absorption systems are given in Perry's Chemical Engineer's Handbook [4] as well as in Wet Scrubber System Study [13].

Example 6

Fellinger has given H_{O_G} for an NH_3-air-H_2O system using Raschig rings and Berl saddles (Sc.D. thesis, M.I.T., 1941). At a water rate of L = 500 lb/hr ft^2 and a gas rate of G = 500 lb/hr ft^2, H_{O_G} = 2.4, for 1-1/2 in. Berl saddles. Estimate H_{O_G} for the same system using 1-in. Tellerette packing at G = 700 lb/hr ft^2 and L = 1,500 lb/hr ft^2. Ammonia is moderately soluble in water, and absorption is mostly gas film controlled.

From Table 24-8,

1-1/2 in. Berl saddle $F = 70$
void fraction $\epsilon = 0.65$

1-in. Tellerette $F = 40$
void fraction $\epsilon = 0.87$

a_p Berl saddle $= 70 \times 0.65^3 = 19.22\ ft^2/ft^3$
a_p 1-in. Tellerette $= 40 \times 0.87^3 = 26.34\ ft^2/ft^3$

Using Eq. (21), 1-1/2 in. Berl saddle equivalent diameter

$$De = \frac{6(1 - \epsilon)}{a_p}$$

$$= \frac{6 \times 0.35}{19.22} = 0.109 \text{ ft}$$

For 1-in. Tellerette,

$$De = \frac{6(1 - \epsilon)}{a_p}$$

$$= \frac{6 \times 0.13}{26.34}$$

$$= 0.0296 \text{ ft}$$

If we consider power for Reynolds number and liquid rate in Eq. (33),

$$\text{HTU}_g \text{ (1-in. Tellerette)} = \left(\frac{0.0296}{0.109} \times \frac{700}{500}\right)^{1/3} \times \left(\frac{1{,}500}{500}\right)^{-1/3} \text{HTU}_g \text{ (1}\tfrac{1}{2}\text{-in. Berl saddle)}$$

$$= 0.502 \times 2.4 = 1.20 \text{ ft}$$

NO_x ABSORPTION

Oxides of nitrogen--NO (nitric oxide) and NO_2 (nitrogen dioxide)--are two major oxides receiving attention as air pollutants. Both are toxic and corrosive. Stationary NO_2 emission originates from the following sources:

1. Electrical power plants,
2. Industrial combustion sources,
3. Stationary industrial combustion engines used in pipe lines and gas plants,
4. Domestic and commercial combustion equipment, and
5. Nitric acid plants.

NO_2 can be absorbed in various solutions, e.g., water or alkali (such as sodium hydroxide, sodium carbonate, and urea).

The following chemical reactions take place during absorption [15].

1. $2NO + O_2 \rightleftarrows 2NO_2$
2. $2NO_2 \rightleftarrows N_2O_4$
3. $2NO_2 + H_2O \rightleftarrows HNO_2 + HNO_3$
4. $N_2O_4 + H_2O \rightleftarrows HNO_2 + HNO_3$
5. $2HNO_2 \rightleftarrows NO + {}^1/{}_2N_2O_4 + H_2O$
6. $3HNO_2 \rightleftarrows 2NO + HNO_3 + H_2O$
7. $NO + {}^3/{}_4O_2 + {}^1/{}_2H_2O \rightleftarrows HNO_3$

The principal chemical reactions are oxidation of nitric oxide, decomposition of nitrous acid, and absorption of nitric dioxide and nitrogen tetroxide. It may be noted that nitric oxide cannot be absorbed, but when nitrogen dioxide is absorbed to form nitrous acid, it decomposes to form insoluble nitric oxide.

The time required for half NO to be oxidized to NO_2 was calculated by First and Viels, and values are given in Table 24-11 [16] and Fig. 24-17. It may also be noted that oxidation of NO to NO_2 is not an easily reversible reaction at ambient temperature; it is slow, and it may be a rate-controlling step.

From Table 24-11 and the reactions shown, the following conclusions can be reached:

1. For high-efficiency removal of NO_2, long retention time is required to oxidize NO $\rightarrow$ NO_2, hence long or multistage scrubbers are required.
2. When the influent concentration of NO is small, it is impossible to reduce the effluent concentration below a few hundred parts per million for practical sizes of scrubbers.
3. Scrubbers will have an efficiency of 50 to 60 percent absorption based on NO_2 + N_2O_4. Theoretically, the maximum amount of NO_2

TABLE 24-11

NO concentration in air (ppm)	Time required for half NO to be oxidized to NO_2 (min)
20,000	0.175
10,000	0.35
1,000	3.5
100	35
10	350
1	3,500

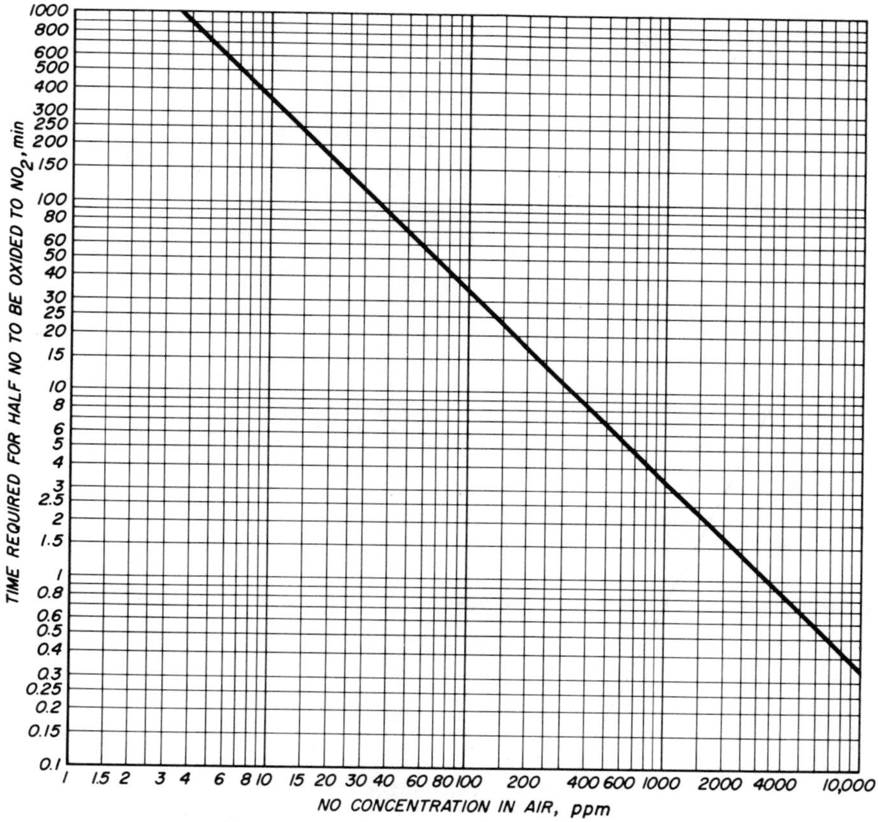

FIG. 24-17. Time required for half NO to be oxidized to NO_2.

+ N_2O_4 that can be removed is 66.66 percent (for every 3 moles of NO_2 absorbed, 1 mole of insoluble NO is formed).

Attempts have been made to increase the oxidation rates of NO by adding strong oxidizing agents such as potassium permanganate, hydrogen peroxide, and chromic acid. It was found that $KMnO_4$ increases efficiency by approximately 20 percent. However, this increase still makes the oxidation time of NO beyond the practical range when concentrations are low.

Alkaline solutions, such as sodium hydroxide, ammonium hydroxide, and soda ash are also used instead of water as a scrubbing liquid.

This change may only have very slight effect on scrubbing efficiency since back pressure of solute from the solvent is seldom a major resistance to NO_x absorption. These solutions are, however, used to reduce fresh water requirements and reduce the waste water volume by permitting recirculation of the scrubbing liquid. Chambers and Sherwood [17] suggested that a gas-phase reaction might occur during the process, and this was confirmed by Peters et al. [18].

NO_2 removal by wet scrubbing can be increased by the following process conditions:

1. Low liquor temperature,
2. Long solute-solvent contact time, and
3. NO_2 concentration at scrubber inlet to be as high (i.e., as undiluted) as possible.

Scrubber Design for Low Concentration of NO_x

Bowman et al. [19] tested a 30-in. diameter, 5-ft high scrubber for treating a small influent concentration of NO_2. Figures 24-18 and 24-19 show the performance of the scrubber and can be used for designing scrubbers with conditions similar to those of the test.

Scrubber Design for High Concentration of NO_x

The overall absorption reaction is given by $3NO_2 + H_2O \rightleftarrows 2HNO_3 + NO$. The equilibrium constant for this reaction is

$$K = \frac{P_{NO} a^2_{HNO_3}}{P^3_{NO_2} a_{H_2O}} \tag{36}$$

where a_{HNO_3} and a_{H_2O} represent the activities in liquids.

Let

$$K = K_1 K_2 \tag{37}$$

where

$$K_1 = \frac{P_{NO}}{P_{NO_2}} \tag{38}$$

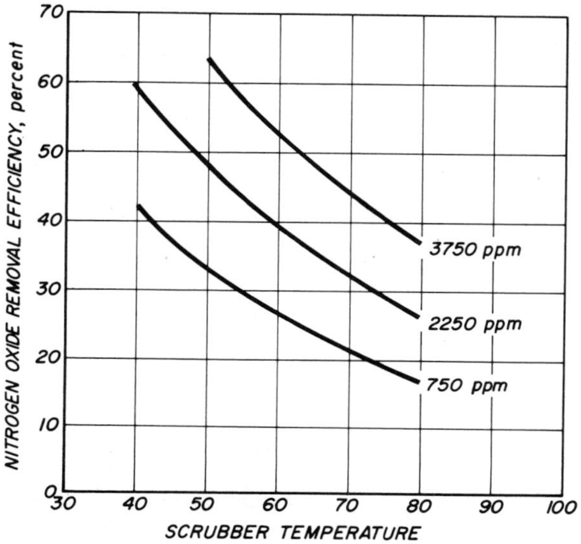

FIG. 24-18. Scrubber performance for NO_2 removal (15 gpm and 330 cfm).

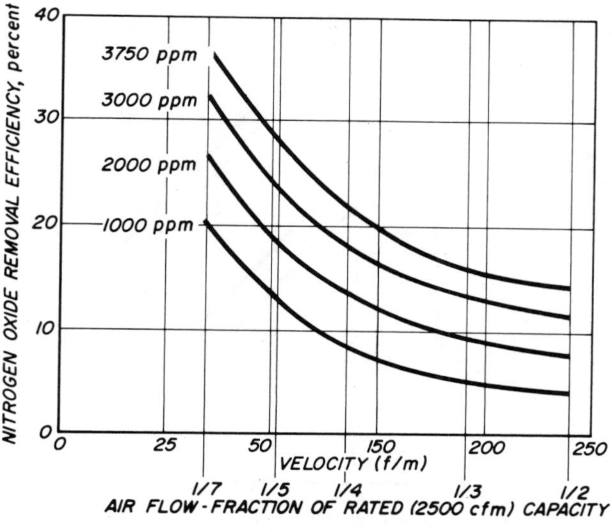

FIG. 24-19. Scrubber performance for NO_2 removal (liquor:H_2O at 15 gpm and 76°F).

and

$$K_2 = \frac{a^2_{HNO_3}}{a_{H_2O}} \tag{39}$$

when the equilibrium constant K_3 (atm^{-1}) between nitrogen dioxide and nitrogen tetroxide, as per the equation $2NO_2 \rightleftarrows N_2O_4$, is given by Wenner [20]

$$\log_{10} K_3 = \frac{2993}{T} - 9.226 \tag{40}$$

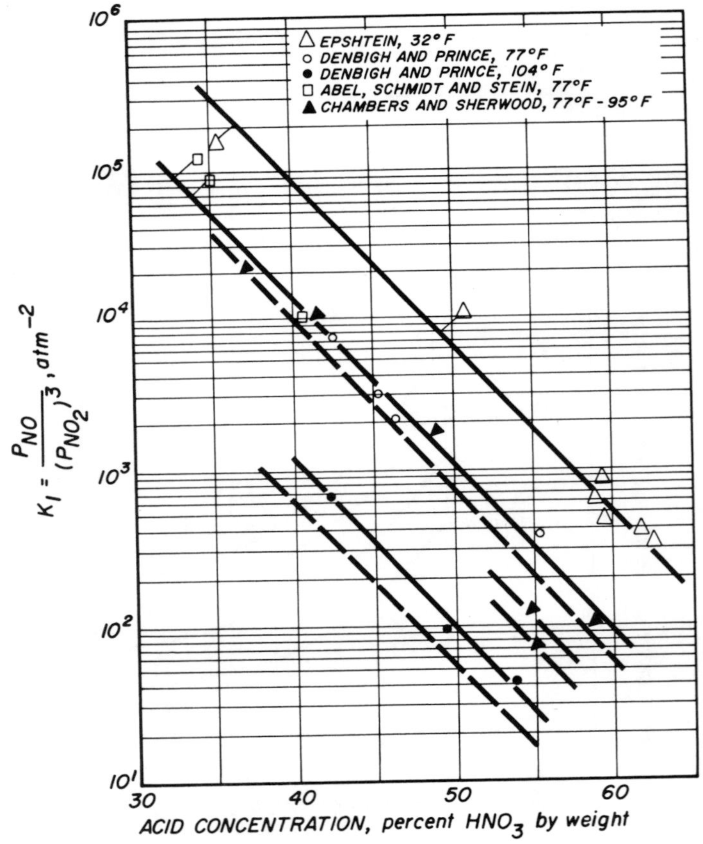

FIG. 24-20. Equilibrium vapor pressures of NO_2 and NO over nitric acid. (From Ref. 20, with permission of McGraw-Hill Book Company.)

where T is in degrees Kelvin. From the above equation, values can be obtained for each equilibrium stage.

In work by Bowman et al. [19] and Krastev and Georgieva [22], it was indicated that gas velocity through the scrubber should be very low, 1 ft/sec or less. The above work and work by First and Viels [16] illustrate that the liquor velocity should be of the order of 2 to 5 gpm/ft^2. From these values, the diameter of the scrubber and the liquid flow rate can be determined.

In many industries, the air pollution control standard for NO_x emission is based on total weight rather than on stack concentration.

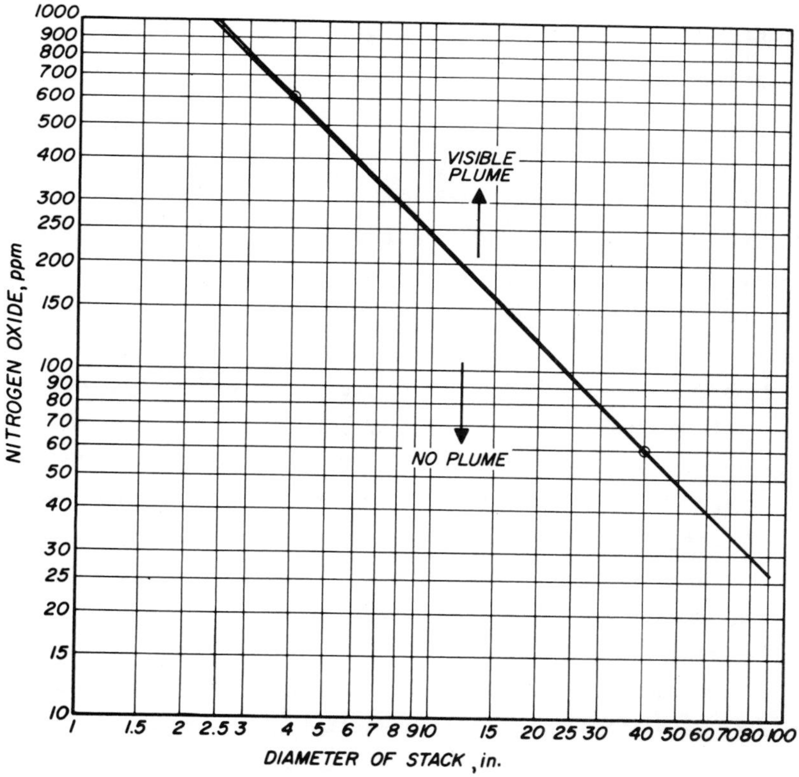

FIG. 24-21. Visible brown plume threshold concentration versus stack diameter. $C_o = 2{,}400/d$, $C_o = NO_2$ threshold concentration (ppm), d = diameter of stack (in.).

Hence, whenever possible, undiluted NO_x can be sent to the scrubber, giving a better driving force in absorption and also permitting the velocity of gas to be scrubbed to be as low as possible. Both of these effects increase the efficiency of scrubbing.

When the gas plume concentration leaving the stack has to be low, dilution of scrubbed gases may be helpful. Hardison [23] has given the following relation between NO_2 visible brown plume threshold concentration in ppm and stack diameter in inches:

$$C_O = \frac{2,400}{d} \tag{41}$$

where C_O is the threshold concentration (ppm), and d is the stack diameter (in.). These data are graphed in Fig. 24-21.

ODOR CONTROL SCRUBBERS

Scrubbers are often used in odor control systems; for example, waste gases from a combined slaughtering and rendering operation have been successfully deodorized in a two-step process. In the first step, gases are filtered. In the second step, they are scrubbed in packed columns. Sea water liquor can give 50 to 80 percent reduction in odor from a fishmeal dryer. However, this reduction may not be high enough. It has been reported that if chlorine is added to the sea water in amounts of 20 ± 5 ppm, the efficiency of the scrubber can be increased to 90 percent [24]. A hypochlorite generator can be used on site to produce sodium hypochlorite, an oxidizing solution, using a brine solution [25]. Figure 24-22 shows a typical odor-abatement scrubber system of this type.

The scrubbing tower is designed in sections. The lower section is the reservoir for the scrubbing solution. A specific level of liquid is maintained in the bottom section. As scrubber liquor overflows the section, it is sent back up to the tower and reused. Minimum liquor level prevents salt and sediment buildup. Fresh sodium hypochlorite solution is also used in a lower section of the scrubber.

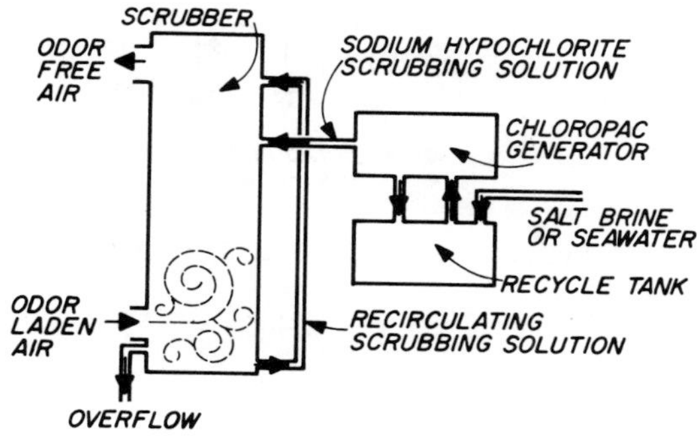

FIG. 24-22. Odor-abatement scrubber system.

PACKED TOWER INTERNALS

Tower internals consist of the following major components: (1) packings, (2) packing support redistributor, (3) liquid distributor, (4) liquid redistributor, (5) packing hold-down plates, (6) retaining plates, and (7) mist eliminator.

PACKINGS

Factors involved in selecting packings were discussed earlier in this chapter.

Packing Support Redistributors

The packing support redistributor supports the packing load. The selection of packing support is based on its (a) physical strength to bear the weight of the packings and (b) capability to provide sufficient free space for both gas and liquid to pass through.

Temperature and chemical corrosive action must be considered in selecting construction materials. Previous practice was to use flat perforated or slotted plates as packing supports. However, these plates have some major disadvantages, particularly very low (25 percent) free space for gas and liquid flows and weak structure. Moreover, the gas and liquid have common openings that create a substan-

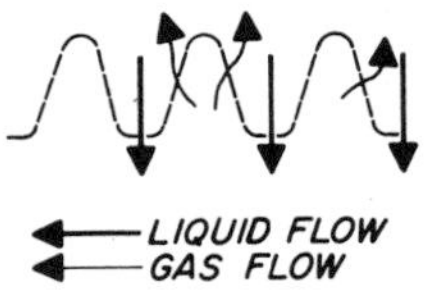

FIG. 24-23. Schematic of gas and liquid flow in gas-injection support plates.

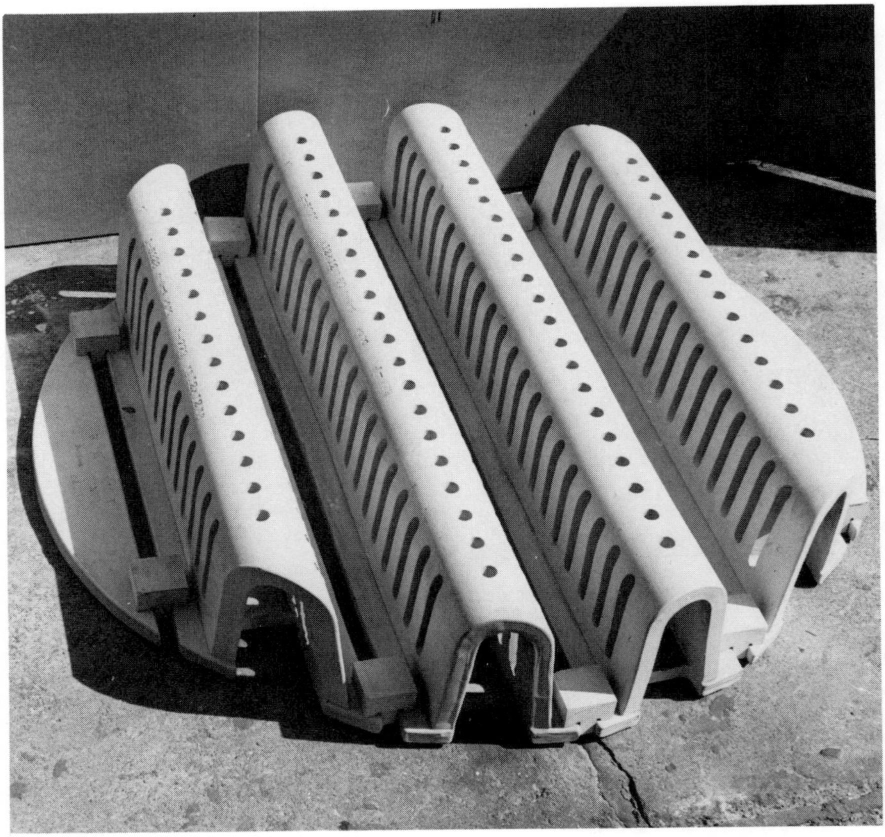

FIG. 24-24. Gas-injection support plate. (Courtesy of Maurice A. Knight Co.)

tial hydrostatic head over the plate. An improvement was made in this design by development of "gas-injection" support plate (Fig. 24-23).

In these plates, gas and liquid are separated by providing gas inlets above the liquid flow level (Fig. 24-24). These support plates can even provide open area greater than the mass sectional area of the tower. The corrugated structure also provides more surface area to hold the weight of packing. These support plates can be made as two- or three-piece assemblies. In highly corrosive applications such as sulfuric acid absorbers, grid bar supports are also used. Packing support plates also function as redistributors.

Liquid Distributors

For efficient diffusion it is necessary to have the packings wet with liquid; the more surface area that is wetted, the higher will be the efficiency of scrubbing. Portions of the bed will remain unirrigated if the liquid is not distributed evenly across the bed. In larger cross-sectional packed beds, therefore, liquid is added through multiple ports. Liquid distributors are used for this purpose.

Liquid distributors are designed with the following considerations:

1. Uniform liquid distribution;
2. High free gas flow area;
3. Resistance to plugging, fouling, etc.;
4. High turndown ratio; and
5. Sectional construction for installation through manways.

Small-diameter towers require relatively more distribution points. For a column with diameter of 2-1/2 ft, 16 distribution ports/ft^2 are recommended, but for 1-1/4 ft diam tower, 32 distribution ports/ft^2 are required.

Liquid distribution is better with smaller packings. In many instances, it is recommended that a few layers of smaller size packing be used at the top, below the liquid distributors.

FIG. 24-25. Spray-type liquid distributor — spray nozzles installed. (Courtesy of Ceilcote Co., Inc.)

Types of Liquid Distributors

Orifice types: 1. Ladder or straight pipe style,

2. Pan style.

Weir type: 1. Trough parting box style,

2. Weir riser style.

Both are gravity types.

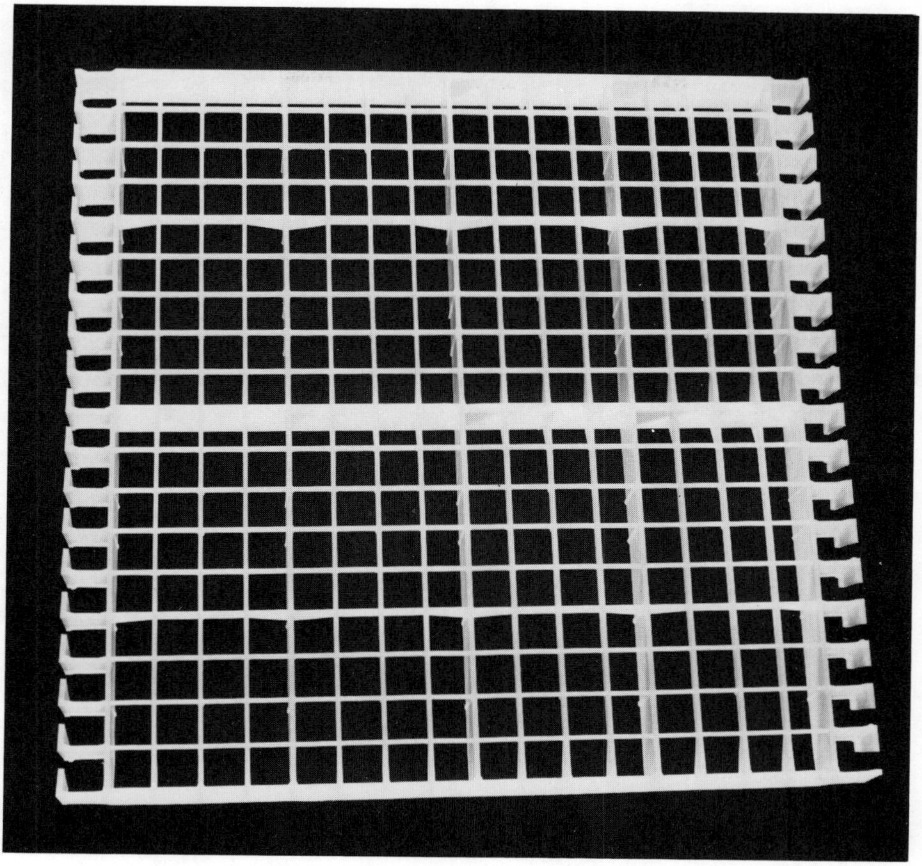

FIG. 24-26. Packing support plates. (Courtesy of Ceilcote Co., Inc.)

Ladder-style orifice liquid distributors provide more area for gas flow. In some instances, nozzles are provided over the orifices. These nozzles can be made of thermoplastic or metal pipes.

Pan-style orifice liquid distributors are generally used for small-diameter towers. Pans are provided with small orifices. A small uniform liquid head is developed that forces the liquid to flow uniformly through the orifices. Pans are made of metal, thermoplastics, FRP, ceramics, and so on.

FIG. 24-27. Trough-parting box-type liquid distributor. (Courtesy of Maurice A. Knight Co.)

In larger-diameter towers, trough-parting box weir distributors are used (Fig. 24-27). Weir-type distributors provide a higher turndown ratio compared to that provided by orifice-type distributors.

Liquid Redistributor

As has been mentioned previously, liquid has a tendency to move toward the tower wall, where void fraction is greater. If the tower is tall, packing in the lower portion of the tower is not effectively wetted, thereby decreasing the scrubbing efficiency. This problem is particularly serious in smaller-diameter packed towers. Wall wipers, a special type of liquid redistributor which pipes the liquid from the wall to the inside section of the tower, are often used to overcome this problem.

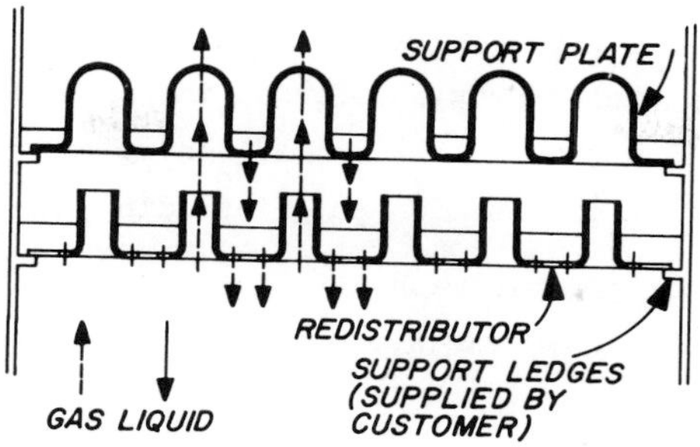

FIG. 24-28. Proper alignment of liquid redistributor and packing support plate.

When packing support plates, particularly gas-injection types, are used, the liquid redistributor must be properly aligned with the support plate (Fig. 24-28).

Packing Holddown Plates (Bed Limiters)

When pressure drop across the bed is high, there is a chance of fluidizing the bed. Upper portions of the bed can be physically lifted up. This may cause an unequal distribution effect in gas and water flow. There is even a possibility of breaking carbon and ceramic packings. These broken pieces can further block free spaces and decrease the mass transfer efficiency if holddown plates are not used on top of the packed bed. These plates can be made of small segments with slots and grids (Fig. 24-29).

The design of holddown plates is based on the following considerations:

1. Low pressure drop,
2. Good liquid distribution, and
3. Throughput capacity equal to at least that of packings.

Retaining Plates

A retaining plate is placed below the liquid distributor. It is not supported by the packing but is fixed to a ring in the scrubber tower.

Mist Eliminator Separators

Mist eliminators are used to remove liquid droplets entrained in the gas stream. Ease of separation depends on the size of the droplets.

Droplets formed from liquid bulk are usually large, up to hundreds of microns in diameter. Droplets formed in condensation or chemical reaction may be less than 1 μm in size. Entrainment removal is possible by a number of methods [26], including the following:

1. Knitted wire or plastic mesh,
2. Swirl vanes or zigzag vanes,
3. Cyclones,
4. Gravity settling chambers,
5. Knockout pots in which gas goes through a 180° reversal,
6. Packed column, and
7. Special design where jets of gas impinge on a target plate.

One of the simplest and most efficient means of mist separation is to use porous blankets of knitted wire or plastic mesh (Fig. 24-29). These pads are light in weight (7 to 12 lb/ft^3). They can be made of any size, shape, and height, and the pressure drop across them is very low. For most processes, pressure drop across these mist eliminators ranges from 0.1 to 1.0 in.wc depending on vapor and liquid loadings and the size of the eliminators. Efficiency of separation generally is very high, 90 percent or better. Construction of the separator vessel for mesh pads is very simple.

Relatively lighter vapor can easily pass through the openings of the porous layers of the wire mesh. Liquid droplets cannot change direction or course as they pass through the mesh openings. As these entrained droplets travel, they impinge directly on the wires. Because of surface tension, liquid droplets adhere to intersections of wire mesh, coalesce into larger droplets, and then, because of their weight, fall and ultimately collect at the bottom of the vessel. Gas leaving the mesh pad is free of moisture droplets.

Knowing the vapor density and density of the liquid entrained and K value [Eq. (43)], optimum velocity can be calculated. Hence, if W_m is lb mole/hr flow rate, the cross-sectional area will be

$$A = \frac{W_m}{3,600} \times \frac{359}{u} \tag{42}$$

$$= 0.09972 \frac{W_m}{u} \tag{42a}$$

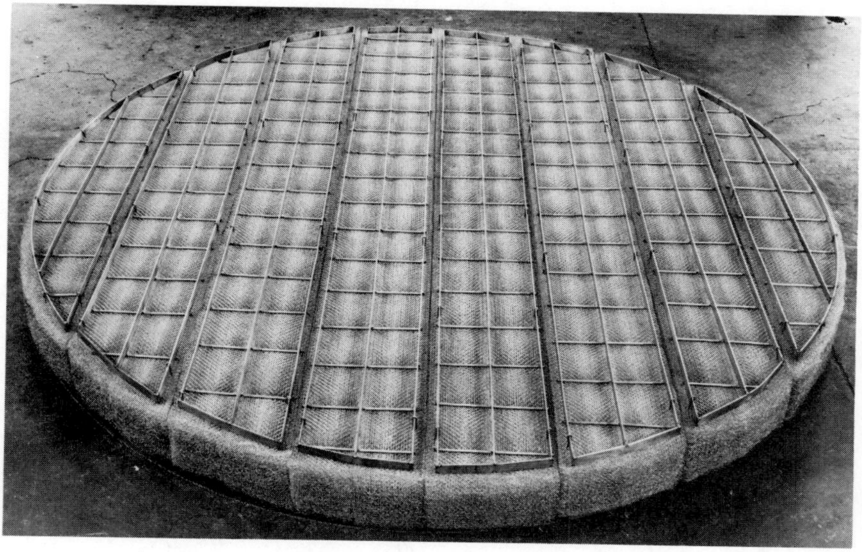

FIG. 24-29. Mist eliminator. (Courtesy of Otto H. York Co., Inc.)

Thickness for most mesh pads is generally 4 to 6 in. If a much finer mist (e.g., 3 to 10 μm) needs to be separated or if high efficiency of separation is desired, the thickness of the mesh pads may vary to 12 in.

Figure 24-30 indicates the general range of vapor velocities and droplet sizes. If vapor velocity is too high, droplets will be carried away in the vapor stream. On the other hand, if the velocity is too low, bubbles will not be formed. If droplets are too small, because of low inertia and high drag, they will be carried away together with the vapor stream.

Design of mist eliminators is relatively simple. The process variables are vapor velocity, vapor density, liquid density, liquid viscosity, surface tension, droplet size, and so on. The linear velocity of the vapor is the most critical design factor. The effect of most of the parameters can be grouped into a single constant K. Together with this constant, only vapor and liquid densities are

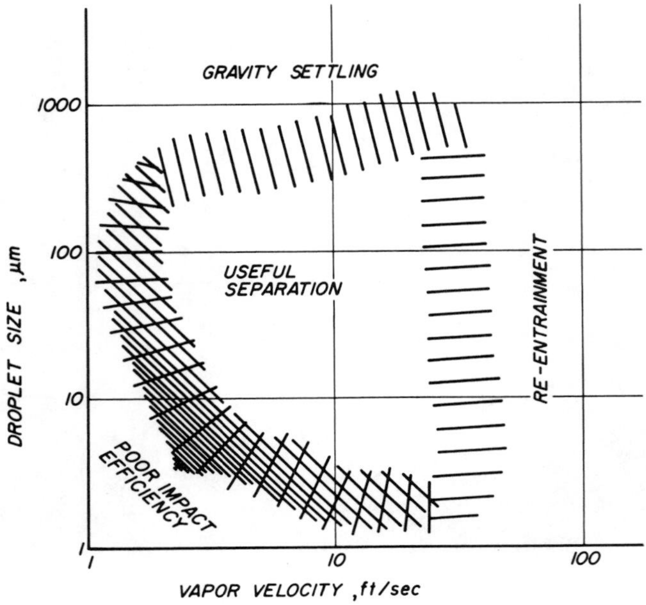

FIG. 24-30.

then required to determine the optimum velocity of vapor passing through the mesh pad. Design velocity for optimum velocity u (ft/sec) is given by

$$u = K\left(\frac{\rho_L - \rho_V}{\rho_V}\right)^{1/2} \tag{43}$$

where K is constant, ρ_L is the density of the liquid (lb/ft^3), and ρ_V is the vapor density (lb/ft^3). Typical values are given in Table 24-12.

Pressure drop through a mesh pad is very low because of the high void fraction of these pads. Hence, the power consumption required to drive vapor through is also very low. Pressure may be higher if liquid loading entrained in the vapor stream is much higher.

The pressure drop across the mist eliminator is given by

$$\Delta P_g = \frac{f_c \ell a_p \rho_G u^2}{g_c \epsilon^3} \tag{44}$$

TABLE 24-12

K	Conditions
0.35	Most oil and gas streams
0.25	Steam at pressures of 1 atm and up
0.20	Most vapors under vacuum
0.15-0.20	Salt, caustic, etc.

Here, ΔP_g is the gas pressure drop across the mesh pad, without entrained load (lb_f/ft^2) [27].

ρ_G = vapor density (lb_m/ft^3)
ϵ = void fraction of mesh pad
g_c = gravitational constant (lb_m ft/lb_f sec^2)
ℓ = mesh pad thickness (ft)
a_p = specific surface area (ft^2/ft^3)
f_c = friction factor, to be estimated from Eqs. (46) and (47)
u = actual superficial velocity (ft/sec)

Values of a_p and ϵ can be obtained from the manufacturer.

Generally, most of the pads are made of 0.06 to 0.011 in. thick wire, and void fraction ranges from 97 to 99 percent. The specific

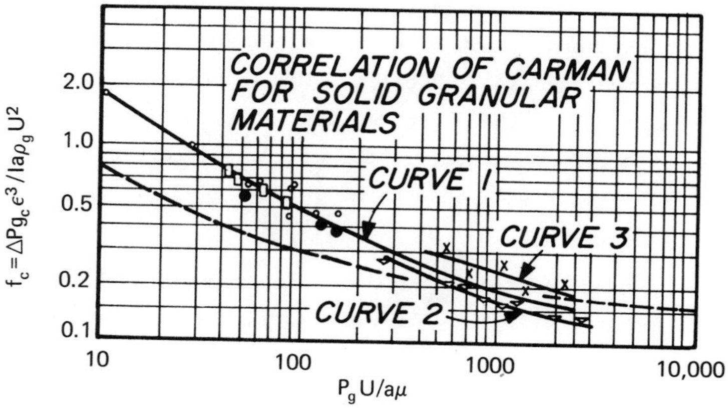

FIG. 24-31.

area of most of the efficient pads is approximately 110 ft^2/ft^3, although some pads have lower specific areas (approximately 45 ft^2/ft^3).

To estimate the liquid load pressure ΔP_L, graphs can be obtained from mesh manufacturers on $u/[(\rho_L - \rho_G)/\rho_G]^{1/2}$ and liquid load as parameters.

$$\Delta P_{total} = \Delta P_{\ell} + \Delta P_G \tag{45}$$

It appears from some graphs in the literature that f_c is a function of a modified Reynolds number. The modified Reynolds number is defined by $\rho_G u/a_p \mu$, where μ is the viscosity of the vapor (lb/ft sec).

For $10 < Re_m < 100$,

$$f_c = 7.5 \ Re^{-0.6} \tag{46}$$

and for $100 < Re_m$

$$f_c = 3.37 \ Re^{-0.43} \tag{47}$$

Venturi Mist Eliminator

From Eq. (43) it can be seen that optimum velocity is a function of gas and liquid densities. Cross-sectional area depends on flow rate. Hence, if the system is to be designed with a wide variation in flow rate or wide variation in vapor densities, the same mist eliminator separator may not be suitable. Venturi mist eliminator separators can be used in such cases. The separator bottom is larger in diameter and the upper section has a small diameter. Therefore, at a certain section vapor will attain a range of optimum velocity and mist will be separated (Fig. 24-32).

Example 7

Design a mist eliminator separator for removing sulfuric acid mist (density 64 lb/ft^3) from air at 25°C and 15 psia. Estimate pressure drop for 3,000-scfm flow across separator. Li-

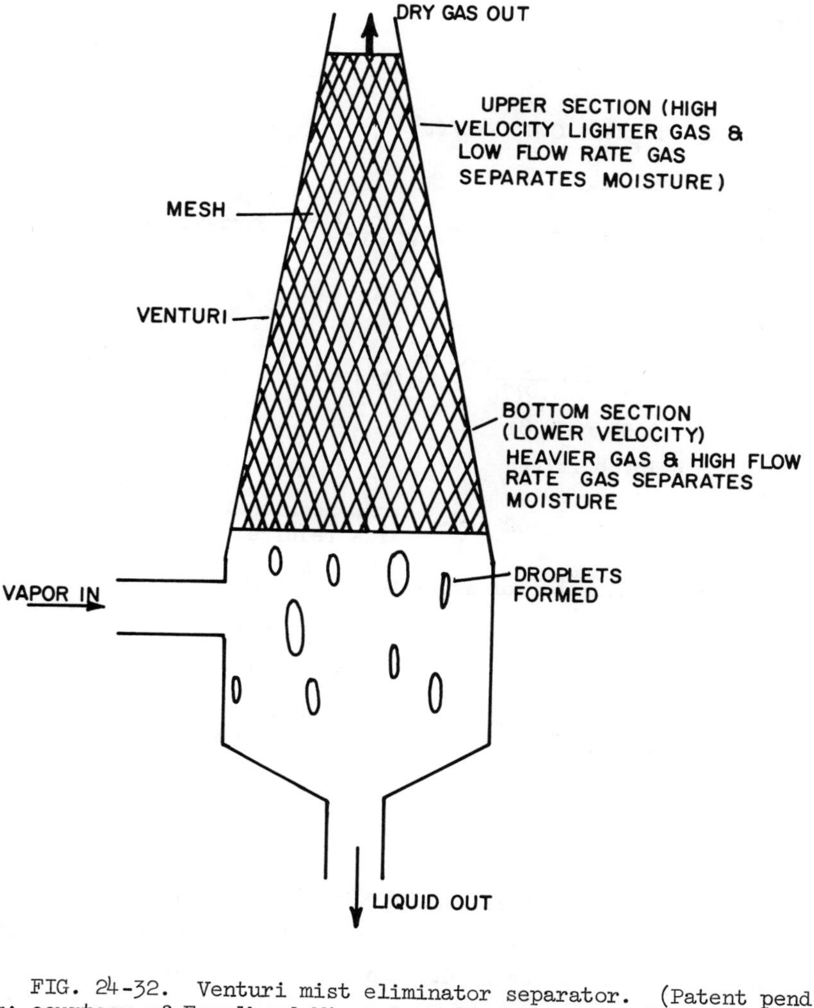

FIG. 24-32. Venturi mist eliminator separator. (Patent pending; courtesy of Engelhard Minerals & Chemicals Corp.)

quid load can be assumed to be very small. Since conventional separators have a limited turndown ratio of approximately 3, a venturi mist eliminator separator is recommended for this application.

Density of gas $\simeq$ 0.077 lb/ft^3

Optimum velocity:

From Eq. (43), $u = K[(\rho_L - \rho_V)/\rho_V]^{1/2}$. From Table 24-12, $K \simeq 0.35$.

Therefore,

$$U = 0.35\left(\frac{64 - 0.077}{0.077}\right)^{1/2} \text{ ft/sec}$$
$$= 0.35 \times 28.81$$
$$= 10.08 \text{ ft/sec}$$

When flow rate is 300 scfm, $\begin{matrix}\text{cross-sectional}\\ \text{area required}\end{matrix} = 0.496 \text{ ft}^2$

Therefore, diameter of separator $= \left(\frac{4}{\pi} \times 0.496\right)^{1/2}$
$= 0.794$ ft
$= 9.53$ in.

When flow rate is 2,000 scfm, $\begin{matrix}\text{cross-sectional}\\ \text{area required}\end{matrix} = \frac{2{,}000}{60 \times 10.08}$

$= 3.30 \text{ ft}^2$

$$D = \left(4 \times \frac{3.30}{\pi}\right)^{1/2} = 2.05 \text{ ft}$$

A venturi mist eliminator with a diameter at the top of approximately 10 in. and a diameter at the bottom of approximately 25 in. can be used. Length of the separator is selected to be 12 in. A mesh pad having a specific area of 100 ft^2/ft^3 and $\epsilon = 0.97$ is selected.

Pressure Drop Calculation:

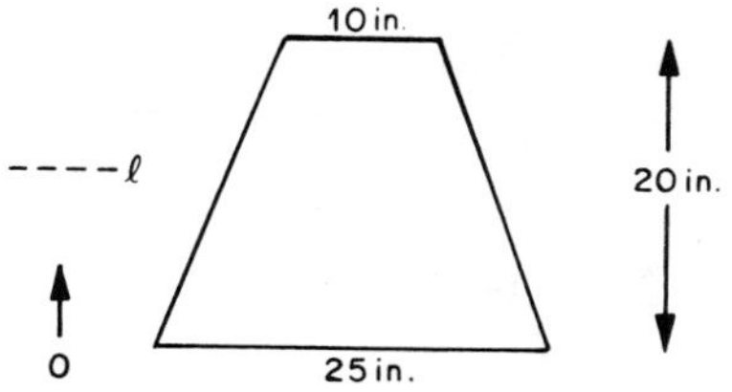

Cross-sectional area at top $= \frac{\pi}{4} \times \left(\frac{10}{12}\right)^2 = 0.545\ \text{ft}^2$

Velocity at top section $= \frac{3{,}000}{60 \times 0.545} = 91.74$ ft/sec

Cross-sectional area at bottom $= \frac{\pi}{4} \times \left(\frac{25}{12}\right)^2 = 3.408\ \text{ft}^2$

Velocity at bottom $= \frac{2{,}000}{60 \times 3.408} = 9.77$ ft/sec

Velocity and diameter of the separator are changing along the entire length.

Reynolds number at the bottom section

$$\text{Re} = \frac{u_\rho}{a_p \mu_g} = \frac{9.77 \times 0.077}{100 \times 0.018 \times 0.000672} = 621.72$$

Reynolds number at the top of separator

$$\text{Re} = \frac{91.74 \times 0.077}{100 \times 0.018 \times 0.000672} = 5839$$

Flow is therefore turbulent.

From Eq. (47), $f_c = 3.37\ \text{Re}^{-0.43}$. Diameter at any height h is

$$D = \frac{25 - \frac{h}{20}(25 - 10)}{12}\ \text{ft}$$

$$\text{Velocity at that section} = \frac{3{,}000 \times 144}{60 \times \frac{\pi}{4}(25 - \frac{3}{4}h)^2}$$

$$= \frac{9167.32}{(25 - \frac{3}{4}h)^2}\ \text{ft/sec}$$

Therefore,

$$\text{Re} = \frac{9167.32 \times 0.077}{100 \times 0.018 \times 0.000672 \times (25 - \frac{3}{4}h)^2}$$

Across small section "dh,"

$$dp = 3.37 \frac{\left[\frac{5.835 \times 10^5}{(25 - \frac{3}{4}h)^2}\right]^{-0.43} \times 100 \times 0.077 \times \frac{9167.32^2}{(25 - \frac{3}{4}h)^4}}{32 \times 0.97^3} dh$$

$$dp = \frac{2.475 \times 10^5}{(25 - \frac{3}{4}h)^{3.14}} dh$$

Integrating, we have

$$\int dp = \Delta P = \int_0^{1.67} \frac{2.475 \times 10^5}{(25 - \frac{3}{4}h)^{3.14}} dh$$

Let $25 - \frac{3}{4}h = \ell$. When

$h = 0, \quad \ell = 25$

$h = 1.67, \ \ell = 23.75$

$$\Delta P = 3.3 \times 10^5 \int_{25}^{23.75} \frac{d\ell}{\ell^{3.14}}$$

$$= -1.54 \times 10^5 \left[\frac{1}{\ell^{2.14}}\right]_{25}^{23.75}$$

$$= 1.54 \times 10^5 [0.001137 - 0.0010195]$$

$$= 1.54 \times 10^5 \times 1.175 \times 10^{-4} \text{ lb/ft}^2$$

$$= 18.095 \text{ lb/ft}^2$$

$$= 0.125 \text{ lb/in.}^2$$

Construction Material

Mesh pads are made of stainless steel (304, 316, 421, 430), Hastelloy, Monel, nickel, and some plastics (e.g., Teflon, polypropylene, PVC, etc.). Table 24-13 indicates applications and also construction materials [28].

Gas outlet and vapor inlet are approximately at a distance of two times the diameter of the mesh pad. These pads are placed between the top and bottom grids and are supported over a retaining annular ring.

TABLE 24-13

Construction Materials for Mist Eliminators

Material	Mesh	Grids	Typical service
Carbon steel	NR	Yes	Petroleum
AISI 304	Yes	Yes	Petroleum, aqueous, mild alkali, mild chemical
AISI 304 ELC	Yes	Yes	
AISI 316	Yes	Yes	Corrosive chemical
AISI 316 ELC	Yes	Yes	H_2SO_4
AISI 430	Yes	Yes	Mesh for magnetic metal pickup in food applications. Mild chemical service
Monel	Yes	Yes	Caustic evaporators
Nickel	Yes	Yes	Caustic evaporators
Carpenter 20	Yes	Yes	Sulfuric acid
Hastelloy	Yes	Yes	Acids, including HCl
Polypropylene	Yes	Yes	Water, acid, alkali 160°F max.
Polyvinyl chloride	NR	Yes	Water, acid, alkali 160°F max.
Teflon	Yes	NR	Hot sulfuric acid, other very corrosive up to 300°F
Metal/Fiberglass	Yes	NA	Mild chemical fine mists
Polyester/Fiberglass	NA	Yes	Acid, salts, not for alkalis

NR = not recommended; NA = not available.

REFERENCES

1. J. O. Hirschfelder, R. R. Bird, and E. L. Spotz: Transactions of the A.S.M.E., 71:921(1949); Chem. Riv., 44:205(1949).
2. Wilke and Lee: Industrial Engineering Chemistry, 47:1253(1955).
3. R. E. Treybal: "Mass Transfer Operations," McGraw-Hill Book Company, New York, 1968, p. 27.
4. J. H. Perry: "Chemical Engineer's Handbook," McGraw-Hill Book Company, New York, 1973, pp. 14-22 to 14-26, 14-36.

5. E. W. Hanf and J. W. McDonald: Chemical Engineering Progress, 71(3):48(1975).

6. Pratt: Transactions of the Institute of Chemical Engineers, 29:282(1950).

7. Sherwood and Holaway: Industrial and Engineering Chemistry, 30:768(1938).

8. Lobo et al.: Transactions of the American Institute of Chemical Engineers, 41:693(1945).

9. M. Leva: Chemical Engineering Progress Symposium Series, 10 (51):52(1954).

10. J. S. Eckert: Chemical Engineering Progress, 66(3):39(1970).

11. F. A. Zenz: Chemical Engineering, 79(25):121(1972).

12. F. A. Zenz: Chemical Engineering, 60(8):176(1953).

13. Wet Scrubber System Study, NTIS Report PB-213 016, July 1972, pp. 5-12, 5-35.

14. M. Leva: "Tower Packings and Packed Tower Design," U.S. Stoneware Company, 1951, p. 55.

15. P. N. Cheremisinoff and K. B. Rao: "Nitrogen Oxide Control for the Chemical Industry--Mass Transfer Techniques," Metrochem 72, American Chemical Society, Fairleigh-Dickinson University, May 13, 1972.

16. M. W. First and F. J. Viels: Paper presented at the 63rd Annual Meeting of the Air Pollution Control Association at St. Louis, Missouri, June 1970.

17. Chambers and Sherwood: Industrial and Engineering Chemistry, 29:1415(1937).

18. M. Peters et al.: A.I.Ch.E. Journal, 1:105(1955).

19. D. H. Bowman, C. J. Kulczak, and J. J. Shulman: Pollution Engineering, 38(Nov. 1974).

20. R. R. Wenner: "Thermochemical Calculations," McGraw-Hill Book Company, New York, 1941.

21. T. K. Sherwood and R. L. Pigford: "Absorption and Extraction," McGraw-Hill Book Company, New York, 1952, p. 372.

22. I. Krastev and A. Georgieva: Khim Ind. (Sofia), 4:147-151 (1968).

23. L. C. Hardison: Journal of the Air Pollution Control Association, 20(6):377(1970).

24. R. M. Bethea, B. N. Murthy, and D. F. Carey: Environmental Science & Technology, 7(6):504(June 1973).

25. Air Pollution Abatement Systems, Engelhard Ind. Bulletin EM-A-236-5M.

26. P. N. Cheremisinoff and R. Young: "Pollution Engineering Practice Handbook," Ann Arbor Science Publishers, Ann Arbor, Mich., 1975, p. 141.

27. O. H. York and E. W. Poppele: Chemical Engineering Progress, 59:6(1963).

28. A.C.S. Industries, "Design Manual."

Chapter 25

PACKED WET SCRUBBERS

J. W. MacDonald
Air Pollution Control Division
The Ceilcote Company
Berea, Ohio

Packed wet scrubbers for air pollution control encompass a group of devices within which an air stream (containing noxious gases and entrained particles of mists, liquids, dusts, or fumes) is passed through a depth of packing material which is irrigated with a scrubbing liquid. This class of equipment contains four basic types, which include several variations and combinations. These basic types are (1) countercurrent-flow packed scrubbers; (2) cocurrent-flow packed scrubbers; (3) vertical and horizontal air washers; and (4) cross-flow packed scrubbers.

Countercurrent packed scrubbers (Fig. 25-1) are used to their greatest advantage in the abatement of noxious or corrosive gases. However, this type is also used to remove entrained liquid, soluble dusts, or mist particles.

In the countercurrent-flow packed scrubber, the gas stream moves upward in direct opposition to the scrubbing liquid stream, which is moving downward through the packed bed. A significant advantage of countercurrent flow is that the gas stream (loaded with contaminants) comes into contact with spent liquor at the bottom of the packed bed, whereas the fresh liquor, coming in at the top of the scrubber, is in contact with the least contaminated gas. This characteristic provides a fairly constant force throughout the packed bed for "driving" the gaseous contaminant into the scrubbing liquid.

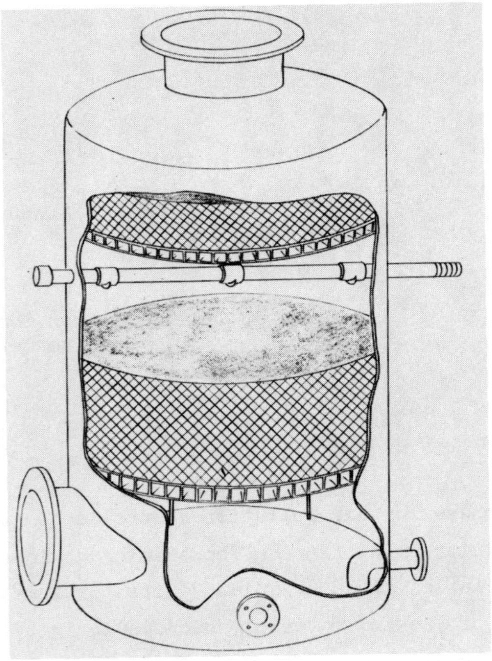

FIG. 25-1. Countercurrent-flow packed scrubber.

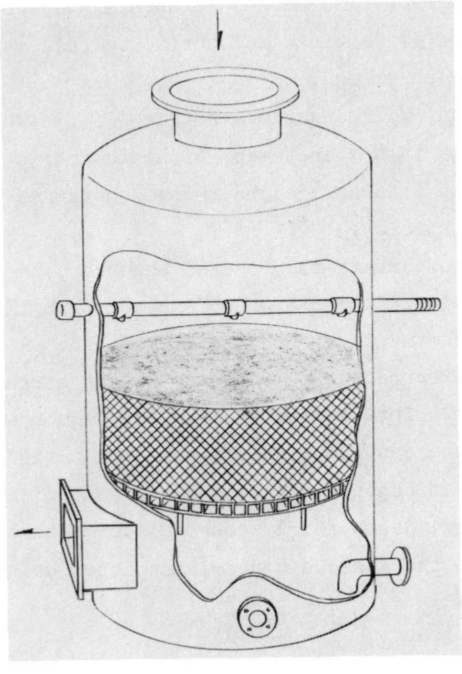

FIG. 25-2. Cocurrent-flow scrubber.

Normally, the capital cost of countercurrent scrubbers is less than that for cross-flow types. Countercurrent-flow scrubbers are usually more expensive to operate than cross-flow scrubbers because of their higher liquid flow rate and higher pressure drop. However, although this is generally true, countercurrent scrubbers are the only scrubbers that are effective in removing low-solubility gases, which are generally regarded as the most difficult air pollution problem for wet scrubbers.

In cocurrent-flow scrubbers (Fig. 25-2) the gas stream and liquid stream move in the same direction, usually vertically and downward through the packed bed. These scrubbers can be operated at high gas rates and at high liquid rates. Cocurrent-flow scrubbers are capable of removing gases of high solubility in water, but are not very efficient for removal of low-solubility gases because the driving force between the gas and liquor diminishes as the gas passes through the scrubber.

These designs are useful where there is limited space for scrubbers, because they can handle very high gas stream velocities. High gas velocities in these scrubbers do not cause flooding, as they would in the countercurrent design, because the gas stream helps to "push" the liquid stream through the packing.

However, this design generally requires a separate entrainment separator. It is for this reason that the cocurrent designs are often used in conjunction with other types of scrubbing units. One such application for removing a soluble gas from an air stream loaded with dust uses two scrubbers in series. The first tower is a cocurrent scrubber which removes the dust particles by impingement and inertial impaction. It also provides some limited gas absorption and in certain applications can provide the necessary preconditioning of the gas stream. After passing through the cocurrent tower, the gas stream enters a countercurrent scrubber where the soluble gas is absorbed in the irrigating liquor.

The vertical air washer (Fig. 25-3) is a low-cost scrubber used primarily for removing liquid particulates from exhaust air streams in metal finishing and plating operations. Liquid entrainment and

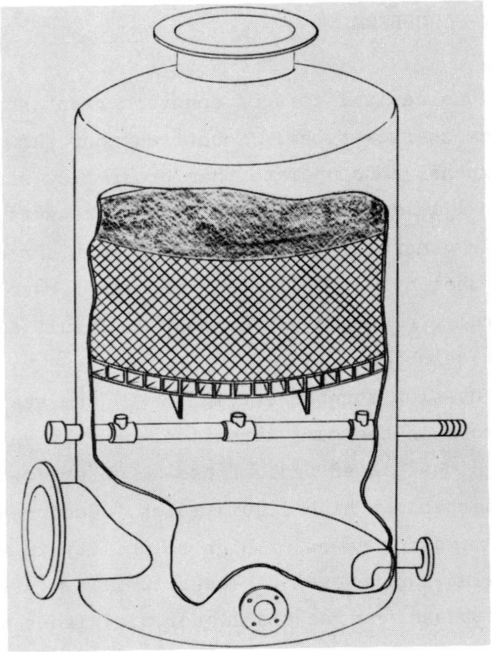

FIG. 25-3. Vertical air washer.

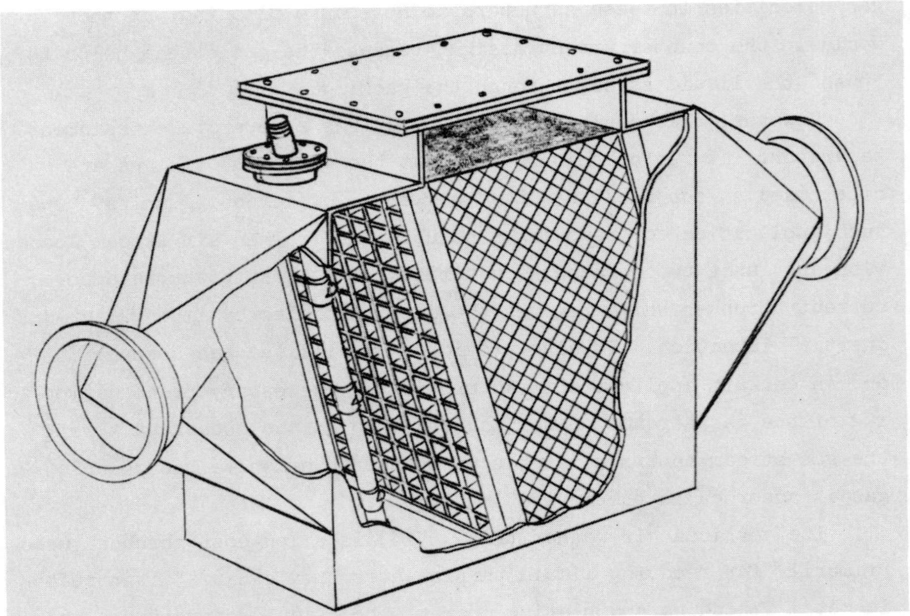

FIG. 25-4. Parallel-flow scrubber.

carryover from these processes have particle sizes of 10 μm and larger.

Using a cocurrent upward irrigation system with shallow packed beds, mesh pad or similar device, the vertical air washer is ideal for removing liquid particulates from anodizing, pickling, etching, cleaning, rinsing, and some plating operations. Its design makes it effective in removing particle sizes down to 10 μm with high collection efficiencies calculated on a weight basis.

Two important features of this scrubber are (1) its low irrigation rate, which can be as low as 1 gpm per 1,000 cfm, and (2) its low pressure drop, which is usually less than 1.0 in.wc. The low pressure drop characteristic and low irrigation rate result in extremely low operating costs because of the savings in horsepower required. This washer design should not be used where gaseous contaminants are present in the air stream.

Air washers built in a horizontal configuration with similar characteristics are normally classified as parallel-flow scrubbers (Fig. 25-4). In the parallel-flow scrubber, the gas stream and scrubbing liquid pass through the bed in the same horizontal direction. As with the vertical air washer, the parallel-flow scrubber has low water consumption, high gas flow capacity, and low pressure drop.

In the cross-flow packed scrubber (Fig. 25-5), the air stream moves horizontally through the packed bed and is irrigated by scrubbing liquor flowing vertically down through the packing. It is useful for removing noxious gases, entrained liquid particles, and dusts. It has been used successfully on such diverse applications as fertilizer plants, pickling, and other metal finishing operations, as well as in chemical plants.

Particulate matter 5 μm and larger is effectively removed by impingement within the cross-flow packed scrubber. Since particulate matter in the air stream is of greater density than the air, it will tend to flow in a straight line when the gas stream is forced to change direction as it encounters the packing. Usually this results in particles striking the packing. The particles will then be washed away from the packing by the scrubbing liquor.

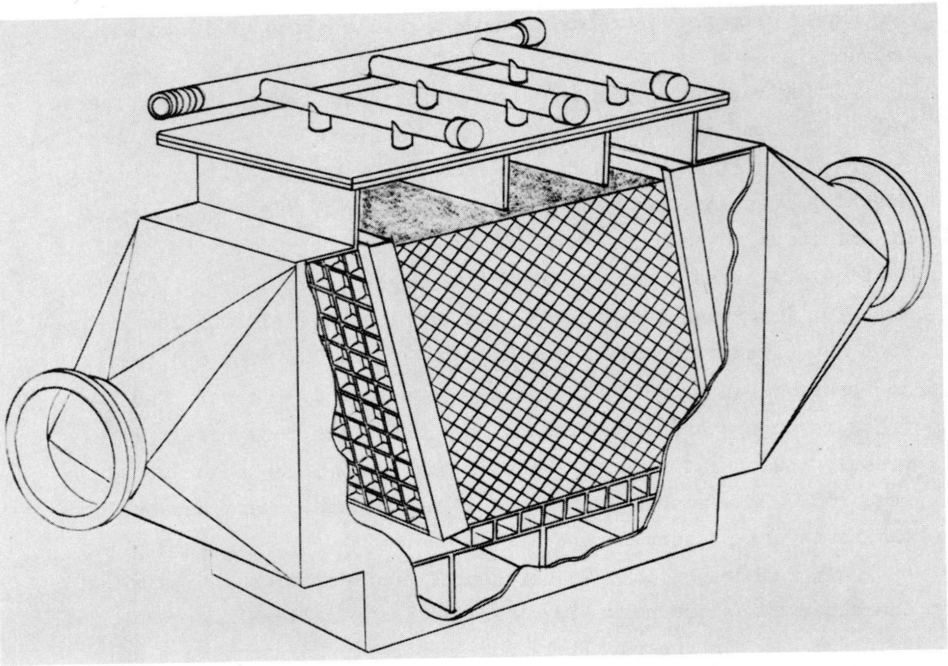

FIG. 25-5. Cross-flow packed scrubber.

In most industrial applications, packed beds are irrigated with scrubbing liquor which carries away the particles that have been removed from the gas stream by impingement. Sufficient flow is provided to keep the concentration of particles in the liquor low enough to prevent plugging of the packed bed and liquor recycle system.

Often two designs (parallel flow and cross flow) are combined into one unit in which the face of the packing is kept washed by spray nozzles located in front of the packed bed (parallel flow), while the main bed of the packing is irrigated by a liquor distribution system located at the top of the packed bed (cross flow). This combination design is particularly effective for pollution control of air streams carrying solid particulate, as well as liquid particulates and/or noxious gases.

In addition to this combination, multiple beds of packing are frequently employed where complex combinations of noxious gases have to be removed. A typical example of this configuration is shown in Fig. 25-6, which depicts a cross-flow scrubber with three packed beds used in the control of odors from high-temperature rendering operations.

The use of two or more packed beds is usually associated with applications requiring odor abatement, multiple gas/solids removal, solids/odor removal, and other complex scrubbing problems.

In an application where two or more gaseous contaminants must be removed, it is often desirable to scrub with chemically different liquids. This can be achieved readily with the cross-flow scrubber because its flow geometry permits the use of two or more packed beds, thus allowing different scrubbing liquids to be used without mixing.

PACKING

The specific packing medium selected for a wet scrubber depends on the nature of the contaminants being handled as well as on the specific geometric mode of contact and the specific scrubbing objectives.

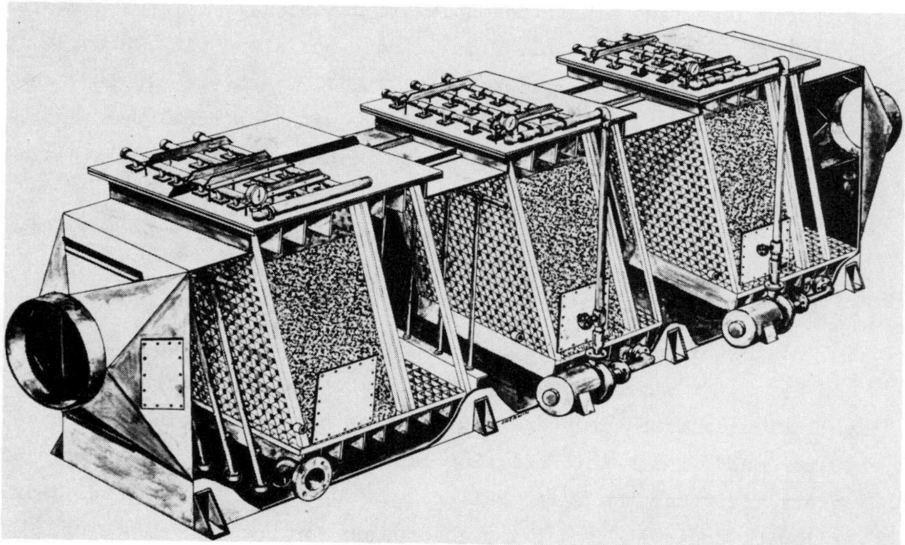

FIG. 25-6. Cross-flow scrubber with three packed beds.

Packing media include stationary or mobile glass spheres, fibrous beds, mobile or stationary plastic spheres, stationary ceramic packing, and stationary plastic packing.

The packing provides a surface across which the liquid flows in such a manner that it presents as large a surface as possible. This permits the gas to be absorbed into the scrubbing liquor, if absorption is the objective. Where the packing is used as a particulate collector, its purpose is to provide a target which will allow the gas stream to move around it while the solid particulate matter carried by that gas stream continues on a straight-line path and hits the target (the packing).

Five relatively new tower packing shapes (Fig. 25-7) are being used more frequently by engineers who design scrubber equipment for removing noxious gases. The packing materials are constructed of high-density polyethylene, polypropylene, or other thermoplastic materials. These provide good resistance to corrosion as compared to metallic packings, and a lighter weight in the packed bed section as compared to the ceramic packings. Following is a brief description of three of these packing shapes.

The Plastic Intalox Saddle® is made of linear polyethylene, polypropylene, and other thermoplastic materials. This packing is available in nominal 1-in., 2-in., and 3-in. sizes with 92 to 94 percent free volume and weights of 4.75, 4.0, and 3.75 lb/ft^3, respectively. The plastic Intalox saddle, with its scalloped edges, is a significant improvement over the ceramic saddle, and is claimed to be superior to ring-type packings.

Pall Rings® are a German development. The Pall ring shape is essentially a modification of the Raschig ring. It is made of polypropylene and other thermoplastics in sizes ranging from 5/8 to 3-1/2 in. nominal diameter. The 1-in. and 2-in. sizes, which weigh 5.5 and 4.2 lb/ft^3, respectively, are the most popular sizes for air pollution control scrubbers. Both sizes have free volumes around 90 percent. The 1-in. Pall ring has much higher pressure drop characteristics than the 2-in. Pall ring. On the other hand, the 1-in. Pall ring achieves higher gas absorption for the same depth of packing.

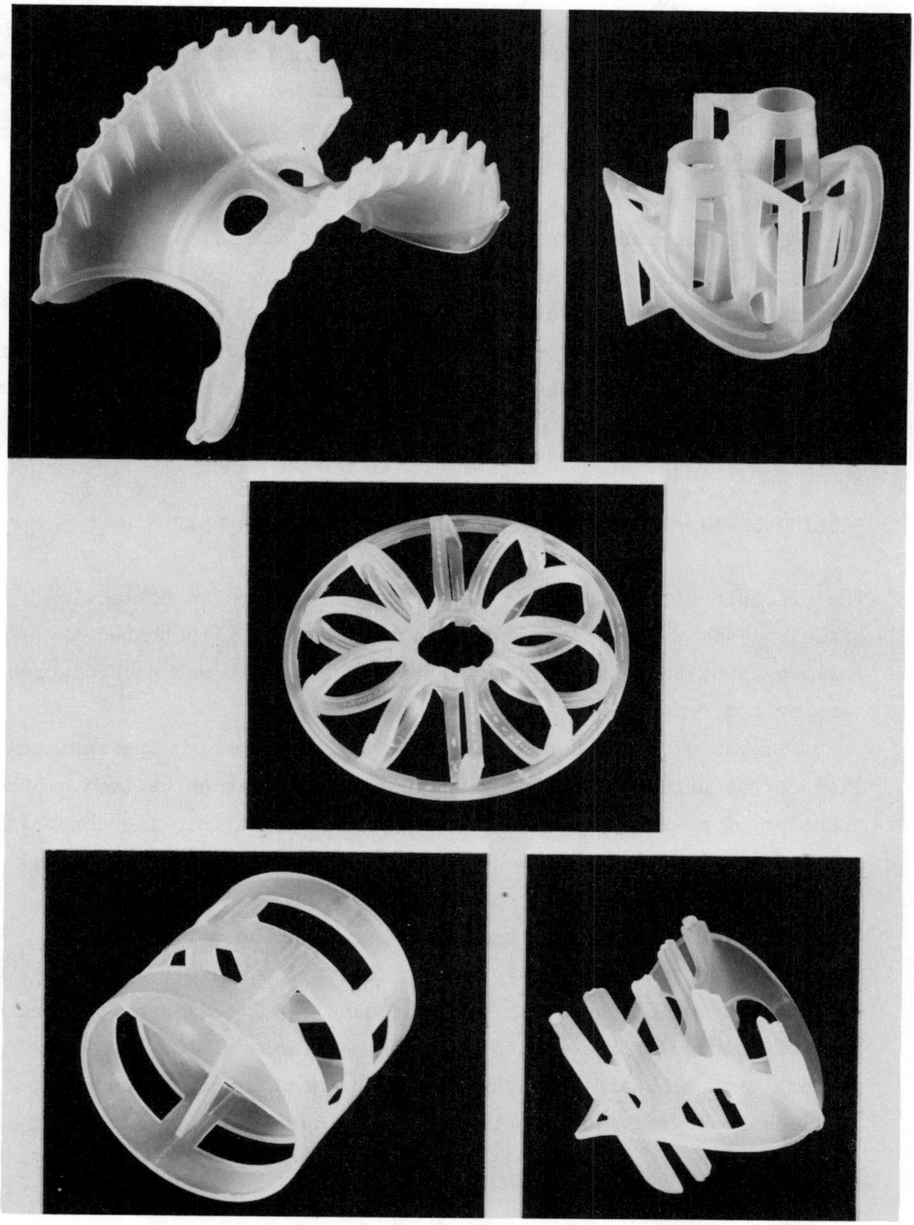

FIG. 25-7. Tower packing shapes. Upper left, 2-in. plastic Intalox saddle®; upper right, 2-in. plastic Heilex®-200; center, 2-in. Tellerette® Type R; lower left, 2-in. plastic Maspac®; lower right, 2-in. plastic Pall® ring.

Tellerette® packing has a rosette-like filamentous shape that has extremely low pressure drop characteristics and high gas absorption capability. The Tellerette is made in 1-in., 2-in., and 3-in. sizes of high-density polyethylene, polypropylene, and other thermoplastics. The 1-in. and 2-in. sizes weigh 7.5 and 3.8 lb/ft^3, respectively.

The filamentous character of Tellerette packing results in the formation of many more interstitial holdup points than can be obtained with other types of packing. These holdup points and the nonwetting polyethylene or polypropylene surface create a dispersion-agglomeration cycle of the scrubbing liquor which results in greater gas absorption capacity per foot of packing.

PACKING DEPTH, SCRUBBING LIQUOR, AND THE TRANSFER UNIT

The priciple of the transfer unit and its relation to collection efficiency as a function of packing depth can be illustrated as follows using the example of a countercurrent-flow wet packed scrubber for the removal of gaseous pollutants.

Removal of gaseous pollutants is achieved through the application of the principles of mass transfer. This can be defined as the transfer of gaseous molecules from the air stream into the scrubbing liquor, where transfer is achieved by a combination of diffusion, physical absorption, and/or chemical reaction.

In wet packed towers, the gaseous collection efficiency is directly dependent on the irrigation rate and the depth of the packed bed. Depending on the depth of the packed bed, the liquor flow rate, and composition of the scrubbing liquor, efficiencies up to 99.99 percent can be obtained in the removal of gaseous contaminants.

The following equation is applied to determine the number of transfer units (N_{OG}) required to achieve a specified scrubbing efficiency.

$$N_{OG} = \ln \frac{Y_1}{Y_2}$$

where

N_{OG} = number of mass transfer units required to achieve a specific scrubbing efficiency

Y_1 = mole fraction of incoming contaminant

Y_2 = desired contaminant concentration of the effluent gas (mole fraction)

Based on the preceding equation, the transfer units required for various scrubbing efficiencies are as follows:

N_{OG}	Percent scrubbing efficiency
0.5	39.0
1	63.0
2	86.5
3	95.0
4	98.2
5	99.3
6	99.75

The above formula can be applied only where the polluting gas dissolved in the scrubbing liquor exerts no vapor pressure to retard further absorption. Thus, N_{OG} is controlled by the inlet and outlet concentrations of only the polluting gas. More complex calculations are required when the scrubbing liquor exerts vapor pressure from absorbed gases.

After determining N_{OG}, the following equation is used to calculate the required packing depth for a specific scrubbing application:

$$Z = (N_{OG})(H_{OG})$$

where

Z = packing depth

N_{OG} = number of transfer units

H_{OG} = height of transfer unit

The height of a transfer unit (H_{OG}) is determined experimentally and depends on the following factors:

1. Type of scrubber packing,

2. Concentration and solubility of the polluting gas,
3. Gas and liquor flow rates,
4. Type of scrubbing liquor, and
5. Liquor and air stream temperatures.

Scrubber packing manufacturers have performed tests on their packings (Tellerettes, Intalox Saddles, Pall Rings, Saddles, etc.) to determine the H_{OG} values for various scrubbing systems such as NH_3-air-water, Cl_2-air-NaOH, and HCl-air-water. H_{OG} values are stated in feet (or inches) of packing for the system at various gas and liquor flow rates and temperatures. For example, Tellerette packing in a 1 percent NH_3-air-water system has an H_{OG} of 0.8 ft, based on a gas rate (G) of 500 lb/hr ft^2 and a liquid rate (ℓ) of 1,000 lb/hr ft^2 at 70°F. In the same system, the H_{OG} is reduced to 0.5 ft when the liquid rate is increased to 4,000 lb/hr ft^2.

TABLE 25-1

Heights of Packing (Feet) to Obtain Listed Efficiency[a]

	Packing size				
Packing efficiency	1 in.	1-1/2 in.	2 in.	3 in.	3-1/2 in.
63.2	1	1.25	1.5	2.25	3
77.7	1.5	2	2.25	3.5	4.25
86.5	2	2.5	3	4.5	5.75
90	2.5	3.25	3.75	5.75	7
95	3	3.75	4.5	6.75	8.5
98	4	5	6	9	11.25
99	4.6	5.75	7	10.25	13
99.5	5.25	6.5	8	12	14.75
99.9	7	8.75	10.5	15.75	19.75
99.99	9.25	11.5	14	21	26

[a]Restricted to specific applications on highly soluble gases or absorption followed by chemical reaction.

Table 25-1 is a generalized table for estimating packing height (or depth). This table is limited to applications with highly soluble gases or low-soluble gases accompanied by a rapid chemical reaction. The table is based on the equation $N_{OG} = \ln (Y_1/Y_2)$, which applies only when the equilibrium line is straight, the mole fraction Y is small, and the slope of the equilibrium line approaches zero. It is also based on a "rule of thumb" that 1-in. size packing yields an H_{OG} (height of a transfer unit) equal to 1 ft, the 1-1/2-in. size equals 1-1/4 ft, the 2-in. size equals 1-1/2 ft, etc. While there are variations in the height versus the type of packing used (approximately plus or minus 25 to 30 percent), the "rule of thumb" for a quick estimate for this purpose can be applied.

PACKING DEPTH AND SOLIDS REMOVAL CAPABILITY

In some industries, cross-flow scrubbers handle inlet concentrations of total solids in the range of 15 to 25 mg/scf. Much higher loadings are actually found in some duct systems ahead of inexpensive pretreatment equipment where the loadings are reduced to the levels indicated. These loadings are, in some cases, twenty to thirty times greater than the loadings which would cause rapid plugging or fouling of a countercurrent packed scrubber.

To obtain satisfactory operating reliability when entrained solids are to be removed, blinding of the front packing support plate and the first few inches of packing must be prevented. This is achieved through the use of front washing sprays and high liquor irrigation rates in those zones most subject to plugging. The front spray, which washes the front packing support plate, operates at a rate of 0.5 to 2.0 gpm/ft^2. The top spray rates are increased substantially in the area immediately behind the front packing support plate.

In most applications where solids are removed, the greatest weight of particulates is removed in the first 12 in. of packing depth. This is shown in Table 25-2, which includes the approximate

TABLE 25-2

(Packing depth 12 in.; air velocity 8 ft/sec)

Particle size	Collection efficiency by particle count (sp. gr = 2.0)	Collection efficiency by particle count (sp. gr = 4.0)
10	99.5	99.9
9	99.3	99.9
8	98.7	99.8
7	94.7	99.7
6	87.0	99.5
5	57.0	95.5

percentage of particles removed by particle size down to 5 μm, based on a particle specific gravity of 2.0 and 4.0 and the use of 1-in. Tellerettes.

Assuming normal particle distribution, the quantity of particles removed on a weight basis in the first 12 in. of packing is often of the order of 97 to 99 percent, because weight is a function of the cube of the particle diameter. Since most of the solids are deposited in this zone, high liquid rates in the range of 4 to 20 gpm/ft^2 are used to provide high washing rates for removal of solids.

APPLICATIONS

Table 25-3 lists several types of wet scrubbers, packed bed as well as other designs. Applications suitable for each type are listed in the table according to the type of pollutant. It can be seen from the table that any one scrubber may range from a rating of excellent to "not recommended," depending on the application. To understand this table fully, it is necessary to define the pollutant categories in it. In general, all pollutants can be broken down into three categories: gaseous, entrained liquids, and solid particulates.

TABLE 25-3

(Key: Normal emissions contain particles of mixed sizes. The chart efficiencies are based on the number of particles collected in all sizes on a percentage basis. Gas absorption efficiencies are based on a percent by weight of the gases collected. E = 95%; G = 85 to 95%; F = 50 to 85%; NR = not recommended.)

Types of wet scrubbers	Gas absorption		Mists under 10 μm	Entrained liquids over 10 μm	Dusts above 5 μm	
	High solubility	Low solubility			Low loading	High loading
Cross flow (HRP)	E	G	G	E	E	NR
Countercurrent packed tower (VCP)	E	E	G	E	G	NR
Wet cyclone (VWC)	G	NR	NR	E	G	G
Air washers (VAW)	NR	NR	F	G	F	NR
Cocurrent packed tower	G	F	F	E	G	NR
Parallel flow	G	F	F	E	G	NR
Spray towers	G	NR	NR	G	F	F
Spray chambers	F	NR	F	F	F	F
Combination scrubbers (VWC and HRP)	E	G	G	E	E	E
Jets	F	NR	F	E	E	G
Venturis	F	NR	F	E	E	E

Noxious gases (such as hydrogen fluoride, ammonia, hydrogen chloride, hydrogen sulfide, and chlorine) are generally removed from air streams by a gas absorption process, sometimes accompanied by a chemical reaction. This process involves the transfer of the noxious gas from the gas phase into a liquid in which it is more or less soluble. In most air pollution control cases, the polluting gases are carried by an air stream (the gas phase). The basic factors controlling this gas absorption process are (1) the degree of solubility (or chemical reaction) of the gas to be removed in the liquid used for scrubbing, and (2) the means of obtaining intimate contact between the gas and liquid streams to facilitate quick absorption. Normally, plant water is used to remove gases of high solubility, e.g., ammonia or hydrogen chloride. In some cases, caustic or acid solutions may be used because they react chemically with less soluble gaseous contaminants. For example, sodium hydroxide scrubbing liquor is used to react with chlorine gases to produce sodium hypochlorite.

In air pollution control equipment designed for removing polluting gases from air streams, a surplus of scrubbing liquor is usually provided. This obviates the possibility of buildup in the concentration of the absorbed gas in the recycled liquor stream to a point where it may reenter the existing gas stream. Packed scrubbers are the most economical equipment available for the removal of noxious gases. They provide intimate contact between the gas stream and scrubbing liquor stream at low initial equipment costs and low operating costs. These scrubbers utilize beds of tower packings to force the gas and liquid streams to come into intimate contact with each other. Although packing is available in different materials, the most popular for air pollution control equipment are constructed of lightweight, corrosion-resistant plastics.

Liquid entrainment is composed of corrosive mists, sprays, and other forms of particulate matter as small dispersoids suspended in the air stream. To a large degree, the separation process for this type of pollutant is determined by the size of the particle. Most mists range in size from 0.1 to 10 μm, while liquid entrainment normally has a size of from 10 to 100 μm. A typical example of mist is sulfuric acid mist from a sulfuric acid plant. Examples of liquid entrainment include sulfuric acid and carryover emanating from a steel pickling line and chromic acid carryover coming from a chrome plating tank. Because of their relative size, mist particles are usually much more difficult to remove from an air stream than the larger liquid entrainment particles.

In many air pollution control problems, small quantities of solid particulates may be present in the exhaust air stream along with the gaseous and liquid pollutants. These solids usually exist as dust (solids over 1 μm in size) or fumes (solids less than 1 μm in size).

Certain types of packed scrubbers are effective in removing solid particulates by impingement or interception, as long as the solid loading in the air stream is not over 25 mg/ft^3. Particle sizes of 5 μm and over can be collected by impingement in packed scrubbers, while a process called nucleation is often necessary to effect removal of dust and fume particles in the submicron range. The nucleation process employs a humidification and cooling cycle which creates water condensation on submicron particles in order to build up their particle size to a level where they can be removed by impingement on the packing.

COLLECTION EFFICIENCY

There are several types of inexpensive air washers on the market today which are claimed to have collection efficiencies of 99 percent and over. Unfortunately, in many cases, the basis on which the high collection efficiency claims are made is not fully described. The low-priced air washers (which use only shallow packed beds or filter pads for the scrubbing medium and low liquor rates) can often give 99 percent collection efficiency only on a weight basis of contaminants removed.

Most of these air washer designs effectively remove entrained liquid particles with diameters of 10 μm or greater. Their ability to collect smaller-diameter particles in the 1- to 8-μm-diameter range, however, is poor. These particles will pass through the air washer and into the atmosphere. Once in the atmosphere, the small-diameter particles will fall out and possibly cause deterioration of plant structures, automobiles, or nearby homes.

To illustrate the difference between collection efficiency on a weight basis and collection efficiency on a particle-size basis, assume that a ventilation air stream coming from a chrome plating tank contains four 25-μm particles, two 10-μm particles, and ten 5-μm particles. On a relative weight basis, the chromic acid mist particles weigh as follows:

Particle size (μm)	Relative weight
25	125
10	8
5	1

Using the relative figures above, the weight of the chromic acid mist particles in the ventilating air stream would be

Four 25-μm particles	500
Two 10-μm particles	16
Ten 5-μm particles	10
Total weight particles	526

If the air washer used was extremely ineffective, it would remove only the four 25-μm particles. However, its efficiency on a weight-removal basis could be calculated as follows:

Weight removed	500
Weight entering	526
Percent collection efficiency	$\frac{500}{526} \times 100\% = 95.2\%$

Should the air washer be effective enough to remove the 25- and 10-μm particles, then its collection efficiency could be reported as follows:

Weight entering	526
Weight removed	516
Percent collection efficiency	$\frac{516}{526} \times 100\% = 98.1\%$

It is important to note in this case, however, that ten 5-μm particles (62.5 percent of the total number of particles) passed through the scrubber and into the atmosphere. This means that efficiency, based on particle size, is only 37.5 percent.

In mist and liquid entrainment removal problems, it is important to have knowledge of the particle size distribution of the contaminants in the air stream. With this information, the manufacturer can design a scrubber unit that will remove the smallest sized particles necessary.

Another aspect of this same subject is the tendency of some users to overspecify. They often ask for a collection efficiency of 99 percent or more, in the belief that equipment with this high level of collection efficiency will comply with existing and future air pollution laws. Also, an efficiency of 99 percent or more is often specified simply because the user is not aware of the true nature of the pollutants in his process exhaust stream, and a 99-percent system provides a good "insurance policy."

This tendency to ask for 99 percent as "insurance against ignorance" leads to overdesigned equipment that is expensive both in terms of first cost and operating costs. On the other hand, lack of proper data can lead to the specifying of equipment that is undersized and cannot provide adequate pollutant removal capability. It must then be replaced or modified.

Many equipment selection errors can be avoided if the time is taken to obtain information about the loading of pollutants in the exhaust stream of a process through in-plant testing. For example, unclean exhaust air may carry 50 lb/day of pollutant and the local air pollution control law may allow 10 lb/day to escape to the atmosphere. Under these conditions, a scrubber operating at 80 percent collection efficiency would be satisfactory. On the other hand, had the law stated that only 0.1 lb/day was permissible, even a 99-percent-efficiency scrubber would be unacceptable. To achieve an acceptable condition, a scrubber with a collection efficiency of 99.8 percent would be required.

With data available on the concentration and amount of noxious gases in the exhaust air stream, the scrubber designer can design equipment that is tailored to the exact pollution control requirements for each job. This becomes particularly important in designing scrubbers for removing noxious gases, since the difference between 95 percent efficiency and 99.9 percent efficiency can mean several feet of packing.

Once the designer knows exactly what is in the air stream, he can select the scrubbing liquor that will contribute to optimum performance. This may consist of fresh water, process water, or an acid or caustic solution. Fiberglass-reinforced plastics would be an obvious material choice for the shell for the last two scrubbing liquids, with thermoplastics being used for the packing, packing support plate, and other internals.

CONSTRUCTION

In recent years there has been an increased acceptance of reinforced plastics as a material of construction for scrubber equipment. Many control problems where noxious gases are being removed involve a corrosive environment. The air stream and its gaseous contaminants are highly corrosive, and/or the scrubbing solution is corrosive. This dictates that the materials of construction for the scrubber be highly resistant to corrosion.

Fiber-reinforced plastics (commonly designated as FRP) are being used more and more to replace rubber- and PVC-lined steel, stainless steel, and other metallic alloys for scrubber bodies. FRP is a material with a combination of polyester or epoxy thermosetting resins and glass fiber reinforcements that produce a finished laminate with excellent corrosion resistance and a high strength-to-weight ratio.

Availability of fiber-reinforced plastic materials and plastic tower packings has enabled scrubber designers to obtain lighter-weight structures. For example, a countercurrent scrubber with a 6-ft diameter and a height of 10 ft-6 in. that utilizes FRP construction and Tellerette packing has a dry weight of 2,000 lb. and an operating wet weight of 5,100 lb. The same scrubber, using stainless steel for the shell and ceramic packing, will have a dry weight of 6,600 lb and an operating wet weight of 9,500 lb. The lighter weight of the plastic scrubber often permits its placement on a rooftop or in some other out-of-the-way location. This is an important space-saving benefit to the plant operator.

Almost any type of equipment can be built from FRP as long as the design engineer is familiar with its characteristics. Special consideration must be given to the location of the scrubber within the system as well as its geometry. Designs for pressure or vacuum location in an exhaust system, type of bottom support, flat sides or cylindrical construction, and other similar items must be critically examined.

SPECIFICATION GUIDE

The key to efficient specification of a wet packed scrubber, or any other pollution control system, lies in a systematic approach to the problem which ensures that all of the required information is collected and organized. The following checklist can be followed by the user-specifier himself if he possesses the skills to develop the data, or it can be used as a guide to evaluate manufacturer's proposals. It must be stressed that the simple phrasing of many of the items in the checklist does not reflect the complexity of the data-development job. In many cases a rigorous mathematical exercise is required to arrive at the proper answer.

1. Determine the nature of the contaminants:
    a. Physical nature and temperature,
    b. Type of process,
    c. Total volume (cfm) required, and
    d. Corrosiveness of contaminants:
        (1) Wet,
        (2) Dry.
2. What efficiency is required:
    a. Local pollution control regulations,

b. EPA requirements, and

c. OSHA requirements.

3. Select the equipment to be used:

a. Is equipment selection correct (Table 25-3)?

b. Can dry collector be used?

c. What is corrosive nature of contaminants?

(1) Can FRP or other plastics be used?

d. Will water treatment be required?

e. Where will makeup water come from?

f. Where is best location for equipment?

(1) Floor space requirement,

(2) Accessibility for maintenance.

4. Determine size of auxiliary equipment:

a. Fan size and horsepower,

b. Pump size and horsepower,

c. Other accessories required.

5. Determine economics:

a. Total first cost of system,

b. Operating cost of system,

c. Installation cost,

d. Annualized cost of operation.

Chapter 26

COUNTERCURRENT CONTACT WET SCRUBBING

Norman D. Phillips
Fuller Company
Catasauqua, Pennsylvania

The basic function of wet scrubbers is to produce contact between the scrubbing liquid and the particulate to be collected. In the many scrubber designs presently available, this contact is achieved in a variety of ways by equipment generally classified by these methods of contact: (1) packed beds--vertical, horizontal or cross-flow; (2) sprays; (3) cyclones; (4) wet orifices; and (5) venturi.

Each type is used in a certain range of applications. Generally, units requiring high-energy inputs have a greater potential for higher collection efficiency of smaller particle sizes. Of course, it is recognized that differences in particle shapes, in hydrophobic or hydrophilic properties of the particulate, and in physical properties of the scrubbing liquid all affect the operating results.

In the development of a new type of scrubber it was reasoned that salient features of several types could be combined to produce a better device. Impingement of particles on liquid films is adequate for larger particle sizes, whereas sprays are required for the small particles. Sprays produced by mechanical means through small orifices, which can either plug or erode and deteriorate, have their deficiencies. Sprays produced by gas atomization in venturis require a high pressure drop of 20 to 100 in.wc because of the high gas velocities required, which range from 100 to 400 fps in the throat. These high velocities are generated by exhaust fans which require high tip speeds, and thus high fan power.

By combining desirable features, it has been possible to develop a compressed air-activated nozzle scrubber.

PRINCIPLE OF OPERATION

This new scrubber operates on the principle of countercurrent contact between a medium-velocity gas stream and a relatively high-velocity air-atomized liquid stream, producing a high-velocity impact zone which creates a very effective turbulent zone for particulate collection. Figure 26-1 shows a schematic diagram for the flow pattern to describe the operation.

Dirty gas enters the top of the contact cylinder in a down-flow mode. The air-atomized liquid is sprayed upward into the contact cylinder where turbulent action causes particulate collection. Recovery of the wetted particles is achieved by impingement on the liquid surface and by additional contact as the gas stream penetrates the curtain of liquid falling from the contact cylinder wall. Entrained water droplets are removed by the spin-vane mist eliminator before the gas stream is discharged--either to the induced draft fan in a suction system or direct to the atmosphere in a pressure system.

With this countercurrent contact action, the nominal gas velocity of 6,000 fpm requires a low pressure drop of 3 to 15 in.wc--thus utilizing a low-speed fan with inherently lower maintenance costs.

The air for water atomization can be supplied by positive-displacement blowers operating at 5 to 10 psig. Collection efficiency can be varied by adjusting the liquid-to-gas ratio (L/G) and/or the air-to-water ratio (A/W) in the nozzle. The total gas flow to the unit can be reduced to 50 percent of the rated value without impairing the collection efficiency and without the use of dampers or adjustable devices located in the contact zone.

The spray nozzle, used for delivering the scrubbing liquor to the contact zone, has a large-area orifice which permits recirculation of high-solids-content slurries (+5 percent by weight) without danger of plugging. The limitation of the solids content is the

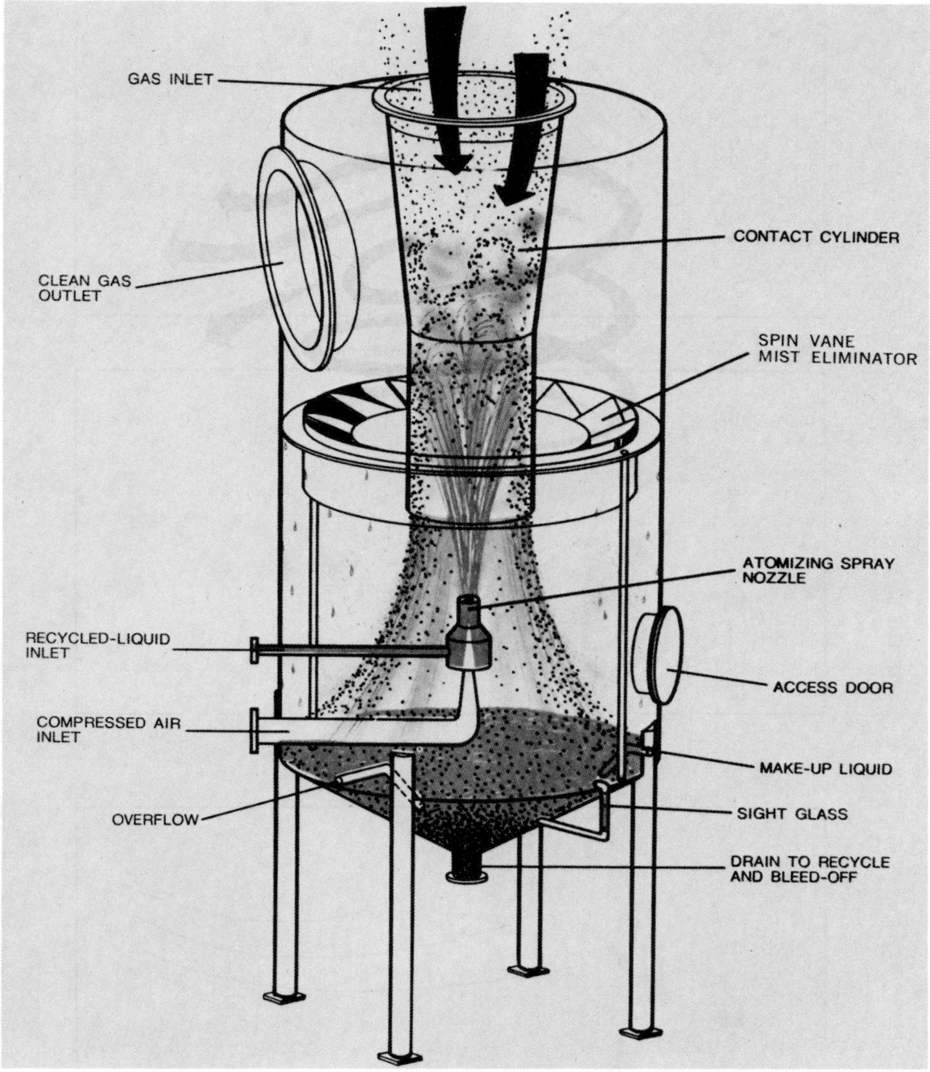

FIG. 26-1. Countercurrent contact wet scrubber.

point at which the viscosity is too high for atomization. This concentration varies with the type of particulate collected; for example, a 30 percent slurry with fly ash was successful, while a

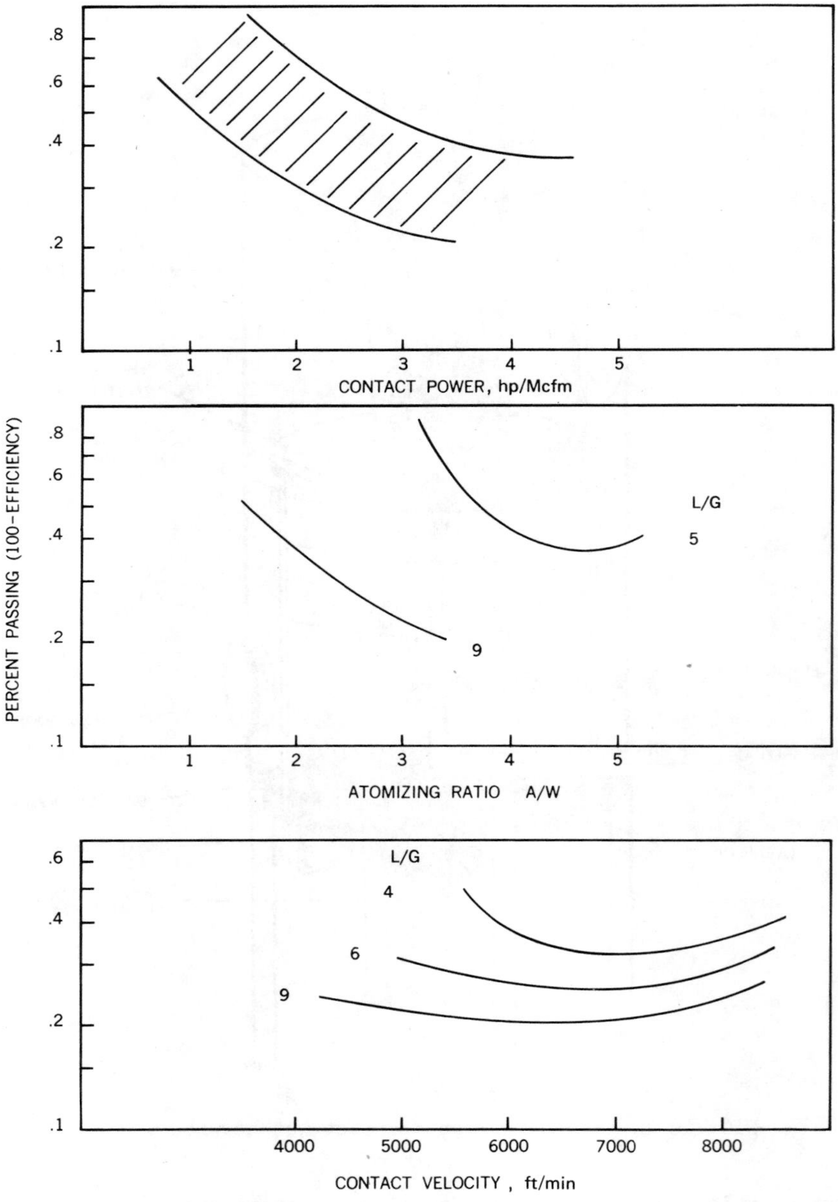

FIG. 26-2. Scrubber performance.

10 percent slurry of fine iron oxide fume was the upper limit for good operation. Since the discharge stream of slurry from the scrubber must be treated, the handling of a higher-solids-content slurry should provide some economy in the waste treatment system.

Figure 26-2 shows the effect of operating parameters on efficiency. The contact horsepower [1] can be calculated directly from pump power, fan air horsepower, and blower air horsepower--all based per 1,000 cfm:

$$P_1 = 0.000582(P_n)\left(\frac{L}{G}\right)$$

$$P_g = 0.1575(\Delta P)$$

$$P_b = 0.00436\left(\frac{A}{W} \times \frac{L}{G}\right)P_n$$

P_1 = pump hp/Mcfm

P_g = fan hp/Mcfm

P_b = blower hp/Mcfm

L/G = gpm/Mcfm

ΔP = in.wc

A/W = scfm/gal

P_n = nozzle pressure, psig

These graphs represent laboratory tests on a prototype unit using coal fly ash previously collected in a utility boiler by electrostatic precipitators. The particle size is nominally 65 percent minus 20 μm and 4 percent minus 1.5 μm, as measured by the Bahco method. The composite curves plotted as percent passing, or 100 percent efficiency, show that the minimum particle loss can be achieved between 3 and 4 hp/Mcfm, with water/gas (L/G) rates of 5 to 9 and air/water (A/W) atomizing ratios of 3 to 5 scfm/gal.

PILOT UNIT

To permit field testing of this concept on pollution sources, a portable pilot plant unit was built. The unit is complete with scrubber, fan, recycle pump, air blower, and control panel. It has a gas capacity of 2,200 scfm outlet volume, and weighs approximately 2 tons. It has been used for several field tests to determine operating efficiency on raw process gases--this being the best way to test a new design. Laboratory tests with dusts cannot always be established to duplicate the various parameters of hot gas, humidity, and raw dust or fume as that being emitted from a process. The use of precollected dusts always presents the problem of possible loss of submicron material during collection or the unknown extent of agglomeration which could affect the results of the scrubber tests. Raw fume tests are the best way to make the complete evaluation necessary for safely predicting performance to meet the air pollution control codes.

The results obtained for pilot tests and typical commercial units operating on lightweight aggregate kilns are as follows:

	Pilot tests		Commercial units	
ACFM, outlet	2180	2450	41,900	73,000
ΔP (in.wc)	8.0	8.5	8.2	4.5
L/G	9.2	12.3	13.0	4.1
A/W	3.1	3.0	3.9	3.1
P_n, psig	4	4.5	9.	5.3
L_p/Mcfm	1.8	2.1	3.3	1.0
Dust loading (gr/scf, dry)				
Inlet	17.0	3.9	6.0	3.7
Outlet	0.04	0.02	0.05	0.04
Efficiency (%)	99.7	99.5	99.2	98.9

The particle size of lightweight aggregate dust is approximately 10 percent minus 1.5 μm and 50 percent minus 10 μm. Scrubber operation will vary with kiln feed which may be various types of shale or clay.

CONCLUSION

This novel contact principle used to achieve particulate collection from industrial gas streams should find broad application--due to decreased horsepower requirements with the fan operating at lower tip speeds. Lower maintenance costs will accrue with the use of lower-speed fans and minimum abrasion produced by the low-velocity gases in the contact zone. The ability to recirculate concentrated slurries should provide a savings in waste treatment facilities. Several of these scrubbers are operating successfully in the process industries such as lightweight aggregate, stone dryers, and in foundries for sand handling and shakeout hoods.

REFERENCE

1. K. T. Semrau: Dust Scrubber Design, Journal of the Air Pollution Control Association, 13:587-594(1963).

Chapter 27

AEROSOL COLLECTION BY FALLING DROPLETS

David Rimberg and Yei-Min Peng

MicroPul Division
United States Filters Corporation
Summit, New Jersey

The collection of aerosol by falling droplets (or other spherical collector) is prevalent in many physical processes. For instance, the suppression of airborne dust by water spray is one of the chief means of combatting the dust hazards in mining; washout of atmospheric aerosol by rain cleanses the atmosphere of natural and man-made pollutants; particulate-laden air from industrial processes is removed by liquid scrubbing air cleaning equipment.

To describe the mechanisms of capture of the aerosol by the spherical collector, it is necessary to characterize the fluid flow surrounding the droplet and then to examine the trajectory of the aerosol particle on which the fluid forces and other forces act. Mathematical models have been developed which describe these conditions, the results usually being expressed as aerosol collection efficiency by a spherical collector.

The physical mechanisms which accomplish the actual collection are achieved by one or more of the following six mechanisms: (1) inertial impaction, (2) interception, (3) Brownian diffusion, (4) electrostatic attraction, (5) gravitational, or (6) thermal attractive (or repulsive) forces.

If the efficiency of aerosol removal by these mechanisms can be calculated, it would appear that it would be possible to predict the performance of air pollution control equipment--specifically the wet scrubber. To date, much experimental effort has been expended examining the mechanisms of inertial impaction and interception [1-2]. The basis of the experimental work performed by these investigators and others [3] was either to allow the spherical collector to remain stationary with respect to oncoming aerosol, or to allow the spherical collector to traverse a quiescent aerosol. In several of these studies, experimental data often indicated that the collection efficiencies were in excess of the calculated efficiency for impaction and interception. In these experiments the removal mechanisms of electrostatic attraction, gravity, and temperature-gradient motion were either eliminated or considered negligible. Hence, Brownian diffusion was suspected as being the unaccounted-for removal mechanism, especially for submicron aerosol particles. Recently, several theoretical studies have appeared which try to predict collection efficiency of aerosol by diffusion adequately.

THEORY OF AEROSOL COLLECTION

The collection of aerosol by falling droplets occurs when the particles come in close proximity to the spherical collector. As this occurs, the collection mechanisms begin to act, resulting in the collection of aerosol by the spherical collector. The three basic mechanisms considered in this development are inertial impaction, interception, and Brownian diffusion. The effect of temperature and external forces (gravitational and electrostatic) are not considered important.

Inertial Impaction

Inertial impaction of aerosol occurs when the trajectory of a moving particle having a mass greatly exceeding the molecules of surrounding gas departs from the path of the gas molecules as the fluid streamlines spread around the spherical obstacle. The departure of the particles from streamline flow may be sufficient to allow the particles to deposit on the curved surface. At high velocities, the streamlines diverge close to the spherical collector (potential flow). At lower velocities, the streamlines begin to diverge a considerable distance upstream of the collector (viscous flow). In actual situations at Reynolds number of 0.2, a 3-percent disturbance occurs at a distance of 100 collector diameters upstream while at Reynolds number 2000 there is practically no fluid disturbance at a distance of two collector diameters. The body of revolution traced by the limiting particle trajectory hitting a sphere is shown in Fig. 27-1.

The collection efficiency due to inertial impaction, E_{im} (often called the target efficiency), is defined as the ratio of the number of particles striking the collector to the number which would strike it if the streamlines were not diverted by the obstacle. (It is assumed that all particles adhere upon impaction.) If the aerosol is uniformly distributed throughout the approaching gas stream, then the collection efficiency E_{im} equals the ratio of the area swept clean to the cross-sectional area of the spherical obstacle, or

$$E_{im} = \frac{\pi Y_{lim}^2}{\pi D_c^2/4} = \left(\frac{Y_{lim}}{D_c/2}\right)^2 \tag{1}$$

where Y_{lim} is the distance from the axis of the limiting streamline for which impaction occurs, and D_c is the diameter of the spherical collector (or droplet).

The impaction efficiency is often correlated with the dimensionless parameter called Stokes number. It is defined as:

$$Stk = \frac{C\rho_g D_p^2 V_{rel}}{9\eta_g D_c} \tag{2}$$

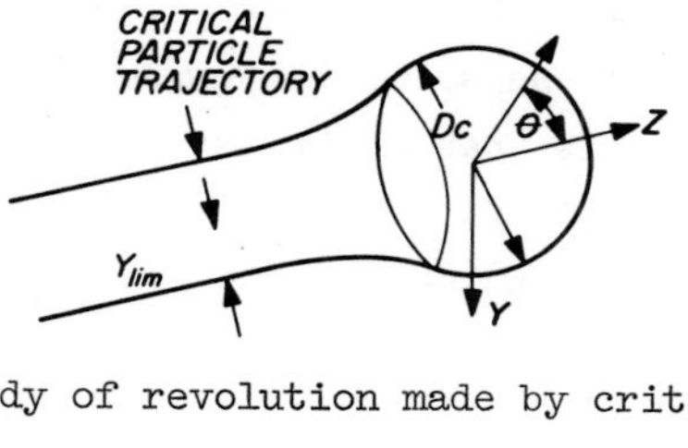

FIG. 27-1. Body of revolution made by critical trajectories of particles.

where

C = Cunningham correction factor for the particle (slip factor)

ρ_g = density of gas

D_p = diameter of particle

V_{rel} = relative velocity of droplet with respect to gas $= (V_\infty^2 + V_{ter}^2)^{1/2}$

V_∞ = free stream gas velocity

V_{ter} = terminal velocity of droplet

η_g = viscosity of gas

D_c = diameter of spherical collector (droplet)

If the particles are larger than 3 μm, the Cunningham correction factor C is almost 1. Therefore, the above equation can be written:

$$Stk = \frac{\rho_g D_p^2 V_{rel}}{9\eta_g D_c} = \frac{2\ Re_c \rho_p D_p^2}{9\rho_g D_c^2} = \frac{L_s}{D_c/2} \qquad (3)$$

where

$$Re_c = \frac{V_{rel}\rho_g D_c}{\eta_g}$$

Written in this manner, the Stokes number becomes the ratio of the particle's stopping distance L_s to the radius of the collector. A particle's stopping distance is the distance it would travel before coming to rest if injected into a still gas at a velocity V_∞ when all the forces on the particle except the drag force are zero.

Since the equations of motion of particles around a spherical collector cannot be solved analytically, numerical methods must be employed. Several of these solutions appear in Table 27-1. The one most often referred to is that of Langmuir and Blodgett [4]. These solutions yield a critical Stokes number, Stk_{cri}, for spheres below which no inertial impaction takes place if $Stk_{cri} = 1/12$.

TABLE 27-1

Expressions for the Collection Efficiency of Aerosols by Spherical Collectors

Collection Mechanism	Efficiency Equation	Conditions	Reference
	Potential flow (high Re_p)		
Impaction	$E_{im} = \left(\frac{Stk}{Stk + 0.5}\right)^2$	$Stk \geq 0.2$ $Stk_{cri} = 1/12$	[4]
Interception	$E_{in} = \left(1 + \frac{D_p}{D_c}\right)^2 - \left(\frac{1}{1 + D_p/D_c}\right) \simeq 3\frac{D_p}{D_c}$		[1]
Impaction Interception	$E_{im,in} = 1 - (1 - E_{im})(1 - E_{in})$		
Diffusion Interception	$E_{diff,in} = 2.18\ Pe^{-1/2} + 2\frac{D_p}{D_c}$	$\frac{D_p}{D_c} \leq 0.3\ Pe^{-1/2}$	
Diffusion Interception Impaction	$E_{diff,in,im} = 1 - (1 - E_{diff})(1 - E_{in})(1 - E_{im})$		
	Viscous Flow (low Re_p)		
Impaction	$E_{im} = \left[1 + \frac{0.75\ \ln(2\ Stk)}{Stk - 1.214}\right]^{-2}$	$Stk > 1.214$	[5]
Interception	$E_{in} = \left(1 + \frac{D_p}{D_c}\right)^2 - \frac{3}{2}\left(1 + \frac{D_p}{D_c}\right) + \frac{1}{2(1 + D_p/D_c)} \simeq \frac{3}{2}\left(\frac{D_p}{D_c}\right)^2$		[6]

TABLE 27-1 (Continued)

Collection Mechanism	Efficiency Equation	Conditions	Reference
Impaction Interception	$E_{im,in} = 1 - (1 - E_{im})(1 - E_{in})$		
Diffusion	$E_{diff} = \frac{2}{(2)^{1/2}}[PeD_c]^{1/2}$		
Diffusion	$E_{diff} = \frac{4}{Pe}(2 + 0.557\ Re_c^{1/2}Sc^{3/8})$		[7]
Impaction Interception Diffusion	$E_{im,in,diff} = (1 - E_{im})(1 - E_{in})(1 - E_{diff})$		
	Transitional flow (finite Re_p)		
Impaction	$E_{im} = \frac{E_{im(vis)} + E_{in(pot)}(Re_c/60)}{1 + Re_c/60}$		[5]
Diffusion	$E_{diff} = 3.18\ Pe^{-2/3}$	Creeping flow; $Re_p \rightarrow 0$	
Diffusion Interception Impaction	$E_{diff,in,im} = 1.68\ Pe^{-2/3}$	Helium $20 < Re_c < 200$	[8]
Diffusion Interception Impaction	$E_{diff,in,im} = 1.53 \times 10^{-[9.9(D_c/2) + 5]\left(\frac{D_p}{2}\right)^{-0.85}}$	Nitrogen $100 < Re_c < 5{,}000$	[9]

Diffusion Interception Impaction	$E_{diff,in,im} = 3.02\, Re_c^{1/6} Pe^{-2/3} + 1.14 \frac{D_p}{D_c} Re_c^{1/3} Pe^{-1/3} + 0.57\left(\frac{D_p}{D_c}\right)^2 Re_c^{1/3}$	$Re_c > 500$	
Diffusion	$E_{diff} = \frac{\pi \mathcal{D}_p D_c Re_c^{1/3} Sc^{1/3}}{Q}$	$Re_c < 3$; $Sc \approx 10^6$	[10]
Diffusion	$E_{diff} = \frac{0.8\pi \mathcal{D}_p D_c Re_c^{1/2} Sc^{1/3}}{Q}$	$600 < Re_c < 2{,}600$ $Sc \approx 10^6$	[11]
Diffusion	$E_{diff} = \frac{0.95\pi \mathcal{D}_p D_c Re_c^{1/2} Sc^{1/3}}{Q}$	$100 < Re_c < 700$ $Sc \approx 10^3$	[12,13]

Interception

In the previous section, the analysis of the inertial impaction mechanisms did not take into account the finite size of the particle. Allowing for the actual size of the particle, disregarding its mass, the removal mechanism known as interception becomes effective. In this situation, the particle follows the streamlines of the gas around the collector. If the particle's center is traveling on a streamline which is closer than the radius of the particle ($D_c/2$), the particle will touch the collector and be intercepted.

Ranz and Wong [1] developed a relationship for interception efficiency when the Stokes number is zero, i.e., a particle following a streamline in potential flow. The collection efficiency by interception is shown in Table 27-1.

Diffusion (Brownian Motion)

Small particles (submicron) suspended in gas experience irregular bombardment by the surrounding gas molecules. The resulting effect on the particle is a zigzag motion, often referred to as Brownian movement. This zigzag motion allows the particle to deviate from its streamlines, thereby enhancing aerosol collection in the vicinity of the collecting obstacle.

Using Fick's law for a stagnant aerosol near a spherical collector, the rate of deposition φ on the collector can be expressed as [6]

$$\varphi = 4\pi \mathcal{D}_p \frac{D_c}{2} C_i \tag{4}$$

where C_i is the aerosol number concentration and $\mathcal{D}_p$ is the particle diffusivity.

The expression most often used to calculate the diffusivity is

$$\mathcal{D}_p = \frac{kT}{3\pi \eta_g D_p} \tag{5}$$

where

k = Boltzmann's constant = 1.38×10^{-16} (erg/K)

T = absolute temperature (K)

η_g = viscosity = 1.83×10^{-4} poise (at 23°C and 760 mmHg)

The criterion most often used for describing the magnitude of particle diffusion is the Peclet number. The larger the Peclet number, the smaller the diffusion collection efficiency. The Peclet number is identified in terms of the particle Reynolds number, Re_c, and the Schmidt number, Sc.

$$Pe = Re_c Sc = \left(\frac{V_{rel} D_c \rho_g}{\eta_g}\right)\left(\frac{\eta_g}{\rho_g \mathcal{D}_p}\right) = \frac{V_{rel} D_c}{\mathcal{D}_p} \quad (6)$$

The Peclet number is a measure of the transport by collective forces compared to the transport by molecular diffusion. When $Pe < 1$, diffusion dominates collection.

The Schmidt number is a dimensionless group which indicates the relative importance of molecular momentum transfer in the gas to diffusion transport of the particle. The importance of Brownian motion of the particle decreases with increasing Sc.

Combined Collection Mechanisms

It is convenient to conceive the collection occurring as a result of the superposition of inertial impaction, interception, and Brownian diffusion. However, these processes are all coupled together, even though one may predominate over the others according to conditions such as particle size, drop size, drop density, hydrodynamic field, etc. Several investigators have considered this possibility and have developed mathematical expressions for the collection efficiency with combined mechanisms.

THEORETICAL COLLECTION FOR A FALLING WATER DROPLET

According to the size of a falling water droplet (or particle), it may or may not attain a constant (terminal) velocity rapidly, or continue to accelerate for a relatively long period of time before coming to terminal velocity. In the simplest case of a rigid

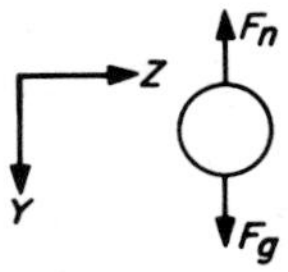

FIG. 27-2. Force diagram for a falling droplet.

spherical particle falling in a gas independently of other particles and free from the influence of walls of the containing vessel, and if the particle is of a size large compared with the mean free path of the gas molecules, then applying the laws of motion to the movement of the particle in a fluid as in Fig. 27-2, the particle moving through the fluid is acted upon by three forces: the gravitational force, $F_g = mg$; the viscous or drag force, $F_\eta = 6\pi r_c \eta_g V_y$; and the inertial force, $F_i = m(dV_y/dt)$. An equilibrium of forces on the particle can therefore be written as

$$\frac{m\, dV_y}{dt_y} = mg - 6\pi \eta r_c V_y \tag{7}$$

where

η = viscosity of the fluid medium
r_c = droplet radius (collector)
V_y = velocity of the droplet
m = mass of the droplet
g = acceleration of gravity
t_y = time

Solution of the above equation yields

$$t_y - t_{yo} = \frac{m}{6\pi r_c \eta} \left(\frac{mg - 6\pi r_c \eta V_{yo}}{mg - 6\pi r_c \eta V_y}\right) \tag{8}$$

and

$$V_y - V_{yo} = \frac{mg}{6\pi r_c \eta} \left[\exp\left(\frac{-6\pi r_c}{m} t_{yo}\right) - \exp\left(\frac{-6\pi r_c}{m} t_y\right) \right] \tag{9}$$

and finally,

$$S_y - S_{yo} = V_{yo}(t_y - t_{oy}) + \frac{mg}{6\pi r_c \eta}\left\{(t_y - t_{oy}) + \frac{m}{6\pi r_c \eta} \times \left[\exp\left(\frac{-6\pi r_c \eta}{m} t_y\right) - \exp\left(\frac{-6\pi r_c \eta}{m} t_{oy}\right)\right]\right\} \quad (10)$$

If the initial velocity of the droplet is considered zero, $V_{yo} = 0$, then

$$t_y = \frac{m}{6\pi r_c \eta} \ln \frac{mg}{mg - 6\pi r_c \eta V_y} \quad (11)$$

$$V_y = \frac{mg}{6\pi r_c \eta}\left[1 - \exp\left(\frac{-6\pi r_c \eta}{m} t\right)\right] \quad (12)$$

and

$$S_y = \frac{mg}{6\pi r_c \eta}\left\{t_y + \frac{m}{6\pi r_c \eta}\left[\exp\left(\frac{-6\pi r_c \eta}{m} t\right) - 1\right]\right\} \quad (13)$$

If air is considered the fluid medium, its density $\rho = 1.19 \times 10^{-3}$ g/cm^3, and viscosity $\eta = 1.83 \times 10^{-4}$ g/cm sec. For a 2,500-μm droplet falling 5 in., it takes 0.1617 sec. It can be shown that if the exponent $6\pi r_c \eta/m$ becomes very small, then the above equation becomes the more familiar Stokes equation,

$$S_y = \frac{mgt_y}{6\pi r_c \eta} \quad (14)$$

Stokes' law is valid so long as there are no inertial effects arising from displacement of air by the sphere, or in other words, when the ratio of the inertial force to the viscous force, the Reynolds number, $Re = D_c V_y \rho_g/\eta$, is very small, about $R < 0.05$. At higher Reynolds numbers, the deviation becomes greater, and Stokes' law tends to overestimate the velocity.

The numerous experimental results for variations of sphere velocities arising from the inertial forces can be coordinated by plotting Re against another dimensionless parameter, the drag coefficient C_D, which is defined as the ratio of the resistance force to the product of the cross section and dynamic pressure.

$$C_D = \frac{(\pi/6)D_c{}^3\rho_c g}{(1/2)\rho_g V_y{}^2(\pi/4)D_c{}^2} \tag{15}$$

In the simple case of Stokes' law,

$$C_D = \frac{24}{Re} \tag{16}$$

Many expressions for the drag coefficient have been published for particles outside the Stokes region. However, a recently published expression [14] for the drag coefficient covers the Reynolds number of immediate interest, $0 \leq Re \leq 5{,}000$:

$$C_D = C_o\left[1 + \frac{\delta_o}{(Re)^{1/2}}\right]^2 \tag{17}$$

where $\delta_o = 9.06$ and $C_o = 24/\delta_o{}^2$. In more general terms, the drag force F_η is written

$$F_\eta = \frac{1}{2}\,\rho V_y^2\, C_D\, \frac{\pi}{4}\, D_c^2 \tag{18}$$

Substituting for C_D and Re, we obtain

$$F_\eta = \frac{C_o\rho\pi D_c^2}{8}\left[1 + \frac{1 + \delta_o(\eta)^{1/2}}{(\rho D_c V_y)^{1/2}}\right]^2 V_y^2 \tag{19}$$

Then, from the equation of motion,

$$m\,\frac{dV_y}{dt} = mg - \frac{C_o\rho\pi D_c^2}{8}\left[1 + \frac{\delta_o(\eta)^{1/2}}{(\rho D_c V_y)^{1/2}}\right]^2 V_y^2 \tag{20}$$

Considering the initial conditions of $t_{yo} = V_{yo} = S_{yo} = 0$,

$$t = \int_0^{V_y} \frac{1}{g - (C_o\rho\pi D_c{}^2/8m)\{1 + [\delta_o(\eta)^{1/2}/(\rho D_c V_y)^{1/2}]\}^2 V_y{}^2}\, dV_y \tag{21}$$

and

$$S_y = \int_0^{V_y} \frac{V_y}{g - (C_o\rho\pi D_c^2/8m)\{1 + [\delta_o(\eta)^{1/2}/(\rho D_c V_y)^{1/2}]\}^2 V_y{}^2}\, dV_y \tag{22}$$

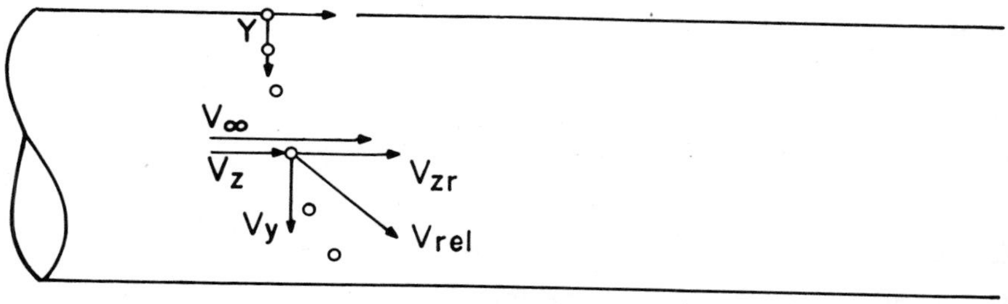

FIG. 27-3. Velocity diagram of a falling water droplet as it traverses a moving air stream. V_z = droplet velocity acquired in the z direction, V_y = droplet falling velocity along the y direction, V_{zr} = relative velocity between gas and droplet along the z axis, V_{rel} = relative velocity between gas and droplet, V_∞ = velocity of free stream.

Slip Correction Factor

The derivation of Stokes' law includes the boundary condition that the relative velocity between the sphere and fluid must go to zero at the surface of the sphere. This means that a thin layer of fluid adheres to the sphere, rather than moving with the fluid. Applying this boundary condition to a gas, such as air, implies that the gas molecules (which must, on the average, have a net velocity in the direction of flow just before they strike the sphere) rebound in such a way that there is no net velocity of molecules adjacent to the sphere surface. This cannot occur in reality, and the gas at the boundary retains some velocity in the direction of flow. As a result, the resistance force acting on a particle in air is less than that calculated from Stokes' law.

This effect, known as slip, is negligible for large particles. It becomes important at a diameter of about 1 μm and less. Using Davies' [15] expression for the slip factor C, we have

$$C = 1 + \frac{2\lambda}{D_p}\left[1.26 + 0.4\,\exp\left(\frac{-1.10\,D_p}{2\lambda}\right)\right] \tag{23}$$

where

D_p = particle diameter
λ = mean free path of air molecules = 6.53×10^{-6} cm (at 23° C and 760 mmHg)

Singlet Droplet Collection Efficiency

As a first approximation to the aerosol collection efficiency by a spherical collector, Strauss' equation, shown below, can be employed. This equation considers the collector to be stationary. Also, to compare theory with practice, potential flow around the collector is used. Hence, to estimate the efficiency of the droplet with the combinations of diffusion impaction and interception, we use

$$E_{comb} = 1 - (1 - E_{diff})(1 - E_{im})(1 - E_{in})$$

where E_{diff} is given by equations listed in the transitional flow section in Table 27-1,

$$E_{im} = \left(\frac{Stk}{Stk + 0.5}\right)^2$$

and

$$E_{in} = \frac{3D_p}{D_c}$$

In the equation for diffusion, the slip factor C must be included when the particle size range being considered is within the slip region. Therefore, the particle diffusivity $\mathcal{D}_p$ must be written as

$$\mathcal{D}_p = \frac{Ckt}{3\pi\eta_g D_p} \qquad (24)$$

where C is given by Eq. (23).

Overall Collection Efficiency

In the previous section, the efficiency calculations did not consider the effect of collection as the droplet sweeps out a volume of gas while falling through its trajectory. Therefore, an overall collection efficiency, E_O, must be obtained, and is given by the expression

$$E_O = 1 - e^{-fE_{comb}} \qquad (25)$$

where f is the fraction of gas volume swept and E_{comb} is the single droplet collection efficiency with the combined removal mechanisms operating.

NOMENCLATURE

C	= Cunningham correction factor for particle (slip factor)
C_D	= drag coefficient
D_c	= diameter of spherical collector (droplet)
D_p	= diameter of particle (aerosol)
E_{im}	= droplet collection efficiency by inertial impaction mechanism
E_{in}	= droplet collection efficiency by interception mechanism
E_{diff}	= droplet collection efficiency by diffusion mechanism
E_{comb}	= droplet collection efficiency by combined mechanisms
Pe	= Peclet number
Q	= gas flow rate in the stream tube having cross section equal to that of the collector
Re_c	= Reynolds number of collector
Sc	= Schmidt number
Stk	= Stokes number
S_y	= falling distance in the y direction
S_z	= traveling distance in the z direction
T	= absolute gas temperature
V_{rel}	= relative velocity of droplet with respect to gas
V_{ter}	= terminal velocity of droplet
V_y	= velocity of droplet in the y direction
V_z	= velocity of droplet in the z direction
V_∞	= free stream gas velocity
C_i	= concentration of particles
f	= fraction of gas volume swept

g = gravitational constant

k = Boltzmann's constant = 1.38×10^{-16} erg/K

m = mass of droplet

r_c = radius of droplet

$\mathcal{D}_p$ = diffusivity of particle

ρ_g = density of gas

η_g = viscosity of gas

φ = rate of deposition of particle on the collector

λ = mean free path of gas

REFERENCES

1. W. Ranz and J. Wong: Industrial and Engineering Chemistry, (1952).
2. Y. Goldshmid and S. Calvert: Small Particle Collection by Supported Liquid Droplets, AIChE Journal, 9:352 (1963).
3. R. Engelmann and W. Slinn: "Precipitation Scavenging," U.S. Atomic Energy Commission, 1970.
4. I. Langmuir and K. Blodgett: U.S. Air Force Technical Report 5418 (1946).
5. I. Langmuir: The Production of Rain by a Chain Reaction in Cumulus Clouds at Temperature Above Freezing, Journal of Meteorology, 5:175 (1948).
6. N. A. Fuchs: "The Mechanics of Aerosols," Pergamon Press, New York, 1964.
7. H. F. Johnston and M. Robert: Industrial & Engineering Chemistry, 41:2417 (1949).
8. V. Hampl et al.: Scavenging of Aerosol Particles by a Falling Water Droplet, Journal of Atmospheric Science, 28(7):1211 (1971).
9. M. Kerker and V. Hampl: Scavenging of Aerosol Particle by a Falling Water Drop and Calculation of Washout Coefficient, Journal of Colloid Interface Science (in press).

10. V. G. Levich: "Physiochemical Hydrodynamics" (Emglish translation), Prentice-Hall, Englewood, N. J., 1962.

11. G. Akselrud: Zhurnal Fizicheskoi Khimii, 27:1445 (1953).

12. F. Garner et al.: AIChE Journal, 4:114 (1958).

13. F. Garner et al.: Chemical Engineering Science, 9:119 (1958).

14. F. F. Abraham: Functional Dependence of Drag Coefficient of a Sphere on Reynolds Number, Physics of Fluids, 13:2191-2195 (1970).

15. C. N. Davies: Definitive Equations for the Fluid Resistance of Spheres, Proceedings of the Physical Society, 57:259 (1945).

16. C. N. Davies: "Air Filtration," Academic Press, London, 1973.

17. H. L. Green and W. R. Lane: "Particulate Clouds: Dusts, Smokes and Mists," Van Nostrand Co., Princeton, N. J., 1964.

18. G. M. Hidy and J. R. Brock: "The Dynamics of Aerocolloidal Systems," Pergamon Press, New York, 1970.

19. E. G. Richardson: "Aerodynamic Capture of Particles," Pergamon Press, New York, 1960.

Chapter 28

VENTURI SCRUBBERS

Jack D. Brady
FMC Corporation
Atlanta, Georgia

L. Karl Legatski
FMC Corporation
Glen Ellyn, Illinois

Venturi scrubbers are applied to air pollution control problems where very high collection efficiencies are required and where most of the particulate matter being removed is smaller than 2 μm in diameter. Because of their physical design, venturi scrubbers utilize fan horsepower more efficiently than any other type of wet scrubber. They are often considered for applications where electrostatic precipitators and fabric filters (baghouses) are also considered. They are always lower in capital cost but higher in energy consumption than these other two devices.

There are a number of applications where venturi scrubbers are the only technically feasible solution to an air pollution problem. If submicron particulate matter is sticky, flammable, or highly corrosive, for example, precipitators and fabric filters cannot be used and venturi scrubbers become a reasonable choice. Venturi scrubbers are also the only ultrahigh-efficiency collectors which can simultaneously remove gaseous and particulate matter from a gas stream without any physical modifications. They are thus often used in applications where acid gases and particulate matter must be removed simultaneously, or where existing air pollution regulations require only particulate removal now but suggest a high probability of including gaseous pollutant controls in the future.

Venturi scrubbers are distinguished from other wet scrubbers by two physical characteristics. First, and most important, is a gas-liquid contacting throat with a constant cross-sectional area over a finite length. In general, the longer the throat, the higher the collection efficiency at a given pressure drop, provided the throat is not so long that frictional losses become significant. For a cylindrical throat, a 3:1 ratio of throat length to diameter is the minimum required to achieve optimum use of fan horsepower. The second feature of a venturi scrubber is the energy recovery section (expander) at the throat discharge which recovers kinetic energy from the mixture of gas and scrubbing liquid drops. The energy recovery section is a constantly expanding duct section starting with minimum cross section at the throat discharge and increasing in area to a point where the scrubbed gas can be discharged at a velocity of less than 100 fps. At less than 100 fps, turbulent losses are minimal and little additional energy recovery is accomplished by slowing the gas down more.

The typical venturi configuration is shown in Fig. 28-1. This is the simplest of all venturi scrubber designs. It includes a converging conical section (the inlet) where the gas is accelerated to throat velocity, a cylindrical throat, and a conical expander where the gas is slowed down and energy is recovered. The scrubber in Fig. 28-1 is described as a "wetted approach" venturi. Liquid is introduced to the scrubber through tangential pipes in the inlet cone. The liquid is distributed at the inlet as a film and flows down the walls to the throat, where it is atomized by the high-velocity gas stream into small liquid droplets which act as the collectors (obstacles) for particulate matter being removed from the gas stream. Once the particulate matter impinges on the liquid drops, it can then be removed by a cyclonic (centrifugal) or chevron (impingement) type of mist eliminator.

A second type of liquid introduction system can also be used. This is the "nonwetted approach" venturi, in which liquid is introduced at the throat rather than on the walls of the converging inlet section. Liquid introduction is accomplished by injecting from the

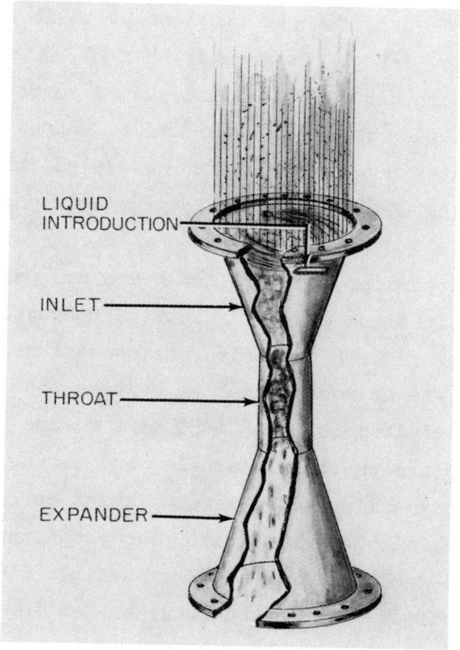

FIG. 28-1. Typical venturi scrubber configuration.

walls of the throat or by injecting from a point somewhere above the throat, using nozzles directed at the throat.

The wetted approach scrubbers are usually more expensive than the nonwetted approach units and, where the inlet gas is already saturated with moisture, or near saturation, the wetted approach is not necessary. Thus, in many applications the nonwetted approach unit is preferred. Where inlet gases are hot and a significant amount of water must be evaporated to bring the gas to the adiabatic saturation temperature, the wetted approach unit is preferred. If a nonwetted approach is used in these applications, a "wet-dry interface" develops at a point where the gas transforms from an unsaturated to a saturated condition. Scale buildup occurs at the wet-dry interface by crystallization and physical deposition of solids suspended or dissolved in the scrubbing liquid. This can eventually cause plugging of the scrubber. The simple round configuration shown in Fig. 28-1 is adequate for gas flows of up to about 40,000 acfm (saturated gas flow rate). At greater than 40,000 acfm, it is virtually impossible to distribute liquid evenly across the cylindrical throat by introduction from the outer walls. Scrubbers built for gas flows of greater than 40,000 acfm require additional liquid distributors including nozzles or a variety of weir and baffle arrangements.

As mentioned earlier, venturi scrubbers are particularly useful in those applications where a sticky or combustible particulate matter is present or where corrosive gas streams are encountered. Because they operate at much higher velocities than baghouses or electrostatic precipitators (150 to 500 fps versus 5 to 15 fps), venturi scrubbers are physically smaller and can be economically manufactured of high-alloy materials to resist corrosive gas streams while the more voluminous precipitators and baghouses cannot. Particulate matter is collected in a noncombustible liquid (usually water) and all surfaces are wetted, thus preventing buildup of solids or any explosion hazard. Performance of venturi scrubbers is independent of particle resistivity, and they can often be used in applications where electrostatic precipitators cannot be used due to the inability to put a static charge on a particle.

Venturi scrubbers are used in applications where pressure drops of between 10 and 100 in.wg are necessary. Below 10 in.wg, low-energy impaction scrubbers are preferred, because the velocities are so low that turbulent losses are not encountered and the expander section is unnecessary. Venturi scrubbers operate at throat velocities varying between about 150 and 500 fps (341 mph). Figure 28-2 shows a throat velocity-versus-differential-pressure curve for a well-designed venturi scrubber operating at a liquid-to-saturated gas ratio (L/G) of 10 gallons per thousand cubic feet (10 gal/Macf) scrubbed. At higher liquid-to-gas ratios, the gas velocity at a given pressure drop is reduced and at lower liquid-to-gas ratios, the velocity is increased. Venturis are operated at liquid-to-gas ratios varying between 3 and 20 gal/Macf, with 7 to 10 gal/Macf being the most common range for efficient operation.

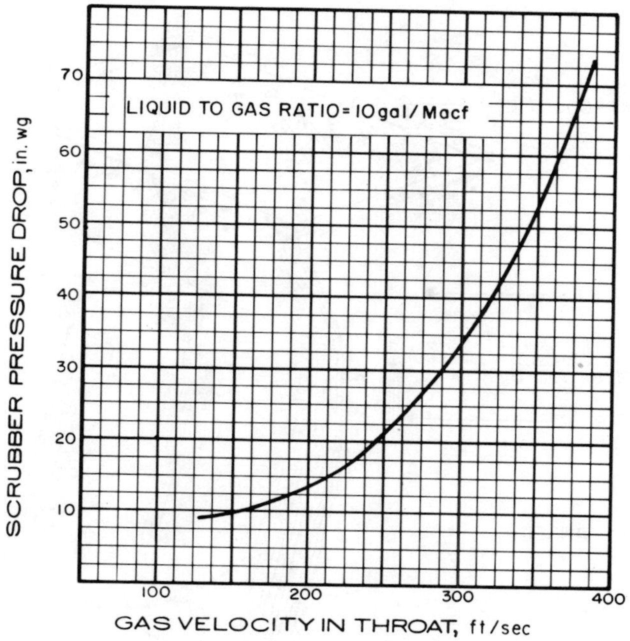

FIG. 28-2. Throat velocities in venturi scrubbers.

Until the early 1960s, venturi scrubbers seldom were built for gas flow rates exceeding 100,000 acfm. This was primarily because of the difficulties encountered in distributing liquid evenly over large throat areas. Since that time, the market has grown rapidly for very large venturi scrubbers, and a number of innovative approaches have been developed for introduction of liquid into the gas stream in a uniform manner over the entire scrubbing throat. It has also become necessary to build variable-throat venturi scrubbers to accommodate changing gas flows while maintaining constant collection efficiencies.

To accommodate the larger gas flow rates, most manufacturers have adopted a rectangular configuration having a long, narrow throat as shown in Fig. 28-3. Some manufacturers have also adopted round configurations having doughnut-shaped throats, with liquid distribution on inserts in the center of the venturi housing. However, as

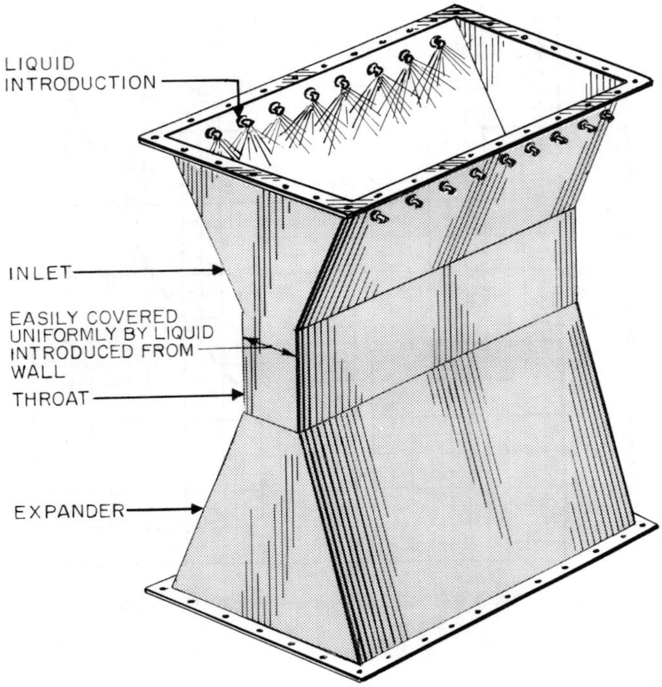

FIG. 28-3. Rectangular venturi scrubber.

can be seen in Fig. 28-4, the venturi design has been compromised in this design by not providing a constant cross-sectional area throat.

Figure 28-5 shows schematically a nonwetted approach, rectangular configuration venturi scrubber with a variable throat which maintains a constant cross-sectional area over the entire range of adjustment. Here, because of the rather narrow throats, liquid distribution does not present a problem. Individual units have been built for up to 350,000 acfm and larger. Figure 28-6 shows a full-scale, nonwetted approach, rectangular configuration, variable-throat venturi scrubber with a cyclonic mist eliminator. This unit was installed on a coal-fired steam boiler to remove sulfur dioxide and flyash from the exhaust gas. Figure 28-7 shows schematically a wetted approach unit with features similar to the unit shown in

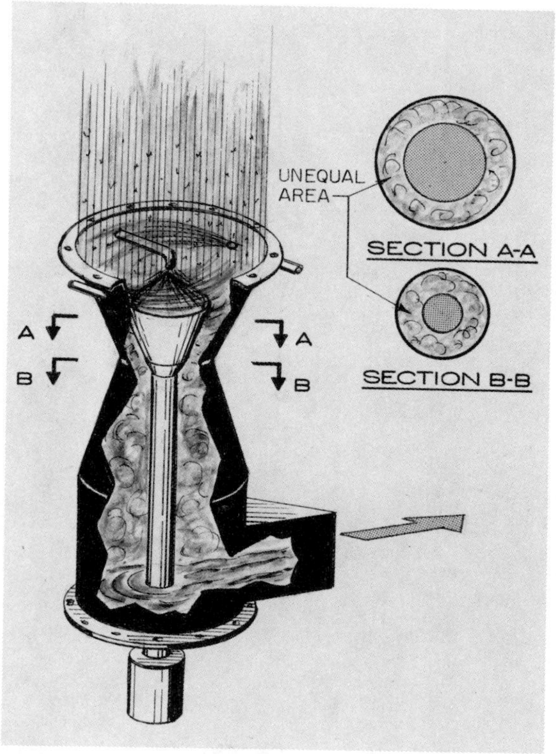

FIG. 28-4. Variable cylindrical venturi scrubber.

Fig. 28-5. The "hidden" liquid inlets are almost universal in large wetted approach units made by all manufacturers. This type of liquid introduction has virtually eliminated the "wet-dry interface" as an operating problem in venturi scrubbers.

PRINCIPLES OF OPERATION

The mechanisms affecting collection of particulates in venturi scrubbers are numerous. The physical phenomena involved are inertia, diffusion, electrostatics, Brownian motion, nucleation and growth, and condensation. All of these affect particulate collection in a venturi scrubber, but it is generally agreed that the predominant phe-

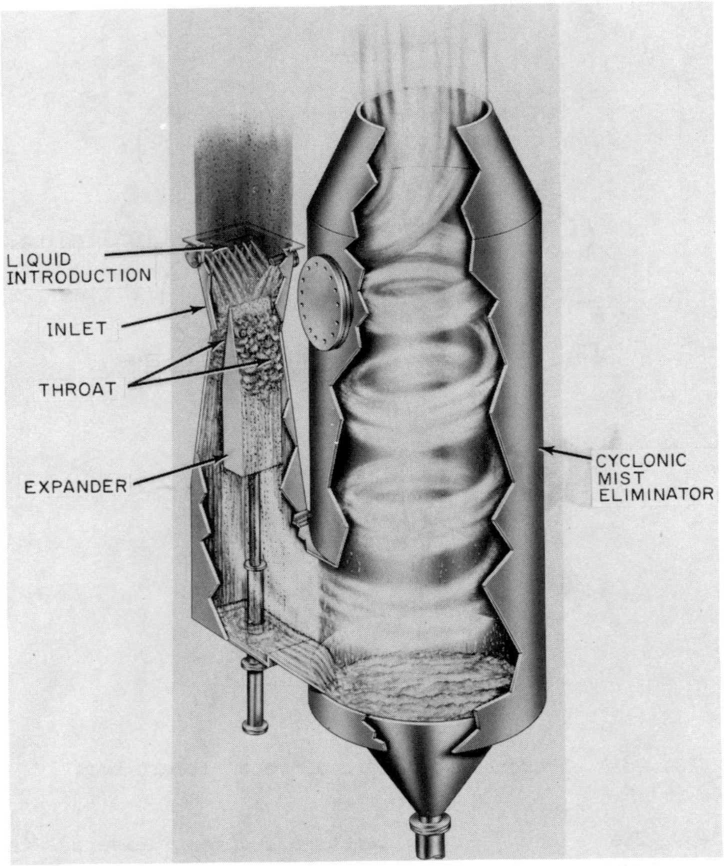

FIG. 28-5. Nonwetted approach, variable rectangular venturi scrubber.

nomenon is inertia. In most wet collectors and many dry collectors, the basic idea is to throw the dust or aerosol particle at an obstacle at a sufficiently high relative velocity so that the particle runs into, or is trapped by, the obstacle. The earliest and most fundamental work in inertial impaction theory was conducted 20 years ago by Ranz and Wong [1]. Ranz and Wong showed that the collection of a particle by an obstacle is a function of what is defined as the inertial impaction parameter K where

FIG. 28-6. Variable venturi scrubber for flyash and sulfur dioxide removal.

$$K = \frac{C\rho U D_p^{2}}{18\mu D_c} \tag{1}$$

where U is the relative velocity, ρ is the particle density, D_p is the particle diameter, μ is the gas viscosity, D_c is the characteristic collector diameter, and C is the Cunningham slip correction factor. The Cunningham slip correction factor is equal to

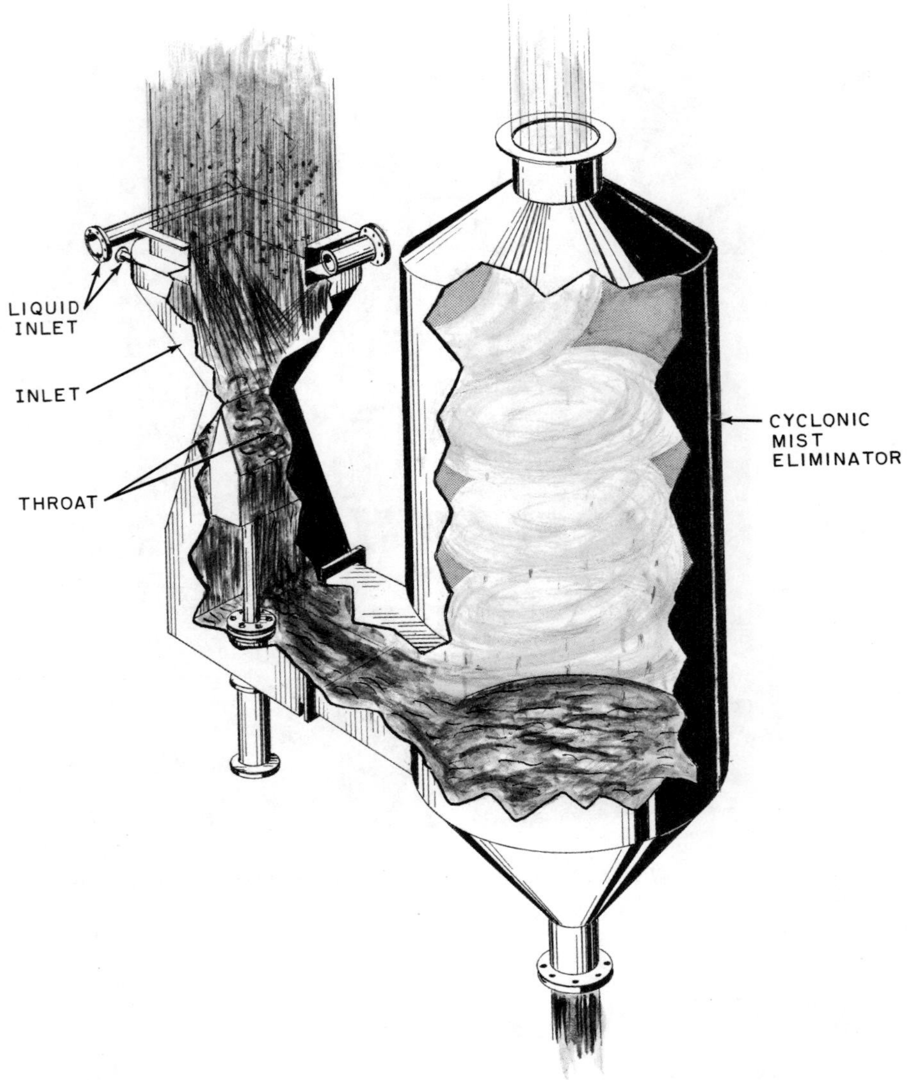

FIG. 28-7. Wetted approach, variable rectangular venturi scrubber.

$$C = 1 + \frac{0.16 \times 10^{-4}}{D_p} \tag{2}$$

This factor corrects for the fact that as particle diameters approach the mean free path of the gas molecules, they tend to "slip" between gas molecules more easily, and are therefore more easily able to cross the bulk flow streamlines and impact on the obstacle. An example might serve to demonstrate the difficulty of capturing a particle in an air stream. If one threw a baseball (collector diameter 3 in.) at a velocity of 100 mph at a droplet of water (density 1 g/cm^3) suspended in air (viscosity 1.8×10^{-4} g/cm sec), he would have a 50 percent chance of hitting a 19-μm-diameter particle. In order to collect a 1-μm-diameter particle (as is common in a venturi scrubber), even if one could throw the particle at the speed of sound (1,128 fps), the collector would have to be no more than 1/16 in. in diameter in order to have a 50 percent chance of capturing the particle.

It is obvious from inspection of the inertial impaction parameter that in order to capture a particle of a given diameter and density in a gas stream of a given viscosity, there are only two variables that can affect the collection efficiency (assuming, as stated earlier, that impaction is the predominant mechanism): (1) the relative velocity between the gas and the collection object, and (2) the characteristic dimension of the collection object. A venturi is a well-known device for accelerating a fluid stream to a high velocity and returning it to its original velocity with a minimum loss of energy. This, of course, is the reason it is used in high-velocity wind tunnels. It is therefore only natural that the venturi was chosen as the most efficient means of contacting a gas and a liquid for particulate collection.

In a venturi scrubber, the collection object is a water drop. Its diameter is a complicated function of velocity, liquid rate, and fluid properties. Normally, a centrifugal fan is installed upstream (forced draft) or downstream (induced draft) from the venturi scrubber. The fan provides the motive force to the gas stream carrying the particulate matter. The gas is accelerated to throat velocity in the conical inlet. It then passes through the throat where the suspended particulate matter encounters the liquid drops (obstacles). Those particles which impact on liquid drops can then be easily separated from the bulk of the gas stream by collecting them in a cyclonic (centrifugal) or chevron (impaction) mist eliminator. Those particles which do not impact on liquid drops "penetrate" the scrubber and are exhausted with the gas stream.

Liquid drops can be created in two different manners in the scrubber. The most common method is simply to allow the high-velocity gas to atomize the liquid exactly as in a perfume spray bottle. This, of course, consumes some energy as fan horsepower. The second method is to atomize the liquid by forcing it through small, high-pressure orifices in spray nozzles. In this case, the energy used to atomize the liquid is provided by pump horsepower. There are no substantial energy savings realized by using either technique, but the high-pressure nozzle technique is limited to those applications where a clean liquid stream is fed to the scrubbing throat.

Due to the rather large volume of liquid which must be circulated to the scrubbing throats, it is often impractical to provide a clear liquid source and this technique often cannot be used. However, the high-pressure atomization technique offers the advantage of simplifying distribution of the liquid evenly across the throat. This becomes a particularly significant consideration when scrubbing throats reach a size where introduction of the liquid along the walls and subsequent atomization of this liquid off the walls does not result in proper distribution of liquid droplets in the center of the throat to provide enough obstacles for the particulate matter to impact on.

The two techniques are sometimes combined to take advantage of the best features of each. Liquid is sometimes introduced in a uniform fashion across the throat via low-pressure spray nozzles.

When these drops enter the throat area and encounter the high-velocity gas stream, they explode into thousands of smaller droplets (atomization).

As the gas exits the scrubbing throat, it carries with it all of the liquid droplets which have now achieved a velocity very nearly that of the gas stream. In the expander section, the gas is slowed down as the cross-sectional area increases. Some of the kinetic energy from the liquid droplets transfers back to the gas stream, resulting in a recovery of part of the energy required to accelerate the gas to throat velocity. This energy regain is what distinguishes a venturi scrubber from any other type of wet scrubber. Once the gas has been slowed down sufficiently to minimize additional turbulent losses, it is then directed to a mist eliminator where, via centrifugal force or an impaction mechanism, it is separated from the gas stream.

For a given particle size and a given throat velocity, one can determine empirically the fraction of particles which will be collected by the venturi scrubber. This represents the "fractional efficiency" for a given particle diameter. If, at a given throat velocity, all empirical fractional efficiency points are plotted versus particle diameter, a collection efficiency curve for a given pressure drop or throat velocity is produced. Typically, collection efficiency curves are plotted for varying pressure drops rather than throat velocities because this provides information on fan requirements. Figure 28-8 shows typical collection efficiency curves at five different pressure drops for particulate matter having a specific gravity of 1.0 in a well-designed venturi scrubber.

The term "pressure drop" refers to the difference in pressure between the gas at the inlet to the venturi scrubber and the gas at the discharge from the venturi scrubber. Figure 28-9 shows a typical pressure profile for a venturi scrubber. As the gas is accelerated, the pressure in the gas stream decreases to its lowest point in the throat. As the gas begins to slow down in the expander section, pressure begins to rise and reaches a level only slightly lower than the pressure at the inlet. The difference between the inlet and

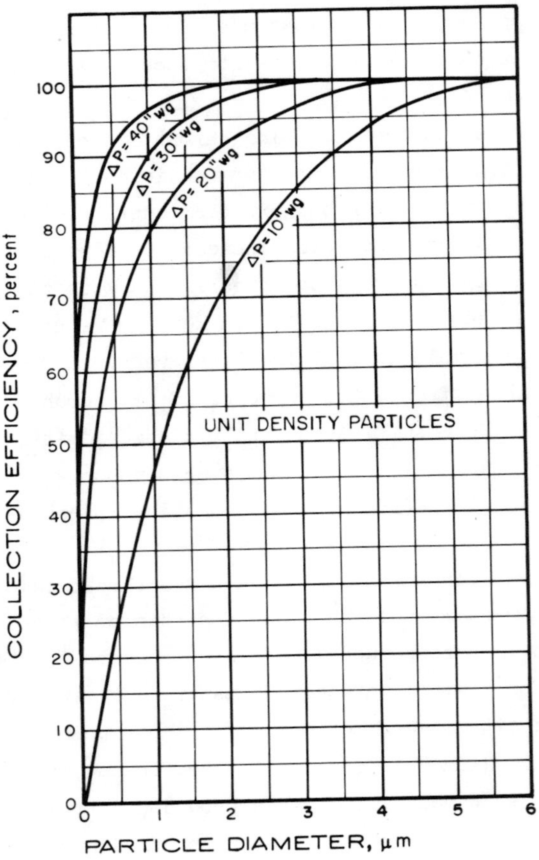

FIG. 28-8. Typical venturi scrubber collection efficiencies.

outlet pressures, or the pressure drop, represents the energy expended in the scrubbing process.

The Calvert equation [2] can be used to predict the pressure drop for a given throat velocity. The Calvert equation is

$$\Delta P = (5 \times 10^{-5})v^2L \tag{3}$$

where L is the liquid rate in gallons per thousand cubic feet at saturated conditions and v is the gas velocity in feet per second. The equation says that the pressure drop is equal to the power

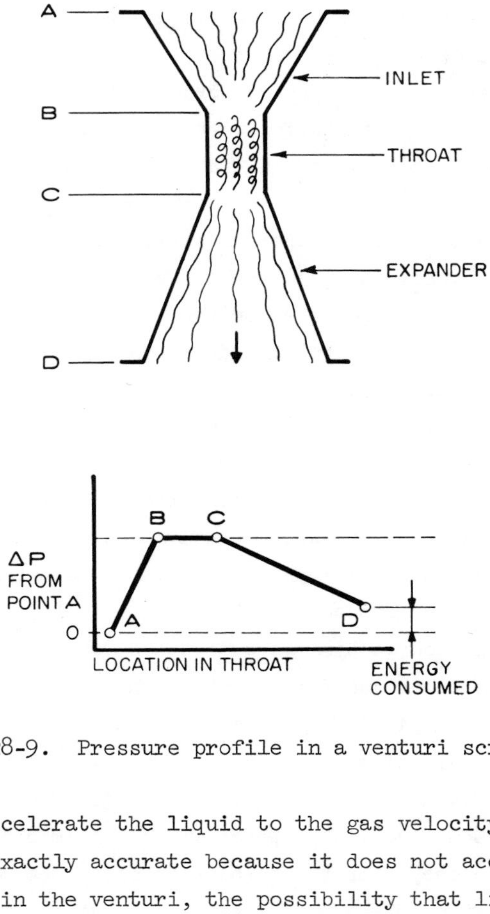

FIG. 28-9. Pressure profile in a venturi scrubber.

required to accelerate the liquid to the gas velocity. This obviously is not exactly accurate because it does not account for frictional losses in the venturi, the possibility that liquid is not accelerated to the full gas velocity, and the fact that the liquid transfers momentum back to gas in the expander section of the venturi. It does, however, predict pressure drop reasonably well, except at high liquid-to-gas ratios. For a liquid-to-gas ratio of 5 gal/Macf, the Calvert equation predicts pressure drop with reasonable accuracy. At liquid rates of 12 gal/Macf, observed pressure drops are usually about 80 percent of those predicted by the Calvert equation. In practice, it has been found that at liquid rates of less than 3 gal/Macf there is inadequate liquid to cover the venturi throat. It has also been found that liquid rates in excess of 10 gal/Macf are seldom justified, in that they do not give any improvement in performance. Most venturi scrubbers are designed for liquid rates between 7 and 10 gal/Macf, and there is virtually no change in performance over this range of liquid rates at a constant pressure drop across the scrubber.

To design a venturi scrubber properly for a given application, it is necessary to rely on previous experience with analogous applications, test a venturi scrubber on the source, or utilize performance curves for a given venturi scrubber and conduct sufficient sampling of the source to determine the characteristics of the particulate matter. If one has a particle size distribution and specific gravity of the particulate matter to be removed from the gas stream, and if collection efficiency curves as a function of particle diameter for given pressure drops are available, the fraction of particulate matter that will pass through the scrubber without being collected (penetrate) can be calculated. To do this, the integral of the performance curve times the size distribution must be calculated. A simple technique for computing this integral is shown in Table 28-1. From a plot of the size distribution, such as the typical size distribution shown in Fig. 28-10, the fraction of the total dust loading in each of the discrete size ranges listed in Table 28-1 is determined. The penetration (one minus the fractional collection efficiency) for the total size distribution is then computed by calculating the sum of the products of the fractions in each size range and the penetration for that size range at that pressure drop. This same calculation is done for a number of pressure drops and a penetration-versus-pressure drop curve is then plotted as in Fig. 28-11. Table 28-2 lists data for a typical scrubber application where this procedure of pressure drop selection has been used.

After this required pressure drop has been computed, the venturi scrubber must be sized. Usually, all dimensions of the scrubber derive from the size of the scrubbing throat itself. Thus, one can

TABLE 28-1

Penetration Calculation

Particle diameter (μm)	Fraction in size range $f(D_p)$		Fractional penetration = $P_t(D_p)$			
			ΔP = 10 in.	ΔP = 20 in.	ΔP = 30 in.	ΔP = 40 in.
> 6	0.063	×	0 = 0	0 = 0	0 = 0	0 = 0
5-6	0.042	×	0.01 = 0.00042	0 = 0	0 = 0	0 = 0
4-5	0.077	×	0.03 = 0.00231	0 = 0	0 = 0	0 = 0
3-4	0.138	×	0.08 = 0.1104	0.02 = 0.00276	0 = 0	0 = 0
2-3	0.245	×	0.20 = 0.049	0.06 = 0.0147	0.012 = 0.00294	0 = 0
1.5-2	0.165	×	0.35 = 0.05775	0.10 = 0.0165	0.035 = 0.005775	0.005 = 0.000825
1.0-1.5	0.173	×	0.45 = 0.07785	0.16 = 0.02768	0.072 = 0.012456	0.025 = 0.004325
0.5-1.0	0.087	×	0.60 = 0.0522	0.27 = 0.02349	0.162 = 0.014094	0.05 = 0.00435
0-0.5	0.010	×	0.87 = 0.0087	0.50 = 0.005	0.35 = 0.0035	0.20 = 0.002
$\int_o^\infty P_t(D_p)f(D_p)\,dD_p$ = total penetration =			0.25927	0.09013	0.038765	0.0115
1 - total penetration = collection =			0.741	0.910	0.961	0.989

Note: Calculations for size distribution in Fig. 28-10.

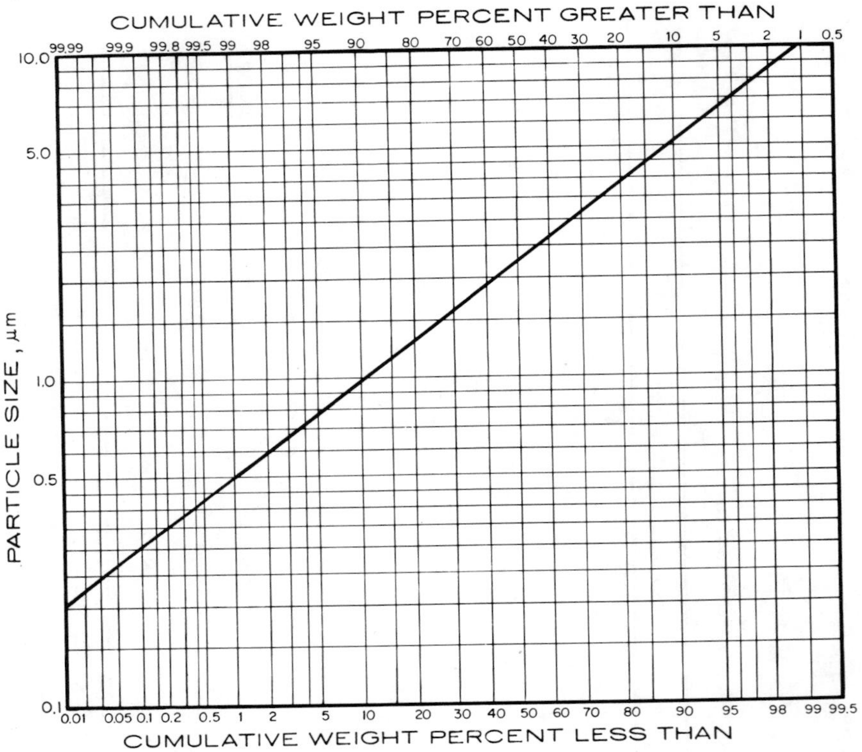

FIG. 28-10. Hypothetical particle size distribution.

TABLE 28-2

Pressure Drop Determination
for Typical Venturi Scrubber Application

Data supplied by purchaser

1. Inlet dust loading to scrubber = 1.83 grains/sdcf
2. Inlet gas flow to scrubber = 183,000 sdcfm
3. Allowable emission = 48.79 lb/hr
4. Size distribution of particulate matter:

Particle diameter (μm)	Weight percent less than
10.0	99.0
5.0	89.5
2.0	43.5

TABLE 28-2 (Continued)

Particle diameter (μm)	Weight percent less than
1.0	9.7
0.5	1.0

Note: This distribution is plotted in Fig. 28-10.

Calculation and graphical solution

1. Determine required collection efficiency:

   a. Inlet loading

   $$1.83 \text{ grains/sdcf} \times 183{,}000 \text{ sdcf/min} \times \frac{1 \text{ lb}}{7{,}000 \text{ grains}} \times \frac{60 \text{ min}}{\text{hr}} = 2{,}870.5 \text{ lb/hr}$$

   b. Collection efficiency

   $$\frac{2{,}870.5 \text{ lb/hr} - 48.79 \text{ lb/hr}}{2{,}870.5 \text{ lb/hr}} \times 100 = 98.3\%$$

   c. Allowable penetration

   100.0 - 98.3 = 1.7% = 0.017 penetration

2. Use Table 28-1 to calculate penetration at 10, 20, 30, and 40 in. water gauge.

   10 in.wg = 0.25927

   20 in.wg = 0.09013

   30 in.wg = 0.038765

   40 in.wg = 0.0115

3. Plot the penetrations from Table 28-1 as a function of pressure drop:

   See Fig. 28-2

4. Read from Fig. 28-2 the pressure drop required for 1.7% penetration:

   1.7% penetration = 36.5 in.wg

---

use either the Calvert equation or empirical velocity-versus-differential pressure curves for a given scrubber to size the throat for a given saturated gas flow rate. Note that it is imperative to use the saturated gas flow rate rather than the hot inlet gas flow rate for sizing the throat. As a hot flue gas stream enters the throat,

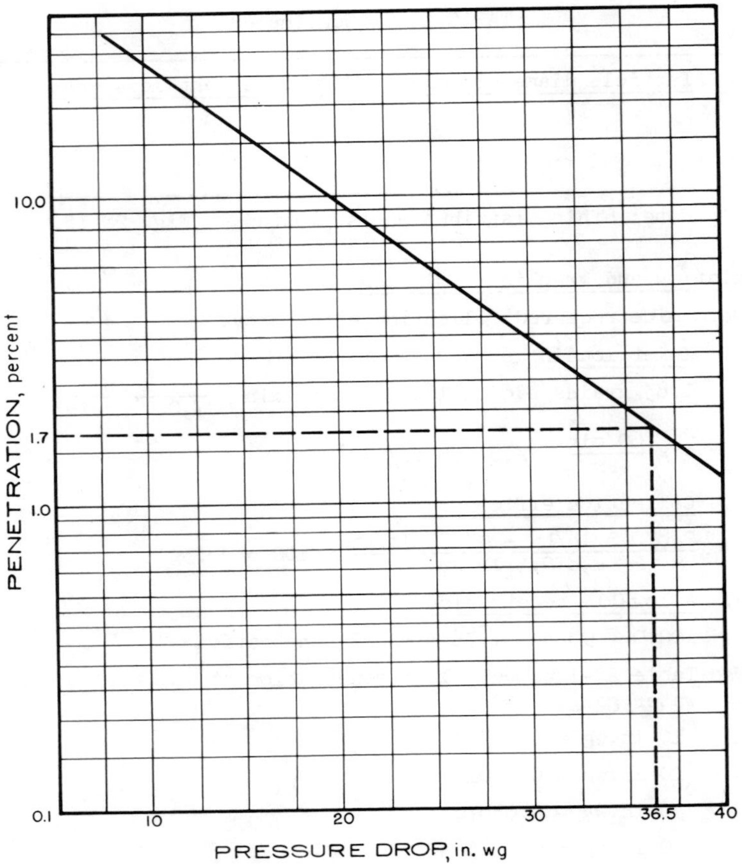

FIG. 28-11. Penetration versus pressure drop.

it is immediately quenched to its saturation temperature and the volume is reduced substantially. If the throat were sized on the unsaturated hot gas volume, it would be far too large for most applications and the required collection efficiency could not be achieved.

A fan must also be selected which will handle the required gas volume. (Note: The fan is sized on the actual gas volume at the fan, not necessarily the saturated gas volume, particularly on the forced draft side.) Finally, pumps, piping, ducting, and tankage must also be designed.

INDUSTRIAL APPLICATIONS

While the predominant application of venturi scrubbers is in the collection of particulate matter from gas streams, they can also be used for gas absorption. They operate as cocurrent contactors (gas and liquid move in the same direction) and are inherently poorer collectors of gases than countercurrent low-energy scrubbers, including packed-bed and tray-type absorbers. However, where gases must be absorbed from gas streams containing small particulate matter, venturi scrubbers are often adequate. They have found widespread application in simultaneous removal of flyash and sulfur dioxide from fossil-fueled power plant exhaust gases.

In addition, venturi scrubbers are used for particulate and gaseous pollutant removal in the industries discussed below. Additional applications are listed in Table 28-3.

Iron and Steel Industry

As in most primary metals industries, air pollution problems are first encountered in the steel industry during mining and processing of the ore. Most of these problems result during size reduction of the ore. As is true of most ore crushing and grinding operations, however, the particle size of the dust emitted by ore crushers is quite large in comparison with most other emissions in the steel industry. Thus, relatively low pressure drops can be used to control these emissions. For a typical iron ore crushing operation, venturi scrubber pressure drops of less than 20 in.wg are used to control particulate emissions. These relatively low pressure drops result in emission reductions of about 99.5 percent or greater.

As the ore moves to the steel mill, the extremely high-temperature processing operations generate much smaller particulate matter.

TABLE 28-3

Typical Venturi Scrubber Applications

Application	Pressure drop (in.wg)	Material of construction
Boilers		
Pulverized coal	15-40	316L stainless steel
Stoker coal	10-12	316L stainless steel
Bark	6-10	Carbon steel
Combination	10-15	316L stainless steel
Recovery	30-40	Carbon steel or 316L stainless steel
Incinerators		
Sewage sludge	18-20	316L stainless steel
Liquid waste	50-55	High nickel alloy
Solid waste		
Municipal	10-20	316L stainless steel
Pathological	10-20	316L stainless steel
Hospital	10-20	High nickel alloy
Kilns and calciners		
Lime	15-25	Carbon steel or stainless steel
Soda ash	20-40	Carbon steel or stainless steel
Potassium chloride	30	Carbon steel or stainless steel
Coal Processing		
Dryers	25	304 stainless steel or 316L stainless steel
Crushers	6-20	Carbon steel
Dryers		
General spray dryer	20-60	Carbon steel or stainless steel
Food spray dryer	20-30	Food-grade stainless steel
Fluid bed dryer	20-30	Carbon steel or stainless steel
Mining		
Crushers	6-20	Carbon steel
Screens	6-20	Carbon steel
Transfer points	6-20	Carbon steel

TABLE 28-3 (Continued)

Application	Pressure drop (in.wg)	Material of Construction
Iron and steel		
Cupolas	30-50	304-316L stainless steel
Arc furnaces	30-50	316L stainless steel
BOFs	40-60	Carbon steel (ceramic lined)
Sand systems	10	Carbon steel
Coke ovens	10	Carbon steel
Blast furnaces	20-30	Carbon steel (ceramic lined)
Open hearths	20-30	Carbon steel (ceramic lined)
Nonferrous metals		
Zinc smelters	20-50	Stainless steel or high nickel
Copper and brass smelters	20-50	Stainless steel or high nickel
Sinter operations	20	Stainless steel or high nickel
Aluminum reduction	50	High nickel, FRP mist eliminator
Phosphorus		
Phosphoric acid		
Wet process	10-30	316L stainless steel
Furnace grade	40-80	316L stainless steel
Asphalt		
Batch plants--dryer	10-15	Stainless steel
Transfer points	6-10	Carbon steel
Glass		
Container	25-60	Stainless steel
Plate	25-60	Stainless steel
Borosilicate	30-60	Stainless steel
Cement		
Wet process kiln	10-15	Carbon steel or stainless steel
Transfer points	6-12	Carbon steel

TABLE 28-3 (Continued)

Application	Pressure drop (in.wg)	Material of construction
Fertilizer		
Dryers	10-20	Stainless steel
Ammoniators	15-30	Stainless steel
Coolers	10-20	Stainless steel

Venturi scrubber applications in steel mills usually require high pressure drops. Electric furnaces, open-hearth furnaces, BOF furnaces, and blast furnaces all require some means of air pollution control. In addition, sinter machines and lime kilns in steel mills also require scrubbing equipment. All of these represent excellent applications for venturi scrubbers.

Venturi scrubbers on open-hearth and blast furnaces typically operate at 20 to 30 in.wg. BOF furnace scrubbers are operated at higher energy levels. In all of these applications, wetted approach venturis are used due to the high inlet gas temperatures and the probability of solids buildup at the wet-dry interface in a non-wetted approach unit. For lime kilns, pressure drops of about 15 in.wg are typically used. In all of these applications, collection efficiencies of greater than 95 percent are achieved.

Electric furnaces often require even higher pressure drops to achieve acceptable outlet loadings. They also require extensive hooding and are highly variable, even over a period of a few minutes, in emission rate and particle size. It is not uncommon to find scrubbers equipped with two-speed fan drives on electric furnaces to expend the high pressure drop only during peak emission periods.

The iron foundry industry uses venturi scrubbers to collect emissions from electric furnaces and from cupola furnaces. Venturi scrubbers are also used to remove particulate matter from actual pouring operations and from the sand shake-out operations after castings have been poured. Often, during pouring and the shake-out operations, odoriferous compounds are emitted which require collection in the gas phase. Particulate matter must be collected simultaneously. The shake-out operations require rather low pressure drop scrubbers with the maximum pressure drop usually being about 12 in.wg Cupola emissions, on the other hand, because they are produced by very high-temperature operations, are quite small and require high pressure drops to collect. Typical cupola installations operate at between 30 and 50 in.wg. Electric furnaces usually require from 30 to 50 in.wg. Pouring operations require 20 to 30 in.wg.

Particulate matter from scrubbers in these iron and steel applications are commonly separated by sedimentation from the liquid stream. Often the collected particulate is returned directly to the process.

Chemical Industry

Venturi scrubbers are commonly used in inorganic chemical plants. They find far fewer applications in the organic chemicals industry or in petroleum plants. They are used for emissions from rotary calciners, spray dryers, materials-handling operations, primary ore reduction operations, and reactor-emissions-containing particulate matter or aerosols and contaminating gases.

For ore reduction and materials-handling operations, venturi scrubbers operating at pressure drops of between 10 and 20 in.wg are common. These usually achieve 99+ percent collection efficiency and, where a soluble product is involved, return the product directly to the chemical process for reuse. In applications where the particulate matter collected is not soluble, thickeners, settling ponds, and a variety of filtering devices are used to separate the particulate matter from the liquid. Again, the particulate matter is often returned to the process.

For calcining and spray drying operations, venturi scrubbers operating between 20 and 60 in.wg are common. Usually, the higher the exhaust temperature from the process, the smaller the particulate matter and the higher the pressure drop required to achieve an adequate collection efficiency. Wetted approach venturis are usually used.

There are also a number of installations where venturi scrubbers are used to remove aerosol droplets produced during the reaction of a gas with a liquid to produce an acid. Sulfuric and hydrochloric acid production often require venturi scrubbers for removal of aerosol mists. Some neutralization processes, due to the heat generated during neutralization, also produce extremely fine aerosols which can be collected in venturi scrubbers. Typical pressure drops for these applications are between 30 and 40 in.wg. Nonwetted approach venturis are used for these applications.

In combustion processes, sometimes used to manufacture an inorganic chemical, extremely fine particulate matter is produced. An example is the combustion of phosphorus to produce P_2O_5, which is then contacted with water to make phosphoric acid. The P_2O_5 particulate matter and the acid mist are almost all less than 1 μm in diameter. Pressure drops ranging from 40 to 80 in.wg are required for venturi scrubbers in this application. Paint pigment is sometimes produced by oxidation in a high-temperature flame and, again, extremely high-pressure-drop scrubbers are required to remove this particulate matter from a gas stream.

Prill towers in fertilizer plants often produce a fine dust which requires venturi scrubbing. In these applications, either wetted approach or nonwetted approach units can be used. Pressure drops required are less than 20 in.wg.

Wet process phosphoric acid plants use venturi scrubbers in the reaction system and on the evaporators to collect hydrogen fluoride, phosphoric acid mist, and silicon tetrafluoride. Wetted approach units are often used.

Electric Power Generation and Industrial Steam Production

One of the most rapidly developing applications for venturi scrubbers is the collection of flyash and sulfur dioxide in flue gases from coal- and oil-fired steam-generating boilers used either by the utility industry to produce electric power or by industrial plants to produce process steam. The pressure drops required to collect the particulate matter are dictated more by the size of the boiler than by the actual operation which produces the particulate matter. Most state regulations are written so that the particulate matter emission limitations for very large sources require higher collection efficiencies than for small sources. Thus, the large utility plant installations often must achieve 99.5 percent or greater collection efficiency, while industrial boilers usually require only 97 to 99 percent collection efficiency. Venturi scrubbers used for these applications range in pressure drop from about 10 in.wg for the smallest industrial boiler to as much as 40 to 45 in.wg for the largest utility boilers. Venturi scrubbers are often used to remove sulfur dioxide by collecting it in a scrubbing solution containing an alkaline reactant. Using lime and limestone slurries, sulfur dioxide collection efficiencies of as high as 90 percent can be attained. Using water-soluble absorbers, such as sodium hydroxide, sodium carbonate, or sodium sulfite, collection efficiencies of greater than 95 percent can be achieved. The venturi scrubber is particularly attractive in this application because it is free of internals which would interfere with flow of the gas and act as collectors for scale or flyash. Both wetted and nonwetted approach venturis are used, depending upon the inlet gas temperature.

Coal-processing plants also use venturi scrubbers extensively for particulate control on coal dryers. These units are normally nonwetted approach venturis operating at pressure drops of between 20 and 30 in.wg.

Pulp and Paper Industry

Venturi scrubbers have found widespread application in the pulp and paper industry in three areas. In black liquor recovery boilers, where an organic-rich slurry is burned to reduce sodium sulfate to sodium sulfide, particulate matter suspended in gases containing hydrogen sulfide and mercaptans creates a significant emission problem. Venturi scrubbers have been used to remove the particulate matter from recovery boilers and to concentrate the feed to the recovery boilers by evaporating water into the gas stream during adiabatic saturation of the off-gas. Often, two venturi scrubbers in series are used, one as an evaporator and one as a particulate scrubber. These typically require about 30 in.wg total. Nonwetted approach units are most commonly used if an evaporator is installed to concentrate the feed liquor to the recovery boiler.

Paper mills also have lime kilns installed at the plant site and venturi scrubbers are used to control their emissions. The kilns are similar to those which have already been mentioned for the steel industry and the conditions under which venturi scrubbers are operated are the same.

Many paper mills utilize wood waste to generate steam in hogged-fuel fired boilers. The particulate matter produced from the wood waste combustion process is typically larger than flyash from coal- or oil-fired plants, and the pressure drops necessary to collect the particulate matter are lower. Venturi scrubbers are operated at pressure drops ranging from about 6 in.wg to as high as 15 in.wg. The 15-in.wg scrubbers are typically used only where the wood waste has been soaked in salt water during some point in its transport history. The salt-water-soaked fuels generate a fine sodium chloride particulate matter which is difficult to scrub at the lower pressure drops. Paper mills in the southern United States have made extensive use of venturi scrubbers for this application. The particulate matter collected is usually filtered or screened and returned to the boiler where a significant amount of additional heat can be generated by its combustion.

Rock Products and Asphalt Production

The gravel, cement, and asphalt industries all utilize venturi scrubbers for emissions from kiln, dryer, and bulk materials handling operations. Rotary kilns used in the cement industry utilize venturi scrubbers for emission control, just as in the lime kilns described earlier in the steel and pulp and paper discussions. Usually venturi scrubbers are installed where wet process kilns are utilized. If dry process kilns are used, precipitators and fabric filters usually find greater application.

In hot-mix asphalt plants, venturi scrubbers are used to control emissions from the rotary dryers. Gravel plants use venturi scrubbers to control emissions from crushing and grinding operations, screening operations, and material transfer points. Gravel quarries are often located in areas where scrubbing water is available and where the waste water from the scrubber can simply be returned to the quarry. These applications are usually low pressure drop applications. The crushing and grinding operations and material transfer points usually require pressure drops of between 6 and 12 in.wg. Cement kiln scrubbers operate at pressure drops of between 15 and 30 in.wg.

Nonferrous Metals Industries

Lead, zinc, and copper smelting operations typically produce very fine sulfuric acid mists, high concentrations of sulfur dioxide, and a variety of particulate matter. Venturi scrubbers have been used in a number of instances to control these emissions. Sulfur dioxide absorption in a venturi can be accomplished at virtually any pressure drop. The fine particulate matter, combined with the sulfuric acid mist, requires a high pressure drop to achieve an adequate collection efficiency. Typically, pressure drops ranging from 20 to 50 in.wg are used. Metal-refining operations also produce fine particulate matter. In almost all cases, this particulate matter can be economically recovered using venturi scrubbers and returned to the processing operation. Secondary metal recovery operations, including combustion of X-ray film to recover silver, incineration of insulation on copper wire to recover the copper, and the recovery of lead from lead-zinc batteries are also good applications for venturi scrubbers. The pressure drop selected for these applications generally depends on the economic return gained by an incremental increase in pressure drop. For example, gold-refining operations almost always utilize higher pressure drop scrubbers than copper smelting operations. This simply reflects the value of the particulate matter recovered. In most of these applications, pressure drops of about 20 in.wg are adequate. However, for the rare metals, pressure drops of up to 50 in.wg are common.

Glass Container, Fiberglass, Plate Glass, and Refractory Furnaces

In any process where silicon oxide is melted or heated to an extremely high temperature and where some chemicals are added to the batch to enhance the melting characteristics or to add desirable physical properties to the mixture, a very fine particulate matter is often produced in the exit gas. This particulate matter is typically composed of the more volatile compounds, including sodium sulfate and sodium carbonate, which are volatilized at furnace temperature and then condensed in the gas exiting the furnace. Because they condense from the gas phase, they create extremely fine particles. These usually require high pressure drops to collect. Venturi scrubbers operating at 30 in.wg to as high as 60 in.wg are used. In some cases, it is possible to control the rate at which the gas is cooled so that the condensation can be controlled and so that the particle size can be increased to a much larger size. This enables the use of a lower pressure drop scrubber for the same level of collection efficiency.

Clay and refractory furnaces often utilize compounds containing fluorides and chlorides. Thus, in addition to the particulate matter emitted, highly acidic gases are emitted and the scrubbers must be built to remove not only the particulate matter but the acidic gases as well.

Numerous other venturi scrubber applications exist. As simultaneous control of gaseous and particulate matter emissions becomes more common, use of venturi scrubbers will become even more widespread. The trend is toward larger and larger units for large emission sources rather than multiple small units.

REFERENCES

1. W. E. Ranz and O. B. Wong: Industrial Hygiene and Occupational Medicine, 5:464 (1952).

2. S. Calvert: Source Control by Liquid Scrubbing, Air Pollution III, Academic Press, New York, 1968, chap. 46.

Chapter 29

JET VENTURI FUME SCRUBBING

W. J. Gilbert

Croll-Reynolds Company
Westfield, New Jersey

The jet venturi fume scrubber is a versatile tool for many dust and gas applications. Its ability to handle both dust and gas simultaneously makes it useful in many problem applications. Since there are no moving parts within the scrubber, it is ideal for handling sticky or abrasive materials.

The principle of operation involves a jet effect created by a water spray nozzle. The water spray nozzle is located on the top of the jet scrubber and creates a hollow cone-shaped spray (Fig. 29-1). This is a relatively narrow-angle spray which contacts the wall of the jet scrubber at a point above the throat. The result is an induced air flow through the scrubber. The gas and liquid enter the throat, where extreme turbulence is encountered, and continue through a diffuser section where partial separation of the gas and liquid occurs. The cocurrent nature of this scrubber requires that a separation device be used to separate the gas completely from the liquid.

The scrubbing mechanism includes the cross-flow effect of the air being entrained through the spray plus the turbulence which occurs at the throat area. Several theories have been put forth to describe the scrubbing action and predict its effectiveness, but to date, they have been unable to come up with a single theory which predicts the results under varying operating conditions. Pilot plant studies or field tests of identical situations are the principal means for obtaining data on this type of unit.

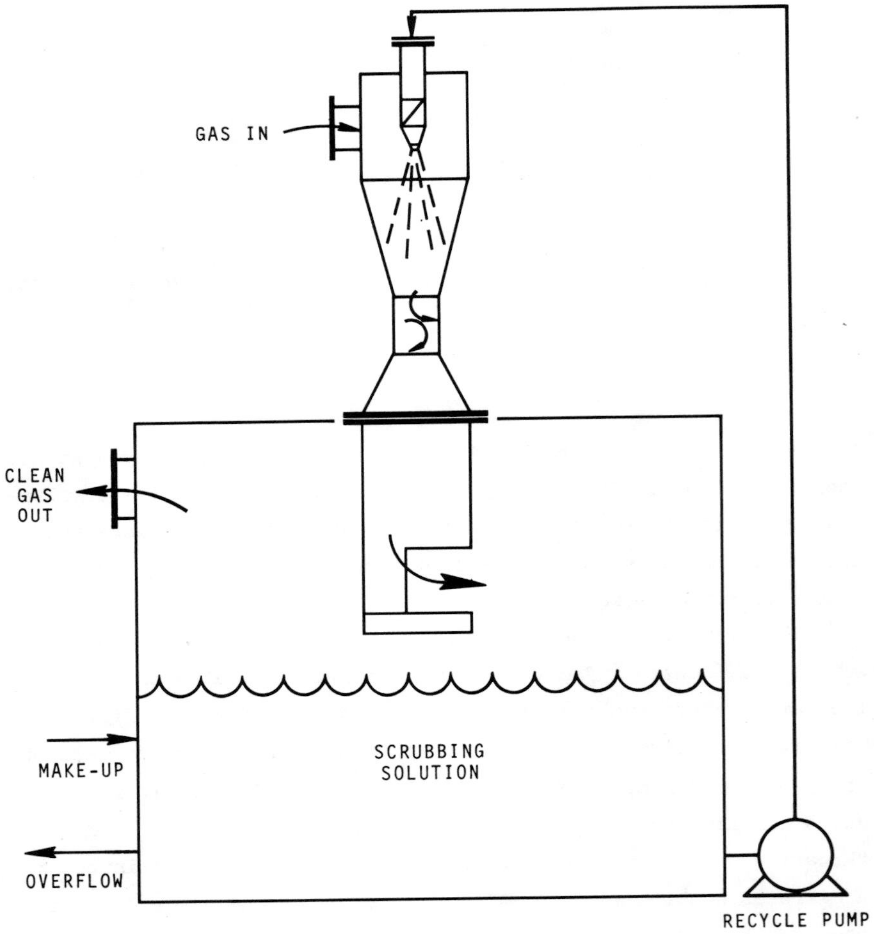

FIG. 29-1. Schematic flow diagram of major components of a jet venturi scrubber.

Because of the relatively high liquid-to-gas ratios normally employed, recycle of the scrubbing liquid is common. A sump is often located in the separator vessel for this purpose. Normally, the scrubber would be set up with a recirculating loop. That is, a recycle pump would take the liquid from the storage area and send it at a pressure of 20 to 80 psig to the liquid nozzle. The liquid entering that nozzle creates the hollow cone spray necessary to draw the air in and thoroughly scrub it.

AIR-HANDLING CAPACITY

Perhaps the most unique feature of the unit is its ability to entrain the gas without the use of fans or blowers of any type. This can be particularly important in applications involving explosive atmospheres or extremely corrosive or abrasive services. It is capable of overcoming its own internal pressure drop, or a slight negative draft can be created. It is not recommended that the unit be used for producing any significant draft in the ductwork. By significant, we refer to vacuums or pressure losses in the ductwork of 2 in. or more of water. If such conditions exist and can be handled by a fan, that would be the preferable approach.

Figure 29-2 shows the relative air-handling capacity at various negative draft conditions. Of course, these data are based strictly on its capacity for air handling and do not reflect the liquid pressure or liquid rates which may be required for a specific application.

Scrubber size	Size factor	Scrubber size	Size factor
22	0.027	18 x 18	5.07
33	0.08	20 x 20	6.31
44	0.25	24 x 24	9.14
66	0.56	30 x 30	14.4
88	1.00	36 x 36	21.0
10 x 10	1.56	42 x 42	29.1
12 x 12	2.25	48 x 48	38.7
14 x 14	3.06	60 x 60	64
16 x 16	4.0	72 x 72	103

If motive pressure is fixed, then the required flow (in gpm) is

$$\text{gpm} = \frac{(\text{hydraulic hp})(1{,}714)}{(\text{psig})} \times \text{size factor}$$

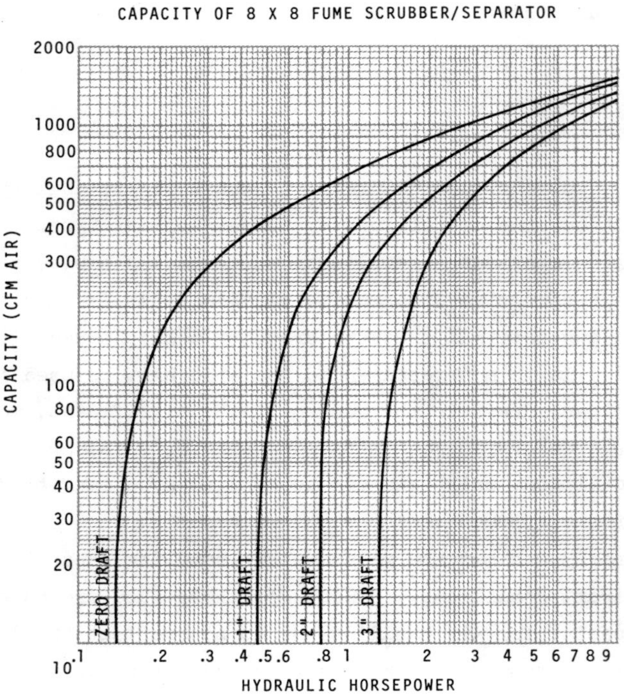

FIG. 29-2. Relative air-handling capacity of jet venturi scrubber at various negative draft conditions.

If available liquid flow is fixed, then the required pressure (in psig) is

$$\text{psig} = \frac{(\text{hydraulic hp})(1{,}714)}{(\text{gpm})} \times \text{size factor}$$

Example 1

A scrubber is needed to handle 800 cfm of contaminated air at 80°F. Pressure drop to the scrubber is 1 in.wc ΔP. Available water supply for scrubbing is 60 psig. Since the pressure drop to the scrubber is 1 in., then 1 in.wc draft will be required. From the capacity curve, 800 cfm can be handled by an 88 Fume Scrubber. The required horsepower is 2.6.

$$\text{gpm} = \frac{(\text{hp})(1{,}714) \times \text{size factor}}{\text{psig}}$$

$$= \frac{(2.6)(1{,}714) \times 1}{60} = 74.2 \text{ gpm}$$

Example 2 (Liquid Flow Given)

A scrubber is required to handle 9,000 acfm of air at 1-1/2 in. draft with 900 gpm of 10 percent NaOH solution. Since the curve for an 88 Fume Scrubber begins tc flatten out at about 1,000 cfm, we need a unit with approximately 10 times the capacity.

$$\text{size factor} = \frac{\text{desired capacity}}{\text{capacity of 88 Fume Scrubber}}$$

Looking at the list of available units, the size factor for a 24 x 24 unit is 9.14.

$$\frac{(\text{desired capacity})}{\text{size factor}} = \begin{array}{c}\text{equivalent required}\\ \text{capacity of 88 Fume Scrubber}\end{array}$$

$$\frac{9{,}000}{9.14} = \begin{array}{c}\text{equivalent required}\\ \text{capacity of 88 Fume Scrubber}\end{array} = 985 \text{ cfm}$$

From the capacity curve, the required horsepower is 4.6 hp.

Since we must use 900 gpm for scrubbing, the required pressure is

$$\text{psig} = \frac{(\text{hydraulic hp})(1{,}714) \times \text{size factor}}{(\text{gpm})}$$

$$= \frac{(4.6)(1{,}714) \times 9.14}{900} = 80 \text{ psig}$$

From the curve it is obvious that we can reduce the motive pressure required by increasing the size of the unit. Taking the same problem as above, we will use a 30 x 30 unit.

$$\begin{array}{c}\text{equivalent required capacity}\\ \text{of 88 Fume Scrubber}\end{array} = \frac{9{,}000}{14.4} = 625 \text{ cfm}$$

From the curve,

$$\text{required hp} = 2.15$$

$$\text{psig} = \frac{(2.15)(1{,}714)14.4}{900} = 59 \text{ (Use 60 psig)}$$

GAS ABSORPTION

Several theories have been promoted for studying the gas absorption capacity of the jet venturi scrubber, including the two-film theory, the penetration theory model, and the surface renewal model. None of these has shown accurate correlation with the experimental data.

Perhaps the easiest rule-of-thumb for the engineer considering installation is to estimate the approximate percentage removal by analogy and allow the manufacturers to give more precise data. Applications can be broken down into relatively high soluble components such as HCl, gases of moderate solubility such as sulfur dioxide, and extremely low-solubility gases such as hydrogen sulfide or chlorine.

By analogy with HTU units for a packed tower scrubber, a jet venturi fume scrubber can achieve approximately three transfer units based on an equivalent countercurrent tower. This assumes that the concentration of the contaminants in the exit solution from the scrubber does not have a significant vapor pressure. This can also be accomplished by the use of a neutralizing solution, as, for instance, sodium hydroxide in the case of hydrochloric acid vapor.

Example 3

With a jet venturi scrubber on HCl gas from a vent stream, we normally expect 95 percent removal at ambient temperatures and low inlet concentrations. This is equal to exactly three transfer units.

Example 4

With water as the scrubbing medium, at ambient conditions, a 70 to 80 percent removal is feasible; with a strong base (e.g., NaOH 10 wt%) 98 percent removal can be achieved. For very low solubility components such as hydrogen sulfide, the unit almost always requires an alkaline medium. Here, contact time is also important, so that even with the presence of an alkaline medium, the equivalent number of transfer units will be less than one. If the chemical reaction goes fast, such as in the case of

lations where condensation occurs. The particle size in this case is determined by the method of cooling.

Experimental data indicate that substantial success can be achieved in reducing smoke emissions to points of 10 percent opacity or less without using excessive amounts of energy. This occurs because of the potential of high liquid-to-gas ratios within the jet scrubber. The same type of ratio cannot be achieved in other units. For example, a typical high-energy venturi scrubber has a liquid-to-gas ratio of 10 to 15 gpm/1,000 cfm. A comparable jet venturi scrubber would have a liquid pressure of perhaps 110 or 120 psig, but a liquid-to-gas ratio of 40 gpm/1,000 cfm. This means a greater reduction in temperature in a very short distance. The result is a quench action which tends to condense out larger particles of smoke which can be collected in the throat of the unit.

EFFICIENCY

One of the questions which always arises in the application of a jet venturi fume scrubber is the energy requirements. Table 29-1 presents a comparison of energy requirements for three types of scrubbers. Articles have been written which indicate that the energy requirements for the jet venturi may be excessive, and in certain applications this may be true. However, what usually happens is that a comparison is made between scrubber systems using fans and a jet scrubber pulling a draft.

The jet scrubber is designed primarily to be a scrubber and not an air-moving device. As soon as additional air-handling capacity requirements are forced on it, its energy requirements tend to rise. In moderate-size applications or for small amounts of draft, this is not a problem and the additional energy requirement is not significant. In fact, in some cases it may be necessary to add a significant amount of energy in order to accomplish a given performance of collection of dust. However, in other cases, comparisons are made on an unequal basis which show a high energy requirement. In Table 29-1 the data show a jet venturi scrubber operating in its normal

TABLE 29-1

Comparative Energy Requirements (1,000 cfm; 90% at 1 μm)

Type of unit	Liquid flow (gpm)	Pressure (psig)	Pump hp	ΔP of gas (in.wg)	Fan hp	Relative hp/1,000 cfm
Wet cyclone	10	60	0.91	8	2.5	3.41
Venturi	10	20	0.233	15	3.94	4.27
Jet venturi	40	70	3.28	0	0	3.28

$$\text{Pump hp} = \frac{(\text{gpm})(\text{psig})}{(1{,}714 \times \text{efficiency})}$$

Assume 50% efficiency.

$$\text{Fan hp} = \frac{(\text{cfm})(\Delta P \text{ in.} H_2O)}{(6{,}356) \times \text{efficiency}}$$

Again assume 50% efficiency.

design range as compared to the energy requirements for alternative systems. All the systems are designed to achieve the same dust efficiency and have the same capacity. The utility costs are based on total horsepower requirements. The actual selection of pumps or fans will determine the relative efficiencies which will be achieved.

SEPARATOR

Separator design is extremely important to the proper function of this unit. Separators may be any of three designs. First would be an impact-type separator, which uses a set of plates or other device which turns the gas flow and allows the liquid to impact on the surface and roll off. Flat plates have been used in the past, although better techniques are now available. These include the use of mesh-type collection devices, which allow part of the air to pass through and eliminate the possibility of solids buildup on the surface. These elements should also be removable so that if plugging due to solids buildup should occur, they can be cleaned. The second arrangement would be the use of a mist eliminator of the mesh type. The gas passes up through the mesh pad and the liquid is separated and returned to the tank. The third arrangement, which is the one least used, is a cyclone separator. Cyclone separators are suitable for single-point designs only. They are not recommended where the gas velocities are likely to vary significantly. Also, the high liquid rate inside the jet venturi scrubber makes the design of these units more difficult.

The separator vessel is also used as the liquid storage capacity for recirculating the liquid. Normally, the material of which this vessel is constructed will be the same as the scrubber. However, in certain cases where the inlet temperature may be high, or where other requirements force the use of exotic materials for the jet scrubber, these may not be necessary for the separator vessel.

In some cases, it is possible to use gravity separation of the liquid. This is especially true of smaller air flows. A horizontal

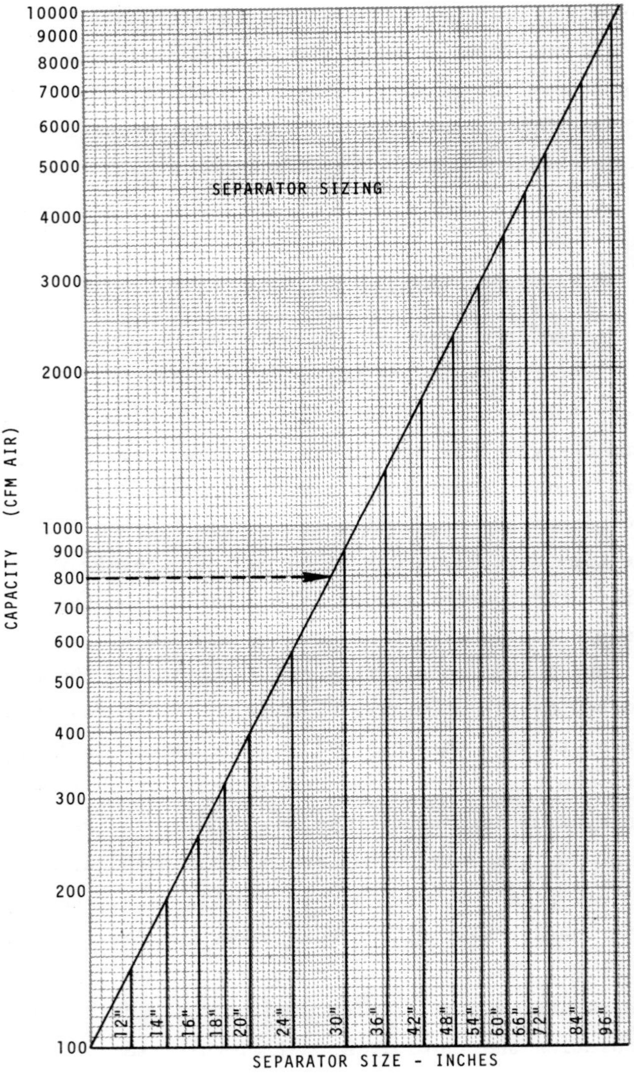

FIG. 29-4. Separator sizing chart for impact-type separator vessels.

velocity of less than 3 fps and a separator residence time of more than 1 sec is usually sufficient. Typical diameters for an impact-

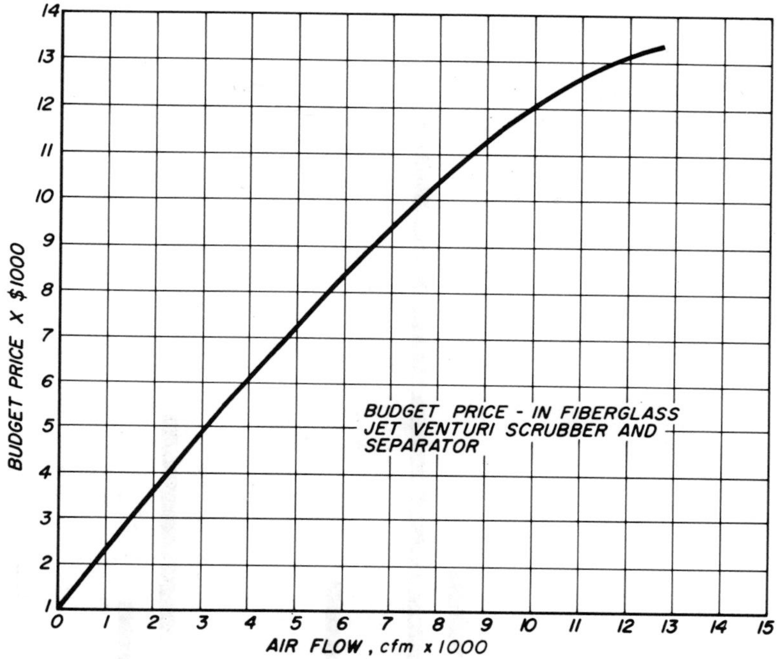

FIG. 29-5. Budget cost data for purchasing a jet venturi scrubber constructed of fiberglass.

type separator vessel are shown in Fig. 29-4. For example, looking at the sizing curve for 800 cfm, a 30-in.-diameter separator should be used.

Figure 29-5 shows budget cost data for purchasing a jet venturi fume scrubber of fiberglass construction. Figure 29-6 shows the relative costs of other materials from which scrubbers can be constructed for comparison. The relative simplicity of design of this unit makes it possible to handle almost any type of fabrication. However, it should be pointed out that the construction material most commonly encountered is fiber glass.

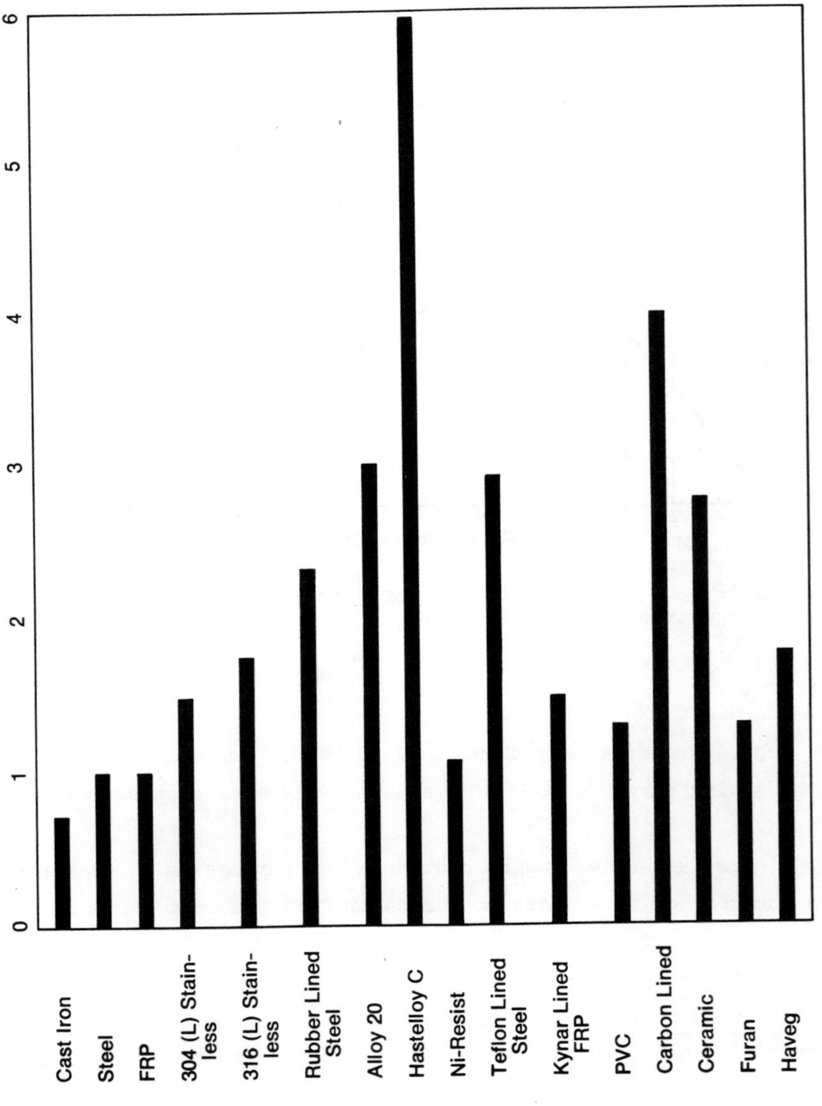

FIG. 29-6. Relative cost of materials used to construct scrubbers.

APPLICATIONS

Table 29-2 lists typical installations of jet venturi scrubbers. Although by no means complete, it indicates the types of applications most commonly encountered with this type of equipment. The list indicates that the applications encountered include those involving extremely corrosive, insoluble, or viscous materials, or gases that form precipitates.

TABLE 29-2

Source of gas by industry	Contaminant	Approximate efficiency (%)
Chemical		
Stack gas	Phthalic anhydride	99
	Benzoic anhydride	99
	SO_2	98
Gaseous	HCl	95
	HF	95
	HBr	95
	I_2	95
	Acetic acid	
	Nitric acid	97
	Ammonia	
	Sulfur dichloride	97
	Dust from misc. sources	97
	SO_3, SO_2	97-98
	Chlorine	
Spray dryers	Odor	
	Dust	
Brushing and grinding	Particulates	98+
Electronics and aerospace industries		
Sanding, grinding, pulverizing, and sandblasting	Dust	

TABLE 29-2 (Continued)

Source of gas by industry	Contaminant	Approximate efficiency (%)
Cleaning and degreasing, painting, etc.	Various solvent fumes	
Rocket propulsion	Hydrazine	
Fertilizer		
Den	SIF_4 and dust	98
	Dust (product handling)	
Fish processing		
Fish metal	Odor and dust	
Drum dryer		
Food industry		
Coffee manufacturing	Caffeine freeze-drying dust	98
Deep fat frying	Grease, fat	
Meat packing	Odor	
Onion production	Odor	
Cabbage cooking	Odor	
Rendering operation	Odor	95
Bouillon manufacturing	HCl	
Hospital		
Viable waste incinerator	Gases, ash	
Metal fabricating		
Plating	HCl	
	Oil mist	
Pharmaceutical industry		
Pharmaceutical manufacturing, i.e., tablet manufacturing	Dust	98+
	Odor	
Petroleum specialties	Ammonia	
	Organic fumes	

TABLE 29-2 (Continued)

Source of gas by industry	Contaminant	Approximate efficiency (%)
Pulp and paper industry		
Multieffect evaporators	H_2S	
	Methyl mercaptan	
	Methyl disulfide	
	Methyl sulfide	
Flyash recovery		
Lime kiln	H_2S dust	
Plywood	Abietic acid and miscellaneous	
Cannery waste	Alkaline liquid waste neutralization with SO_2 from stack	
Textile industry		
Dryer ovens (silk-screen)	Plasticizer-type fumes	95+
	Formaldehyde	85
Aging application	Acetic acid	95
Truck, trailer repair, and cleaning facilities	HCl, SO_2, other waste gases and dust	

A typical example of difficult scrubbing which is handled by this type of unit is phthalic or maleic anhydride. Phthalic anhydride (the gas sublimes) tends to form a solid directly from the air stream. Since this solidification can occur on the scrubber walls if they are cool, problems can be encountered with plugging. Special modifications of the standard unit have been used to avoid this problem. The solution is normally recirculated to a high concentration and then returned to the process. Figure 29-7 shows a flow diagram of a typical anhydride scrubbing system.

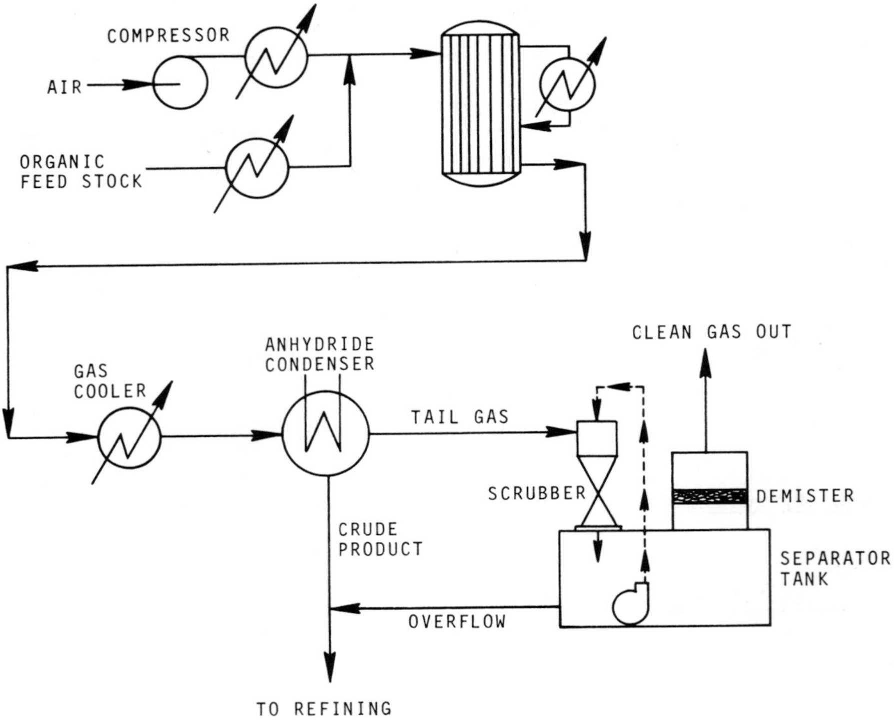

FIG. 29-7. Flow diagram for anhydride scrubbing system.

SPECIAL APPLICATIONS

The unit can also be combined with other types of collection devices such as packed towers to achieve a combination of results. For example, dust collection can be achieved in a jet venturi scrubber with partial absorption of a soluble gas. A packed tower can be added to the end of the system to assure complete elimination of the undesirable gas component. The jet scrubber by itself may not be capable of sufficient efficiencies to satisfy state air pollution control requirements on a moderately soluble or insoluble gas. The combination of the two will satisfy this requirement. The packed tower by itself would run into difficulties handling the dust, which would tend to plug or clog the packing. Such combinations may result in slightly higher initial cost, but the overall savings in operating and maintenance problems more than justify this expense.

Modifications of the standard jet scrubber having smaller throat diameters and using special-design spray nozzles for liquid pressures of 120 to 170 psig are often used on difficult dust or smoke applications. In some cases where extremely high efficiencies are required, these units are used in series with electrostatic precipitators. The jet scrubber is used to eliminate the maintenance problems often encountered with these applications, since it has no moving parts and can readily handle sticky materials. It would eliminate approximately 85 to 95 percent by weight of such condensable smoke or fumes and allows the precipitator to receive a saturated air stream containing the residual submicron smoke. Such an arrangement requires control of the moisture content of the air to avoid condensation within the precipitator, but once established, functions at relatively low cost with a minimum of maintenance problems.

Chapter 30

THE IONIZING WET SCRUBBER

Warren Klugman and S. V. Sheppard
The Ceilcote Company
Berea, Ohio

The Ionizing Wet Scrubber (IWS) is a device that employs the principles of electrostatic particle charging in combination with a wet packed cross-flow scrubber to increase dramatically the collection efficiency of the scrubber action. Developed and patented by Ceilcote Company, the device requires relatively small quantities of electrical energy for its operation. It exhibits a high degree of collection efficiency as it simultaneously removes corrosive, noxious gases and submicron liquid and solid particulate from industrial effluent air streams. Both micron- and submicron-size particles are removed with equal efficiency.

The IWS (Fig. 30-1) is a compact system. It is modularly constructed for easy field installation. Any number of IWS stages can be installed in series to increase collection efficiency. Being a lightweight, low-energy system, it is relatively inexpensive to install and to operate.

The following general characteristics are exhibited by the IWS:

Particle Collection. Both pilot tests and on-stream installation results verify that particles of virtually any size or composition are collected by the IWS. It exhibits consistently high efficiency in collecting particles as small as 0.05 μm and captures both organic and inorganic particles that have either high or low resistivity.

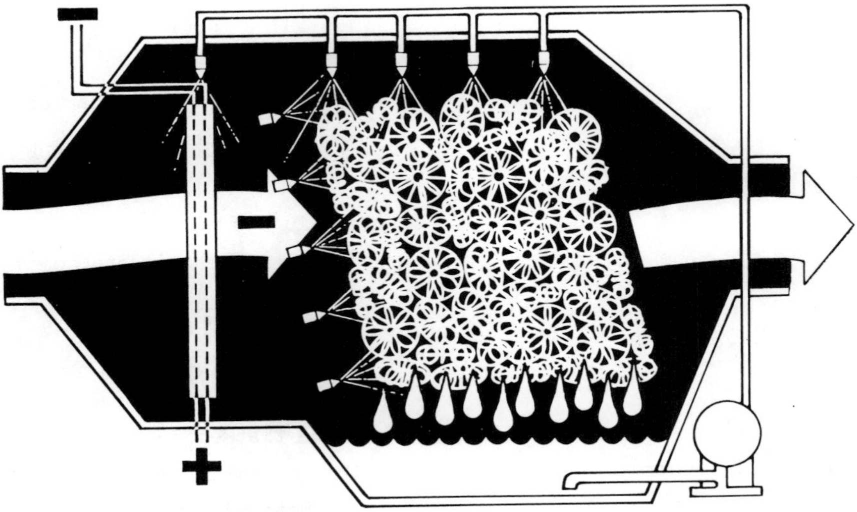

FIG. 30-1. The Ionizing Wet Scrubber.

Energy. Pressure drop through a single-stage IWS is only 1-1/2 to 2 in.wc. Energy for particle charging is low--approximately 0.2 to 0.4 KVA per 1,000 cfm. The actual pressure drop experienced is affected principally by three factors: (1) the depth of packing required to achieve the desired gas removal; (2) the gas velocity through the packed bed; and (3) the liquid recycle rate required for gas absorption and solids flushing. Regardless of the number of stages utilized, however, total energy consumption remains low.

Corrosion Resistance. The shell and most internal parts of the IWS are fabricated of thermoplastics and fiberglass polyester materials. This predominance of plastic construction assures practically corrosion-free operation in the presence of acid gases such as HCl, HF, Cl_2, NH_3, SO_2, and SO_3.

The use of plastic material throughout the IWS is made possible by its unique operating principles. The only metal associated with the IWS is in the few electrical conductors employed within the ionizing section. And these are fabricated of materials that will provide maximum corrosion resistance to the application environment.

Such extensive use of plastics is not feasible in conventional electrostatic precipitators, which require considerably more metal for the larger number of electrical conductors (for particle charging), collecting plates, and support members.

Gas Absorption. The IWS utilizes Tellerette® packing as its collection surface to achieve particulate removal. Scrubbing liquid droplets that are formed also act as collection surfaces. The IWS simultaneously absorbs gases at no additional expense. The noxious gases are removed concurrently with particles through physical absorption and/or absorption that is accompanied by chemical reaction, as is common in a wet packed scrubber.

Fractional Collector. Pilot tests indicate that the IWS essentially acts as a fractional collector. The percentage of particulate removed varies little with load over a wide range. As particulate load increases, the percentage removed remains nearly constant. Unlike electrostatic precipitators, the IWS shows little sensitivity to decreasing particle size. The collection efficiency for fine particles is nearly as great as for coarse particles.

A significant advantage of a fractional collector is that where a higher collection efficiency is required than is possible with one stage, a second stage can be added in series with the first. Both stages have the same collection efficiency, so particulate contained in the gas stream exiting from the first stage is removed by the second stage with the same percentage of removal.

Turndown. Collection efficiency remains constant (or actually improves) with turndown from 100 to 0 percent load. This feature makes the IWS highly desirable for application to systems whose load fluctuates, such as incinerators. High-energy venturis, however, are severely limited in their ability to handle a fluctuating load, particularly as it decreases, because of the difficulty in maintaining a constant pressure drop across the venturi throat. To achieve stability and fairly constant efficiency, elaborate controls must be incorporated.

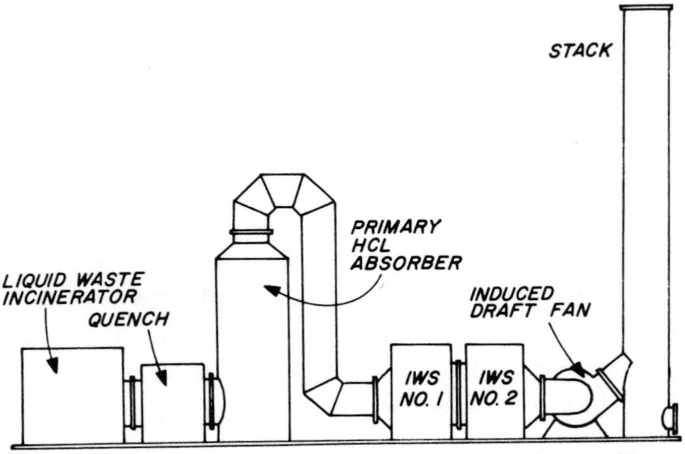

FIG. 30-2. Equipment schematic.

THEORY OF OPERATION

The Ionizing Wet Scrubber utilizes a high-voltage ionizer section to charge particulate in the gas stream before the particles enter the wet packed scrubber. As the charged particles flow through the scrubber, they pass very close to a surface of either a Tellerette or a scrubbing liquid droplet. The particles become attracted to, and attached to, one of these surfaces by image force (to be described below). They are eventually washed out of the scrubber in the exit liquor. Noxious gases are absorbed in the same scrubbing liquor.

Charging of particles is accomplished within a very short, but intensely charged, ionizing section that utilizes high-voltage d-c power. Discharge electrodes are of negative polarity, while wetted plates act as positively grounded electrodes. The wetted plates are continuously flushed with water to prevent any buildup of solid particles. Thus, particles cannot build up a resistive layer on the collecting surfaces, and optimum electrical properties are maintained. This condition or characteristic is far superior to that of a conventional electrostatic precipitator.

Electric corona discharge from the electrodes produces an ion emission. These ions intercept the fine contaminant particles in the gas stream and give up their charge to the particles. As the gas stream leaves the ionizing section, the entrained contaminant particles carry an electrostatic charge. The gas stream then flows through the Tellerette-packed cross-flow wet scrubber where removal in the gas stream is achieved either by impaction or by attraction of the charged particles to a neutral surface. Particles of 3 to 5 μm and larger are collected through inertial impaction. They are virtually certain to impinge upon a packing surface within the first few inches of packing, be captured, and be flushed away.

Efficiency by inertial impaction is a function of target size, and the Tellerette presents a smaller target area than do other packing materials. The filament size, filament cross-section shape, larger number of filament coils, and number of packing units per cubic foot all contribute to make the Tellerette packing an excellent impingement target and a highly efficient collector.

The Tellerette is also designed specifically to create water droplets, as opposed to other packing materials which are traditionally designed to produce large surface films. This Tellerette characteristic produces a greatly increased number of targets in the form of water drops. Tests have shown that the Tellerette-type filamentous shape is far more effective than other packing materials tested.

Particles smaller than 3 to 5 μm are forced to follow a tortuous path through the packing. The probability is high that an actual impact will take place, or that particles will come into proximity with either a packing surface or with a liquid droplet from the scrubbing liquor.

IMAGE FORCE

Particles that are charged and come within the boundary layer of a neutral surface will be attracted to that surface. This phenomenon is known as image force attraction. The force of attraction between a charged particle and a neutral surface is the same as that force that would exist between that same particle and another particle of equal size, but opposite polarity, when each particle is located on either side of a neutral surface and equidistant from it. This is known as the "mirror image" analogy.

While the contaminant particle in the IWS carries a negative charge as it enters the packed wet scrubber, the packing and scrubbing liquid (as well as the entire structure of the scrubber) are maintained electrically neutral to create the conditions for image force attraction. Therefore, when a contaminant particle carrying a charge comes close to the neutral surface of a packing element or water drop (probably within about 1 mm in this case), there is sufficient image force attraction to overcome the inertial and viscous drag of the particle in the gas stream so that the particle moves to that surface. It is then flushed away by the scrubbing liquid.

The design of the Tellerette shape provides the greatest probability that particles will pass sufficiently close to a surface so that the image force will be effective.

The image force described should be distinguished from both coulomb forces and space charge. The principle of coulomb force, as used in conventional electrostatic precipitators, is that charged particles pass through an electrostatic field where a force acts to attract them to the walls of the device. This is said to be analogous to the action of a conductor exposed to a magnetic field. (It should be noted that in the IWS there is no imposed electrostatic field in the collection zone.) The principle of space charge is that charged contaminant particles and water droplets are all charged one polarity. They form a space charge in which all elements tend to repel each other. Some fraction of them will be forced to the container's wall and be collected. As droplets are collected, space charge concentration falls. Neither of these phenomena seems to operate to any significant degree in the IWS.

While inertial impaction and image force are the two most significant phenomena involved in the collection of contaminant particles, agglomeration, another phenomenon that occurs in all particle collectors, also enhances particle removal.

Since all packing surfaces, as well as all water drop surfaces, constitute collection surfaces, the total collection surface area provided by the IWS is probably several orders of magnitude greater than that available in a conventional electrostatic precipitator. With all of these collection surfaces in the IWS, the distance between any two is minute. Thus, particles need be "drifted" only a short distance before they become attached to a surface. This distance is usually within the boundary layer of a surface. In a typical electrostatic precipitator, particles must be "drifted" from 2 to 6 in.

APPLICATIONS

The low energy-consuming IWS can be applied to meet a wide range of applications. It offers the greatest advantages in applications requiring the collection of submicron particulate, especially where corrosive conditions exist. Submicron particulate is most commonly formed through condensation of vaporized liquids and solids. It occurs in almost all high-temperature operations. Many of these operations also produce off-gases that are corrosive.

Some potential applications for the IWS are listed in Table 30-1.

TABLE 30-1

Process	Contaminants
Petroleum coke calciners	SO_3 mist, SO_2, carbon particles, smoke, inorganic fume
Liquid and solid waste incinerators, chemical plants	HCl gas, SiO_2 fume, SO_3 mist, organic mists, smoke, inorganic fumes

TABLE 30-1 (Continued)

Process	Contaminants
Municipal solid waste incinerator plants	HCl gas, smoke, organic mists, odoriferous gases, inorganic fumes
Copper, lead, and zinc smelters with sulfuric acid plants	Dust, inorganic fumes, SO_3 mist
Aluminum refining and processing operations	Smoke, Al_2, Cl_3 fumes, HCl, dust, SO_2, SO_3, HF, carbon particles, Al_2O_3 fumes, fluoride fumes
Dryers--feed and grain mills	Dust, vapors, smoke, odoriferous gases
Phosphate fertilizer manufacture	HF gas, fluoride fumes, nonfluoride fumes, P_2O_5 fume, dust
Food and coffee processing	Dust, smoke, odors, aldehydes, organic acid mist, NO_x
Zinc galvanizing lines	NH_4Cl fumes, AnO fumes, oil vapors
Ferroalloy production	SiO_2 fumes, inorganic fumes, carbon particles
Black liquor recovery furnaces	Na_2SO_4 and Na_2CO_3 fumes, H_2S gas, SO_2 gas, mercaptans

EXAMPLE OF COMMERCIAL INSTALLATION

In 1971, Ceilcote Company designed a venturi system that would be installed at a plant to remove SiO_2 particulate from the effluent of an incinerator that was to be installed to burn liquid wastes including chlorinated silane and siloxy solvent.

Before proceeding with any fabrication, design confirmation tests were performed at Ceilcote's pilot installation. Samples of liquid wastes from the plant were incinerated and the gas stream analyzed. These tests indicated that particle size was 0.05 to 0.2

μm. This turned out to be significantly smaller than anticipated when the request was made for a venturi design.

As a result, the use of a venturi was ruled out, because this type of equipment is generally impractical for removing submicron particulate. To be effective on particles in this size range, a venturi would require an inordinate amount of power, and even then its ability to remove any particles less than 0.1 μm would be questionable.

The recommendation was made that an Ionizing Wet Scrubber be installed rather than a venturi. A 15,000-scfm IWS was fabricated and installed. The success achieved in removing fine particulate from the incinerator gas stream verified the soundness of the recommendation. The IWS operated continuously at the facility on a 24-hr-per-day basis for over 1 year. During that period, it required virtually no maintenance. Based on this success, when a second and larger incinerator was required for this same plant, IWS was specified to remove pollutants from the gas stream (Fig. 30-3).

The hot gas stream output of the second incinerator is quenched and cooled to 175°F (79°C). The quenched output (50,000 acfm) is then conducted to the IWS. It first passes through a packed counterflow wet scrubber (Fig. 30-4), where the gas stream is cooled further and the major part of the HCl is absorbed. This is necessary because of the unusually high HCl content. Output of the HCl scrubber (25,000 acfm at 1,000°F saturated) enters the IWS. Submicron SiO_2 particulate is captured, and the remaining HCl is absorbed. The new, larger IWS was designed to accept an input of 22,500 scfm and to remove per hour a maximum of 1,185 lb of HCl and 350 lb of submicron SiO_2.

One of the difficulties anticipated in this installation was that even where the stack effluents would be within code requirements, it was felt that emissions would still be visible because of the extremely high light-scattering characteristics of particles

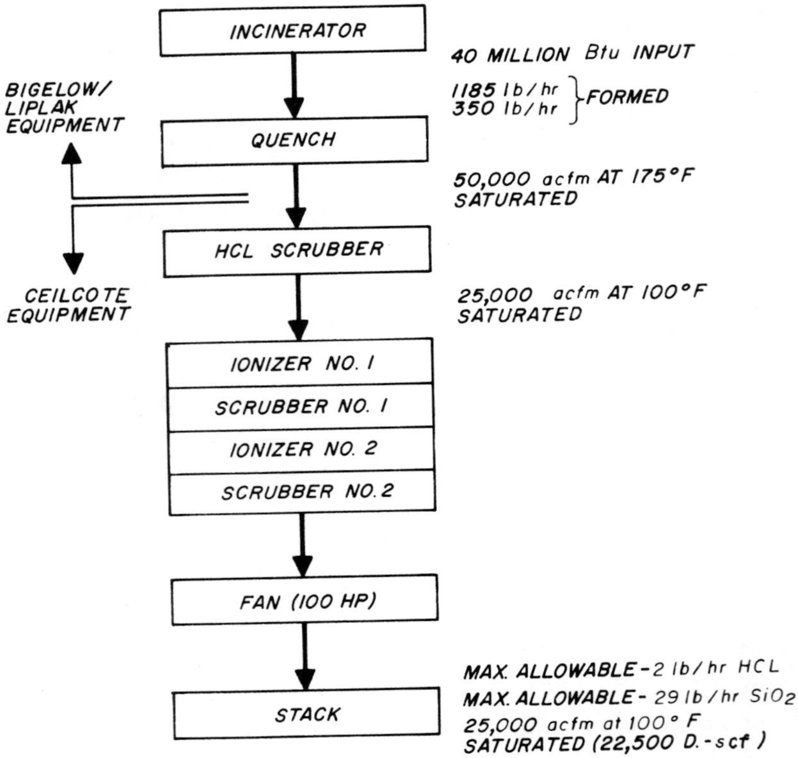

FIG. 30-3.

0.25 μm and smaller. In an effort to reduce the anticipated stack exhaust gas opacity that would result from the high particulate load and submicron particulate size, a two-stage IWS was specified. The second stage of the IWS was identical to the first and was connected in series with it. From the second-stage IWS output, the exhaust gases were drawn through an FRP-induced draft fan and directed out of a 100-ft-high stack.

All components of the system were of reinforced plastic modular design. The modules were fabricated at the Ceilcote plant, shipped as complete sections, and assembled in the field. The second IWS has also operated on a 24-hr-per-day basis and has been on-stream continuously since June 1974.

FIG. 30-4. Packed counterflow wet scrubber.

Results

It was stipulated that emission from the IWS could not exceed 2 lb/hr of HCl and 29 lb/hr of SiO_2. At this level, stack gas opacity would be low, but the effluent still would be slightly visible. However, once the IWS was on-stream, the reduction in stack gas opacity was dramatic. The stack was visually clear. The IWS was performing far better than expected. Performance tests were conducted that proved the IWS was well within the maximum allowable limits requested, and the outlet load was exceptionally low.

In an effort to determine the operating limits of the IWS, materials were burned at rates which produced a theoretical SiO_2 generation four times that of the system's designed maximum capacity. Despite this huge increase in load, a visually clear stack was maintained. Even materials such as methyltrichlorosilane, which when incinerated produces a theoretical SiO_2 content of approximately 40 percent of the input weight, have been incinerated without creating any visible stack effluents.

APPLICATION TESTING

Tests were conducted on diverse applications that involved exhaust gases from a primary aluminum reduction facility using the Soderberg cell; an elemental silicon furnace; a meat-processing smoke house; a petroleum coke calcining kiln; a corn gluten dryer; and incineration of chlorinated hydrocarbons, methyltrichlorosilane, siloxy and silane solvent wastes, and municipal sludge after primary treatment.

Figure 30-5 shows results obtained from a typical pilot test in which the gas stream contained a mixture of dioctylphthalate and kerosene. Samples were taken at the equipment outlet only--first with the power to the ionizing section turned off and then with the power on. In this way the electrostatic influence on the wet scrubber's collection capability could be evaluated.

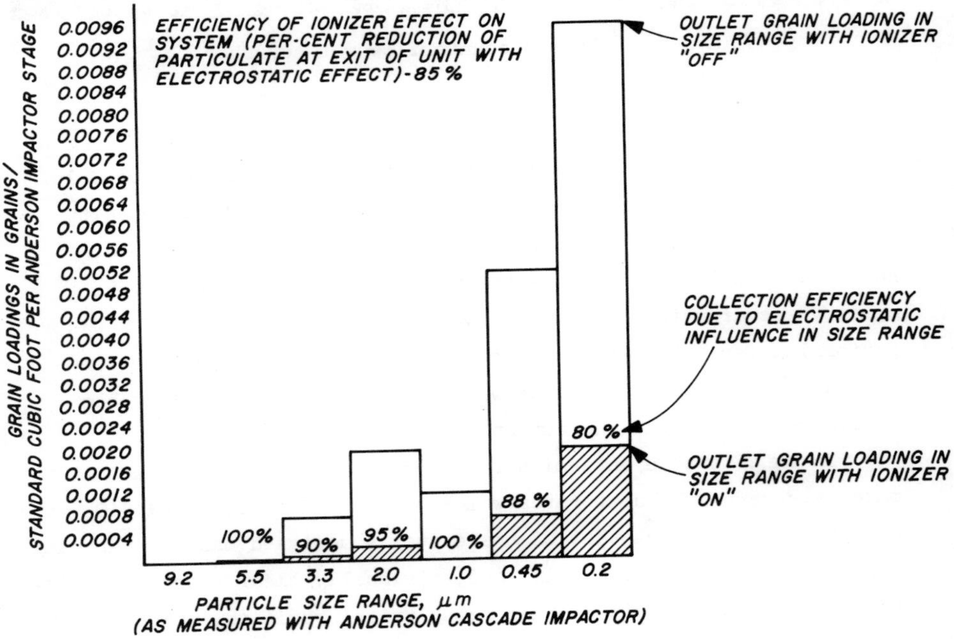

FIG. 30-5. Pilot test results for IWS test media: 50 percent dioctylphthalate; 50 percent kerosene. Total outlet grain loadings: ionizer off--0.0175 gr/ft^3, ionizer on--0.0027 gr/ft^3.

The samples taken are shown in the bar graph as grain loading versus various particle size ranges from 0.2 to 9.2 μm. Each size range has two grain loading figures--one for each operating condition (ionizer on or off). The increase in collection efficiency due to the electrostatic influence is indicated for each particle size range.

Chapter 31

A SURVEY OF LIME/LIMESTONE SCRUBBING FOR SO_2 REMOVAL

Robert T. Fellman
Ebasco Services, Inc.
New York, New York

Paul N. Cheremisinoff
New Jersey Institute of Technology
Newark, New Jersey

The earliest experience with lime or limestone wet scrubbing took place in England during the 1930s, where closed-loop lime and limestone SO_2 scrubbers were installed on full-scale boilers [1]. Although the problem of SO_2 removal from flue gases appeared to lend itself to a simple application of wet scrubbing using lime or limestone slurries, such serious problems as equipment corrosion, erosion, scaling, and plugging soon emerged as the primary difficulties inherent with these systems.

The London Power Company installed on a large boiler a test unit in the early 1930s. A test program was carried out which resulted in the construction of a large gas-washing installation at the Battersea Power Station on the bank of the Thames River in London. The Battersea scrubbers were revised several times, with the final mode of operation being the use of a chalk ($CaCO_3$) slurry in a once-through scrubbing cycle. The pH of the slurry effluent was too low for release to the Thames River, so the effluent was diluted with condenser water to reduce the pH. The Battersea process was installed on one additional generating unit--but this became the upper limit of application, because the capacity of the Thames River to absorb the sulfite effluents had been exceeded.

Attention then focused on the development of a noneffluent process, which could avoid the discharge of waste products to the receiving water body. Imperial Chemical Industries and James Howden and Co. developed what came to be called the ICI-Howden process, wherein a slurry of lime or chalk was circulated through a grid packed scrubber. A sidestream of the slurry was filtered to remove the product as a solid.

The ICI-Howden method was successfully utilized on London Power's Fulham Station, although maintenance costs were quite high due to scrubber fouling from particulate overloading and excessive scaling. The unit was shut down during World War II, because steam plumes from the stacks were thought to be guiding enemy aircraft. The unit was not put back on-line after the war.

Interest in wet scrubbing and stack gas cleaning for pollution control subsided after World War II, but was revived in the 1960s. However, in the United States during the 1950s, the Tennessee Valley Authority (TVA) carried out small, limited pilot plant tests.

A scrubber installation, removing SO_2 from about 1,700,000 scfm of waste gas from an iron ore sintering plant, has been operating in Russia since 1964.

A scrubber in Japan has been operating for about 5 years, during which the tail gas from a 480 ton/day sulfuric acid plant is scrubbed with a lime slurry. The calcium sulfite by-product is converted to gypsum (calcium sulfate), and then sold [2].

The first major pilot plant work is probably that of Universal Oil Products, beginning in 1965. A limestone slurry, circulated through a mobile-bed scrubber, yielded good SO_2 removal.

Boiler injection followed by wet scrubbing was tested by Combustion Engineering, Inc., on a pilot plant basis in 1966 and 1967. Two installations have been made by Union Electric Company, St. Louis, Mo., and Kansas Power and Light Co., Lawrence, Kansas.

The interest generated by the Clean Air Act Amendments of 1970 and the proliferation of air pollution guidelines promulgated by state governments to comply with federal EPA requirements, have resulted in an enormous research effort in SO_2 removal technologies. One research effort in lime and limestone scrubbing technology has been spearheaded by the TVA, EPA, and Bechtel Corporation at the TVA's Shawnee Power Plant pilot testing facility. The results of this work are the prime reference for this chapter on the chemistry of SO_2 removal using lime and limestone.

CHEMISTRY OF LIME AND LIMESTONE SCRUBBING

Flue gas desulfurization using lime or limestone wet slurries is more an art than a science. On the equipment market, at present, there are lime/limestone systems available using turbulent contact absorbers, venturis, spray towers, fluidized beds, and spray trays as contacting mechanisms. For each of these systems, there are volumes of data available, complete pilot plant testing histories, endorsements, and--in rare cases--performance guarantees.

One of the most confusing aspects of lime and limestone scrubbing is the chemistry of the process, and the manner by which the chemistry can be controlled to optimize the performance of the process. Understanding the chemistry of lime/limestone wet scrubbing is especially useful in the following areas:

1. Control of pH to avoid hard scale formation,
2. Control of pH to avoid plugging (soft scale) formation,
3. Control of pH and liquid/gas ratios to optimize removal of SO_2 and avoid carbon dioxide scrubbing,
4. Understanding the nature of the sludge discharged from the system, and
5. Controlling and optimizing reagent utilization.

The following assumptions have been made about the flue gas to be treated:

1. The flue gas is from a utility boiler, firing the coal shown in Table 31-1 and having the flue gas composition shown in Table 31-2.

TABLE 31-1

Western Kentucky Coal Composition

Constituent	Percent by weight
Volatile matter	32.9
Fixed carbon	46.4
Sulfur	3.3
Moisture	5.4
Ash	12.0
	100.0
H	4.5
N	1.3
C	66.0
O	7.5
S	3.3
Ash	12.0
Moisture	5.4
	100.0

TABLE 31-2

Composition of Stack Gas Evolved in Burning Coal from Western Kentucky

Constituent	Formula	Percent by volume
Nitrogen	N_2	74.6
Carbon dioxide	CO_2	12.6
Oxygen	O_2	4.9
Water vapor	H_2O	7.8
Sulfur oxides		
Sulfur dioxide	SO_2	0.22
Sulfur trioxide	SO_3	0.001
Nitrogen oxides	NO_x ($x = 1, 2$)	0.04
Particulate matter		3.6 gr/scf

2. The flue gas has been passed through an electrostatic precipitator with an efficiency of 99.4 percent removal. (This assumption eliminates the need to discuss the effects of flyash carryover.)

3. The flue gas has been quenched to its wet bulb temperature (about 130°F).

The discussion which follows introduces the principal reactants, defines their properties, shows how they interact, and discusses the consequences of their interactions.

PROPERTIES OF SULFUR DIOXIDE, LIME, AND LIMESTONE

Sulfur Dioxide as the Prevalent Sulfur Oxide in Flue Gas

A variety of sulfur compounds are formed solely from oxygen and sulfur [3]. The most common of these are SO_4, S_2O_7, SO_3, SO_2, S_2O_3, and SO.

Most of these compounds are unstable, or are formed by peculiar reactions, thus attention needs to be focused only on the two gaseous, stable oxides: sulfur dioxide (SO_2) and sulfur trioxide (SO_3).

Sulfur dioxide can be formed by the reaction of elementary sulfur with oxygen, acidifying a solution containing sulfite ion, air oxidation of pyrites (iron sulfides), or the oxidation of elementary and organic sulfur in coal.

The chemical reaction between elementary sulfur and oxygen forming SO_2 is given below (H_f is the enthalpy of formation):

$$S + O_2 \rightarrow SO_2 \quad H_f = -71 \text{ kcal/mole} \tag{1}$$

The reaction is exothermic and highly favorable thermodynamically.

Sulfur trioxide (SO_3) may be formed from sulfur dioxide and oxygen, according to:

$$SO_2 + \frac{1}{2}O_2 \rightarrow SO_3 \quad H_f = -23 \text{ kcal/mole} \tag{2}$$

This reaction is also exothermic and favorable thermodynamically at low temperatures (0 to 1,000°F). However, the reaction shown in

Eq. (2) is not kinetically rapid; it proceeds at an infinitesimal rate at low temperatures. Increasing the temperature will increase the rate at which SO_2 is converted to SO_3. However, the thermodynamic equilibrium for the reaction shifts steadily toward SO_2 as the temperature is increased. This behavior is illustrated in Table 31-3.

Although the equilibrium for Eq. (1) also shifts away from SO_2 formation as the temperature is increased, the shift is inconsequential at the temperatures normally observed in coal-fired boilers. Thus, due to the low thermodynamic instability of SO_3 at high temperatures, the high thermodynamic stability of SO_2 at high temperatures, and the extremely slow rate of conversion of SO_2 to SO_3 at low temperatures, the problem of flue gas desulfurization is largely a question of removing the sulfur dioxide rather than the sulfur trioxide from flue gases exiting from coal-fired boilers.

Table 31-2 shows a typical flue gas composition for the gases evolved from a boiler burning coal mined in western Kentucky [2]. Of particular interest is the proportion of SO_2 in the gas stream as compared with SO_3.

TABLE 31-3

Effect of Temperature on the Equilibrium of

$$SO_2(g) + \frac{1}{2}O_2(g) \rightleftarrows SO_3(g)$$

Temperature (°F)	752	962	1184	1396	1608
Equilibrium constant	2.3×10^3	4.0×10^2	7.0×10^1	2.0×10^1	7.0×10^0
Percent SO_3 (mole)	96	88	66	40	20
Percent SO_2 (mole)	2	6	17	30	40
Percent O_2 (mole)	2	6	17	30	40

Chemical and Physical Properties of SO_2 Exploited in Wet Scrubbing

In lime and limestone wet scrubbing of flue gases, the most important chemical property of SO_2 to be exploited is the fact that SO_2 is an acidic anhydride. This means that, in combination with water, SO_2 forms an acid, namely sulfurous acid. Equation (3) shows this reaction:

$$SO_2\ (\text{gas}) + H_2O\ (\text{liq.}) \rightleftarrows \underset{\text{sulfurous acid}}{H_2SO_3} \tag{3}$$

Sulfurous acid is a weak dibasic acid which ionizes according to the following reactions:

$$H_2SO_3 \rightleftarrows H^+ + HSO_3^- \quad K_{A1}\ (T = 18°C) = 1.54 \times 10^{-2} \tag{4}$$

$$H_2SO_3 \rightleftarrows H^+ + SO_3^{2-} \quad K_{A2}\ (T = 18°C) = 1.02 \times 10^{-7} \tag{5}$$

Table 31-4 lists some common acids and their relative strengths as acids compared with sulfurous acid. As seen in this table, sulfurous acid is a stronger acid than carbonic acid, which means that sulfurous acid can decompose salts of carbonic acid. One such salt of carbonic acid is calcium carbonate (limestone), the decomposition of which by sulfurous acid is fundamental to the removal of SO_2 from a flue gas stream using limestone as the absorbent.

The solubility of SO_2 in pure water as a function of temperature is shown in Fig. 31-1. Solubility of SO_2 in water decreases with increasing temperature and increases as the partial pressure of SO_2 over the solution increases. This behavior is characteristic of most gases.

Lime, with the chemical formula CaO (calcium oxide), may be formed by heating (calcining) calcium carbonate at approximately 800° C, according to the following reaction:

$$CaCO_3 \xrightarrow{T = 800° C} CaO + CO_2 \tag{6}$$

Lime is a basic anhydride which, when combined with water, yields calcium hydroxide (slaked lime), $Ca(OH)_2$. This dibasic alkali dissolves in pure water, dissociating according to the following ionic formula:

TABLE 31-4

Comparison of Acidities of Various Common Acids
(Assuming 0.1 Mole Parent Acid Dissolved in 1 Liter of H_2O)

Parent acid	Formula	Dissociation species	Estimated pH of solution
Nitric acid	HNO_3	NO_3^-	1.0
Hydrochloric acid	HCl	Cl^-	1.0
Sulfuric acid	H_2SO_4	HSO_4^-, SO_4^{2-}	1.0
Acetic acid	CH_3COOH	CH_3COO^-	2.9
Carbonic acid	H_2CO_3	HCO_3^-, CO_3^{2-}	3.7
Sulfurous acid	H_2SO_3	HSO_3, SO_3^{2-}	1.4
Nitrous acid	HNO_2	NO_2^-	2.2
Oxalic acid	$(COOH)_2$	$C_2O_4H^-$, $C_2O_4^{2-}$	1.1

$$Ca(OH)_2 \rightleftarrows Ca^{2+} + 2OH^- \quad (7)$$

It is rather insoluble (0.185 g in 100 g of H_2O at 0° C) and unlike most soluble metallic hydroxides, its solubility decreases as the temperature increases.

Limestone consists of a mixture of calcium carbonate and siliceous compounds. Although limestone is quite plentiful, being found in every one of the United States and nearly all foreign countries, it is estimated that only 2 percent of the deposits are of "chemical grade" (i.e., containing 95 percent or more $CaCO_3$). Dolomitic limestone, mixtures of calcium and magnesium carbonates, are plentiful, but are usually unacceptable as absorbent reagents in SO_2 wet scrubbing processes because of their relative inertness.

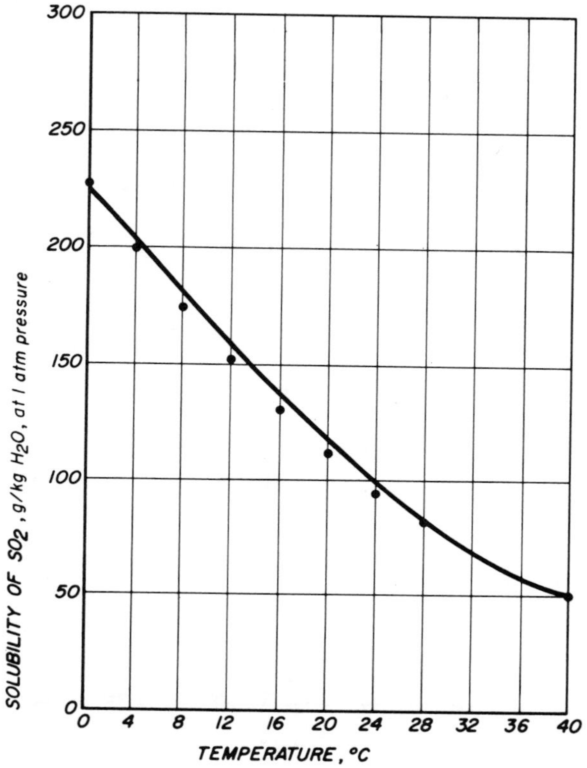

FIG. 31-1. Solubility of SO_2 in pure water as a function of temperature.

Calcium carbonate, the calcium salt of carbonic acid, is quite insoluble (0.00153 g per 100 g of H_2O at 0° C), and its solubility increases only slightly as the temperature is increased.

The low solubility of calcium carbonate is one of the chief drawbacks against its use in wet scrubbing processes for SO_2 removal. Figure 31-2 compares the solubility of calcium hydroxide, calcium carbonate, calcium sulfite, and calcium sulfate, which are associated with the scrubbing of SO_2 from flue gases.

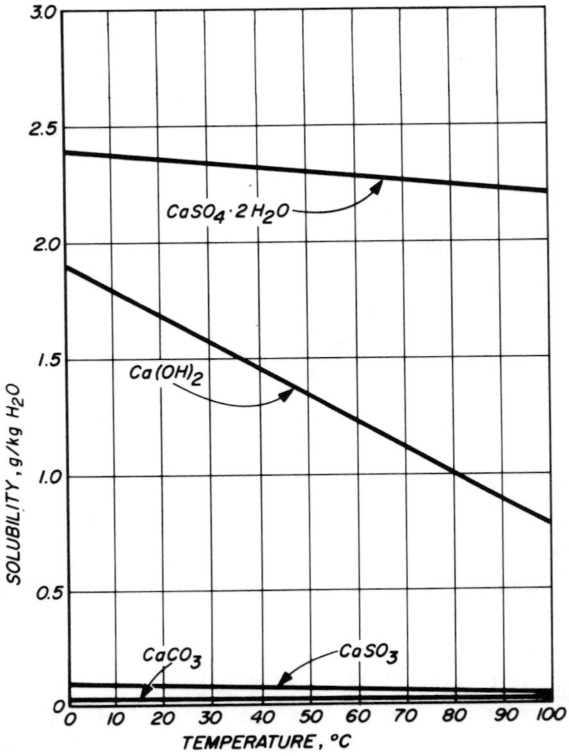

FIG. 31-2. Solubilities of calcium hydroxide, calcium carbonate, calcium sulfite, and calcium sulfate as a function of temperature.

Analysis of Chemical Interactions Among SO_2, Lime, and Limestone

Equations (8) and (9) show the overall reactions for wet scrubbing of SO_2 with limestone and lime, respectively. Scrubbing with either limestone or lime results in the formation of calcium sulfite dihydrate.

$$CaCO_3 + SO_2 + 2H_2O \rightarrow \underset{\text{calcium sulfite dihydrate}}{CaSO_3 \cdot 2H_2O} + CO_2 \quad (8)$$

$$CaO + SO_2 + 2H_2O \rightarrow CaSO_3 \cdot 2H_2O \quad (9)$$

Thermodynamic calculations indicate that the reaction of lime with SO_2 should be more favorable by several orders of magnitude than the reaction of limestone with SO_2 (Table 31-5), but that both reactions are highly favorable.

A detailed mechanism for wet scrubbing with limestone has been worked out by W. H. Boll of Babcock and Wilcox. The postulated reaction mechanism is as follows:

TABLE 31-5

Thermodynamic Calculations for Reactions Between Lime/Limestone and Sulfur Dioxide

Reactions:

$$CaO + 2H_2O + SO_2 \rightarrow CaSO_3 \cdot 2H_2O \quad \text{(Lime)}$$

$$CaCO_3 + 2H_2O + SO_2 \rightarrow CaSO_3 \cdot 2H_2O + CO_2\uparrow \quad \text{(Limestone)}$$

Thermodynamics:

Compound	Free energy of formation at 25°C (kcal/mole)
CaO	-144.4
$CaCO_3$	-269.5
H_2O	-56.7
SO_2	-71.8
$CaSO_3 \cdot 2H_2O$	-374.1
CO_2	-94.3

ΔG for lime reaction:

$$\Delta G = -374.1 - [-144.4 - 2(56.7) - 71.8] = -44.5 \text{ kcal/mole}$$

ΔG for limestone reaction:

$$\Delta G = -374.1 - 94.3 - [-269.5 - 2(56.7) - 71.8] = -13.4 \text{ kcal/mole}$$

Equilibrium constants:

Lime: $\log K_{eq} = 32.8$ Limestone: $\log K_{eq} = 10.9$

Mechanism I

$$SO_2(g) + H_2O \rightarrow H_2SO_3 \tag{10}$$

$$H_2SO_3 \rightarrow H^+ + HSO_3^- \tag{11}$$

$$H^+ + CaCO_3 \rightarrow Ca^{2+} + HCO_3^- \tag{12}$$

$$Ca^{2+} + HSO_3^- + 2H_2O \rightarrow CaSO_3 \cdot 2H_2O + H^+ \tag{13}$$

$$H^+ + HCO_3^- \rightarrow H_2CO_3 \tag{14}$$

$$H_2CO_3 \rightarrow CO_2 + H_2O \tag{15}$$

This series of reactions was developed for a scrubbing system in which $CaCO_3$ is the absorbent rather than CaO. It was assumed in the early British work that if lime were used instead of limestone, the lime would convert to $CaCO_3$ by reacting with the CO_2 in the flue gas, as shown in Eq. (16):

$$CaO + CO_2 \rightarrow CaCO_3 \tag{16}$$

If this were the case, then the Boll mechanism would be applicable to both lime and limestone scrubbing processes.

Although analysis of the by-products of SO_2 scrubbing with lime indicates that some $CaCO_3$ is formed, this does not prove conclusively that all lime reacts with CO_2 to form calcium carbonate. An alternative mechanism may be postulated for SO_2 scrubbing with lime as follows:

Mechanism II

$$SO_2(g) + H_2O \rightleftarrows H_2SO_3 \tag{17}$$

$$H_2SO_3 \rightarrow H^+ + HSO_3^- \tag{18}$$

$$CaO + H_2O \rightarrow Ca(OH)_2 \tag{19}$$

$$Ca(OH)_2 \rightarrow Ca^{2+} + 2OH^- \tag{20}$$

$$Ca^{2+} + HSO_3^- + 2H_2O \rightarrow CaSO_3 \cdot 2H_2O\downarrow + H^+ \tag{21}$$

$$2H^+ + 2OH^- \rightarrow 2H_2O \tag{22}$$

It should be noted that summation of all the terms in Mechanism I yields Eq. (9), the overall reaction for the limestone system.

Summation of the equations composing Mechanism II yields Eq. (8), the overall reaction for the lime system.

These mechanisms describe the chemical pathways leading from the reactants to the products which must be followed by the respective systems. As such, it should be possible to make certain predictions about the performance of a scrubber using either lime or limestone.

Both mechanisms contain one critical step: the formation of the calcium ion [Eq. (12) in Mechanism I and Eq. (20) in Mechanism II]. This is critical because it is the calcium ion which, upon combination with the bisulfite ion, removes the SO_2 from the solution. This critical step points up a very important distinction between lime and limestone systems:

1. In a limestone system, formation of the calcium ion is dependent on the hydrogen ion (H^+) concentration as well as on the presence of limestone ($CaCO_3$).
2. In a lime system, the formation of the calcium ion is dependent only on the presence of lime (CaO).

This means that a limestone system will operate at a lower pH than will a lime system. Test results from the EPA pilot plant facility at Research Triangle Park [4] indicate that the optimal operating pH for a limestone system would be between 5.8 and 6.2, whereas for lime systems the optimal pH is approximately 8.0.

The difference in optimum pH affects the operation of a scrubber by introducing two problems:

1. Operation of a lime/limestone scrubber at a low pH promotes the formation of calcium sulfate hard scale.
2. Operation of a lime/limestone scrubber at a high pH promotes the formation of calcium sulfite soft pluggage.

Figure 31-3 shows the relationship between the solubilities of calcium sulfite and calcium sulfate and pH. The reader may wonder why a discussion of calcium sulfate is being introduced, because calcium sulfate does not appear in the mechanisms postulated for lime

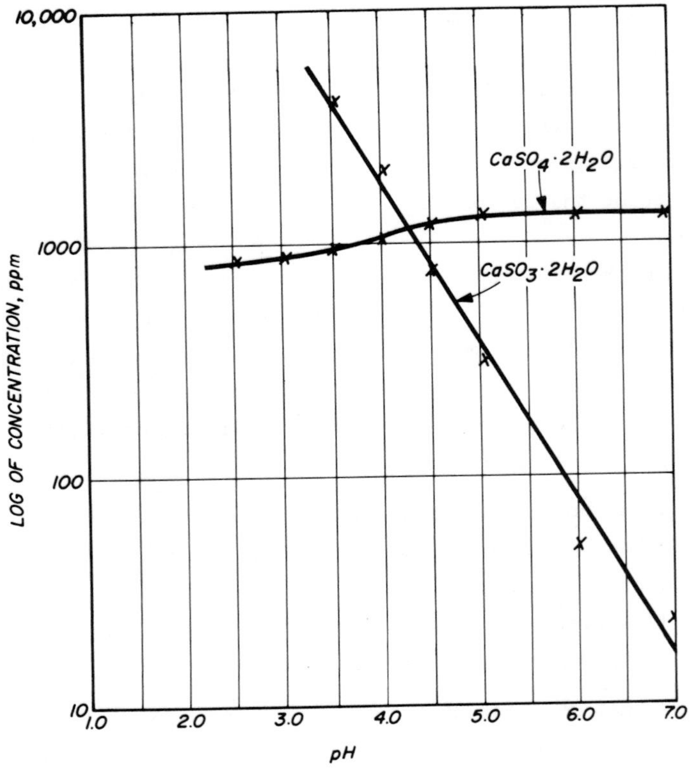

FIG. 31-3. Effect of pH on solubility of $CaSO_3 \cdot 2H_2O$ and $CaSO_3 \cdot 2H_2O$ at 122° F.

and limestone wet scrubbing. However, its inclusion in the discussion is relevant to our topic.

Analogous to the oxidation of SO_2 to form SO_3 as previously described, sulfite ion undergoes oxidation by dissolved oxygen in the absorbent slurry to form sulfate ion, according to Eq. (23):

$$\underset{\text{sulfite ion}}{2SO_3^{2-}} + \underset{\text{dissolved } O_2}{(O_2)} \rightarrow \underset{\text{sulfate ion}}{2SO_4^{2-}} \qquad (23)$$

This reaction occurs in aqueous solution and is favorable thermodynamically, although the rate of the reaction is slow. Calcium sulfate, upon precipitation from solution, forms a hard, stubborn scale on equipment component surfaces. The scale must be removed mechanically and can be done only with considerable effort.

Figure 31-3 shows that the solubility of calcium sulfite increases greatly as the pH decreases; calcium sulfate concentration decreases slightly as pH decreases. The rate at which sulfite is converted to sulfate is expressed by Eq. (24):

$$d/dt(SO_4^{2-}) = K_2(O_2)(SO_3^{2-}) \tag{24}$$

This equation is a second-order rate equation, showing that the rate of sulfate formation is proportional to the product of the concentrations of dissolved oxygen and sulfite ion. Due to the excess oxygen in the flue gas, the concentration of O_2 dissolved in the slurry should be constant. Thus, the formation of sulfate would depend only on the concentration of sulfite. Since sulfite solubility is greater at lower pH, sulfate production should also be enhanced at lower pH.

Furthermore, the solubility of calcium sulfate decreases as pH decreases; thus calcium sulfate is more likely to precipitate out of solution to form hard scale at lower pH's. To avoid hard scale formation, the pH should therefore be kept high.

According to A. V. Slack [2], other factors affecting the degree of sulfite oxidation in scrubbing flue gases have been reported, including:

1. Scrubber contacting mechanism,
2. Scrubber temperature,
3. Oxygen/sulfur dioxide ratio in the gas,
4. Amount of nitrogen dioxide in the flue gas, and
5. Presence of oxidation inhibitors and catalysts, such as manganese and iron in the flyash entrained in the flue gas stream.

Figure 31-3 shows that the solubility of calcium sulfite decreases rapidly as the pH increases. Although this behavior was exploited successfully in order to prevent oxidation of sulfite to sulfate with subsequent hard scale formation, this low sulfite solubility at high pH promotes a phenomenon known as soft pluggage.

Soft pluggage is formations of calcium sulfite on the internal components of a scrubber. They are characterized by large, leaflike masses resembling very delicate corals. They are soft and easily altered mechanically; thus maintenance of equipment that is pluggage-prone requires much less effort than does maintenance of equipment covered with hard scale.

The best way to control soft pluggage is to institute proper pH control on the scrubber operation. Experience has shown that pluggage can be "melted" off components simply by lowering the pH in the scrubber, thus promoting higher solubility of calcium sulfite. When the pH is lowered, the pluggage dissolves away.

Data from full-scale operation of an SO_2 scrubber in Japan, built by Chemico and using lime as the reagent, indicate that an inlet pH of 8.0 should not be exceeded in order to avoid pluggage. Pluggage is not a problem on limestone systems, since the pH of operation rarely exceeds 6.0. EPA/TVA pilot plant test data [4] indicate that for limestone systems, control of scrubber effluent pH to less than 6.2 is sufficient to avoid soft pluggage.

One of the problems associated with SO_2 removal utilizing lime as the reagent is the capture of carbon dioxide from the gas stream [according to Eq. (16)]. As should be noted, CO_2 is present in typical flue gases in concentrations from 50 to 100 times higher than sulfur dioxide.

The formation of calcium carbonate reduces the efficiency of the scrubber because of the lower reactivity of calcium carbonate.

Table 31-6 summarizes a series of tests conducted by the EPA/TVA [4] to show the effects of elevated pH levels on the removal of CO_2 from the flue gas. It is clear from the table that recarbonation is significant at an inlet pH of 9 and above.

For limestone systems, recarbonation would obviously not be a problem.

Reagent Reactivity and Kinetics

Very little work has been done in the way of a systematic study of the factors that affect reaction rates. The main work on kinetics

TABLE 31-6

Recarbonation of Lime in a Scrubber as a Function of Scrubber Inlet pH (Scrubber Type: TCA)

	Scrubber inlet pH			
	6	7	9	10
CO_2 in scrubber effluent (mg/liter)	42	43	47	68
CO_2 in scrubber inlet (mg/liter)	29	26	25	8
CO_2 precipitated as CO_3^{2-} in the recirculation tank	13	17	22	60
Lime utilization (stoichiometric ratio)	1.02	1.03	1.12	1.24

has been done in the past few years, most of it under EPA contracts. Other work, carried out under limited-scope objectives and cost constraints, appears to have led to inadequate exploration of the several variables that may affect kinetics.

Some of the most promising work was done by R. H. Boll of Babcock and Wilcox, who postulated a kinetic model for lime and limestone systems and was able to simulate mathematically the interaction of lime/limestone systems. Boll's model investigated the effects of: (1) SO_2 partial pressure in the flue gas, (2) liquid holdup in the scrubber (L/G ratio), (3) diffusion resistance through the liquid/solid interface, and (4) contact surface area.

The model was calibrated by measurements from a pilot plant testing program utilizing a turbulent contact absorber. Some of the findings from these model tests and results of pilot plant programs relating to the kinetics of SO_2 removal are as follows:

1. Inlet gas temperature: the hotter the inlet gas temperature, the lower the rate of SO_2 removal.
2. Scrubber slurry temperature: the lower the slurry temperature, the greater the rate of SO_2 removal.
3. Inlet SO_2 concentration: the higher the inlet concentration, the greater the rate of removal.

4. Limestone type: the smaller the particle size, the greater the rate of removal.
5. Effect of dolomitic content in limestone: dolomitic ($MgCO_3$) content in limestone lowers the rate of removal.
6. Solubility of reagent: limestone is less reactive than lime because limestone is much less soluble than lime.
7. Special precipitation effects: $CaSO_3$ may precipitate on the surface of lime or limestone particles (see Fig. 31-3), thus "blinding" the reaction.

Reagent Reactivity and the L/G Ratio

The reaction between SO_2 and lime or limestone occurs in solution as reactions between ionized species. As previously noted, neither lime nor limestone is particularly soluble. Thus, the concentration of calcium ion, which is essential to the scrubbing operation (as shown in Mechanisms I and II), is likely to be low.

Solubility of salts and bases can be enhanced beyond the saturation limit if constant agitation of the solid solute in contact with the aqueous solution is accomplished. This turbulent mixing process promotes supersaturation by exposing a maximum solid surface area per volume of slurry.

For lime and limestone scrubbers, one measure of the degree of surface contact is the liquid-to-gas ratio, usually in gallons per 1,000 cfm. The L/G ratio, as this parameter is called, must be high in systems where the reactivity of the reagent is small. TVA/EPA test data have shown that for reliable operation of limestone systems the L/G ratio must exceed 65. For a lime system, however, an L/G ratio of 35 appears to be adequate because of the greater reactivity.

Stoichiometric Ratio, Reagent Requirements, and Reagent Reactivity

Another operating parameter which is governed by the reactivity of the reagent is the stoichiometric ratio. This parameter is defined for a sulfur dioxide removal system as the weight of reagent experimentally required to remove 1 lb of SO_2 from the gas stream divided by the weight of reagent theoretically required to do the same. Stoichiometric ratios, for comparative purposes, must apply to the same reduction in sulfur dioxide concentration measured across the inlet to the outlet of the system. Theoretically, 0.85 lb of lime and 1.52 lb of limestone (100 percent pure in both cases) would be required to remove 1 lb of SO_2 from a flue gas stream.

Due to the large difference in the reactivity of lime and limestone, the stoichiometric ratios for typical lime systems range from 1.05 to 1.15, while those for limestone range from 1.25 to 1.6. These ratios correspond to removals of approximately 90 percent of the SO_2. As can be seen from these stoichiometric ratios and the theoretical weights required for removal, limestone systems require approximately 2.3 times the weight of reagent to perform the same SO_2 removal as would a lime system.

Requirement for Recirculation Tank in Scrubber Design

One of the most important features of a lime or limestone system is the design of the retention or recirculation tank. The recirculation tank is a basin which catches the effluent from a scrubber and provides additional time for the reactions to occur.

It is in the recirculation tank that precipitation of calcium sulfite, calcium sulfate, unreacted calcium carbonate, and other silicate inerts occurs. It is very important that precipitation shall occur in the recirculation tank instead of in the scrubber, because solids deposition within the scrubber would plug and foul the system.

The reactions for lime and limestone systems which occur in the recirculation tank are summarized in Eqs. (25) through (28).

For lime systems:

$$CaO + Ca(HSO_3)_2 + 3H_2O \rightarrow 2CaSO_3 \cdot 2H_2O\downarrow \qquad (25)$$

$$CaO + Ca(HSO_3)_2 + 3H_2O \rightarrow 2CaSO_4 \cdot 2H_2O\downarrow \qquad (26)$$

For limestone systems:

$$CaCO_3 + Ca(HSO_3)_2 + 3H_2O \rightarrow 2CaSO_3 \cdot 2H_2O\downarrow + CO_2 \tag{27}$$

$$CaCO_3 + Ca(HSO_3)_2 + O_2 + 3H_2O \rightarrow 2CaSO_4 \cdot 2H_2O\downarrow + CO_2 \tag{28}$$

There is evidence that calcium sulfate and calcium sulfite precipitate simultaneously in a "solid solution" consisting of about 22.5 mole percent calcium sulfate [4]. This simultaneous precipitation in the recirculation tank of sulfate and sulfite is vital for scale-free scrubber operation.

EPA/TVA pilot plant data show that recirculation tank residence time should be about 10 min for a limestone system and about 5 min for a lime system [4]. This difference in the required residence time is due to the greater reactivity of lime as compared with limestone.

Waste Products from Lime and Limestone Scrubbing

The products crystallized from the slurry in the recirculation tank are the waste products from the scrubbing system. Typically, this waste sludge is conveyed to a thickener for primary dewatering, followed by secondary dewatering through vacuum filtration. After vacuum filtration, the percentage of water in the sludge cake is approximately 35 percent.

Typical chemical compositions of dessicated waste sludges are shown in Table 31-7 for lime and limestone scrubbing systems. The percentages of sulfur in a typical pound of dessicated sludge from lime and limestone systems are 17.0 and 13.6 percent, respectively.

An appreciation of the magnitude of the disposal problem caused by this sludge can be gained from the following example. Using the coal analysis shown in Table 31-1 and assuming a 1,000-MW generating station, scrubbing with a lime system would generate approximately 1,800 tons/day of dry wastes. Assuming a sludge of 35 percent solids content, 5,200 tons/day of wet sludge would be generated. For the American Electric Power System, which has approximately 12,000 MW of coal-fired capacity, the sludge generation could be well over 60,000 tons/day.

TABLE 31-7

Typical Lime and Limestone Scrubber Sludge Compositions

Compound	Percent dry weight
Limestone systems:	
$CaCO_3$	33
$CaSO_3 \cdot 2H_2O$	58
$CaSO_4 \cdot 2H_2O$	9
	100
Lime systems:	
$CaCO_3$	5
$CaSO_3 \cdot 2H_2O$	73
$CaSO_4 \cdot 2H_2O$	11
$Ca(OH)_2$	11
	100

The material in this chapter has been presented to convey to the reader the fact that sulfur dioxide scrubbing is not just a matter of choosing a venturi or packed tower or turbulent contact absorber, but rather is a complex interplay of chemical thermodynamics and kinetics. The data regarding pH control, recirculation tank residence times, L/G ratio requirements, stoichiometric ratios, and reagent requirements are based on the pilot plant test findings performed for the EPA and TVA at the TVA Shawnee Power Station and have a wide range of applications irrespective of scrubber contacting mode.

REFERENCES

1. A. V. Slack: An Introduction to Stack Gas Cleaning Technology, Proceedings of the Conference, "Sulfur in Utility Fuels: The Growing Dilemma," Chicago, Illinois, October 25-26, 1972, p. 161.

2. A. V. Slack: Sulfur Dioxide Removal from Waste Gases, Noyes Data Corporation, 1971.

3. D. H. Andrews and R. J. Kokes: Fundamental Chemistry, John Wiley & Sons, New York, 1963.

4. M. Epstein et al.: Preliminary Report of Test Results from the EPA Alkali Scrubbing Test Facilities at the TVA Power Plant and at Research Triangle Park, Bechtel Corporation, EPA and TVA, December 1973.

Chapter 32

WET SCRUBBING TECHNIQUES ON LIME SLUDGE KILNS

Ken Schifftner

Environmental Elements Company
Division of Koppers Inc.
Ramsey, New Jersey

The lime sludge kiln is used extensively in the pulp and paper industry, especially the kraft process, for the production of quicklime. Quicklime (CaO) is an instrumental part of the chemical recovery system of a pulp mill.

Lime mud, the underflow from the white liquor clarifier, is pumped at approximately 10 percent solids to a lime mud washing system where impurities and considerable soda (expressed as Na_2O) are removed. The residual mud is stored in slurry form in an agitated lime mud storage tank in order to provide uniform feed to the subsequent lime mud filter. The lime mud vacuum filter is a large, slowly rotating drum in which the solids content of the mud is increased to over 65 percent solids for introduction into the kiln. Its role in the preparation of the lime mud is of pivotal importance, because the moisture content of the mud feed can considerably affect kiln production, outlet humidity, and thermal efficiency. Every additional pound of water not removed by the vacuum filter means increased fuel costs and, in all probability, higher maintenance costs through lower thermal efficiency of the kiln system.

The filtered lime is mixed with fresh limestone ($CaCO_3$) to replace chemicals lost in the recovery process through the green liquor (smelt) discharge from the smelt tank vent and from other points of loss. This mixture, containing CaO, $CaCO_3$, water, dissolved sodium compounds, sodium sulfate and carbonate (in the kraft process), and trace unwashed odorous compounds such as H_2S, methyl mercaptan, dimethyl sulfide, and others, is introduced into the feed (higher) end of the kiln. The rotation of the refractory-lined kiln and gravity move the lime mixture to the discharge (lower) end, where temperatures as high as 1,800 to 2,000°F are maintained by a burner. Its fuel, either gas, oil, or coal, produces products of combustion and particulate effluent such as flyash. The hot gases rise through the slowly turning kiln, passing through a chain mix section where the lime mud contact surface is increased, to the feed end. This counterflow motion, draft, kiln rotation, and feed determine the production capacity of the equipment.

The kiln may also do double service as a combustion chamber for noncondensible odorous gases from the evaporator or digester areas. Odorous sulfur compounds are thus oxidized to SO_2 and SO_x compounds. This fairly new technique has provided one of many inputs for the design of the collecting equipment.

The next-to-last device through which the gases pass is the feed-end housing, a hopperlike box with an internal baffle and possibly dust suppression sprays. This chamber is the remnant of the original air pollution control device used on lime sludge kilns, the knock-out chamber. It still serves as a knock-out chamber by permitting the dust-laden gas stream to approach its settling velocity. This chamber removes large (> 25 μm) particles. Following the settling chamber is the scrubbing system, the final treatment applied before the process air is released to the atmosphere.

So far, we have addressed ourselves to the kiln equipment. Equally important are the chemical processes involved. They are as follows:

1. The calcining of calcium carbonate to calcium oxide,

$$CaCO_3 \overset{\Delta}{\rightarrow} CaO + CO_2$$

at temperatures in excess of 1,600°F,

2. The incineration of fossil fuels in the kiln burner,

3. The incineration of noncondensible sulfur compounds (if the kiln is so equipped), and
4. The superheating of water vapor present in the lime mud feed and the primary combustion air.

Thus, the calcining of the lime at elevated temperatures using fossil fuels in a combustion process contributes to the dust and fume emission that a good air pollution control device must accommodate. It must be operable 24 hr/day, 7 days/week with little maintenance and the minimum of cost to the owner. The combination of incineration and the drying of the lime mud sludge provides a humid dusty gas to the scrubber, a gas that may also contain odorous and submicron particulate.

Early scrubbing equipment was designed around the foregoing processes and the requirements of law and economics. The simple knock-out chamber collector was sufficient for a world of cheap fuel and cheap lime. Twenty to twenty-five years ago knock-out chambers consisted of concrete or steel baffle chambers with spray nozzles or weir-type "waterfall" liquid walls through which the dust-laden gases passed. Initial costs dictated the simplest emission control equipment, and that is what was provided. Efficiencies were only 60 to 80 percent. The low-initial-costs ethic most likely permitted the rapid expansion of the pulp and paper industry.

Soon, however, corporate process analysts were concerned about the dollars being exhausted from the kraft process, the smelt tank vent, the recovery stack, and the lime kiln. The chemicals exhausted into the air from low-initial-cost equipment reduces company profits and the ability to improve the process.

This need precipitated the adaptation of high-efficiency impingement-type scrubbers, which had been used for gas absorption in many chemical processes. Basically, the gas is directed by a bottom-inlet plenum to a horizontal perforated plate. Directly above and opposite each of these perforations is a baffle. The top surface of the perforated plate is constantly flushed with the scrubbing liquid. The bubbling of the liquid surface, the double 90° change of direction of the gas stream, the liquid gas film contact, and the inertial impaction on the baffle permitted efficiencies of 90 percent with low liquid pressures, no spray nozzles, and low (6 to 8 in.) pressure loss.

Limitations of this design included most principally the inability to run at high-recycle liquid solids content and its attendant problems. Scrubbing liquid solids were typically 1 to 2 percent solids maximum to keep the perforated plate clean and prevent uneven distribution with resulting lower scrubbing efficiency. Because this underflow (scrubber effluent) was typically sent to the mud filter feed, excess energy was expended in removing that water. As kiln sizes increased for additional production capacity, this underflow became significant. The need to operate at higher solids content was important, but federal and state laws merely required efficiencies easily attained by these low-energy scrubbers.

Their low cost and adequate efficiency recommended impingement devices for the job. But they suffered from restricted efficiency in a world of changing laws and economies. A lesson difficult to learn in the 1950s and 1960s was that certain air pollution control equipment had very limited efficiencies and that as the theoretical 100 percent efficiency target was approached, more and more energy was required. Devices such as the plate scrubber described in the previous paragraphs were inertial absorbers, that is, they collected particulate through inertial impaction and absorbed fume through long contact. Theoretically, the addition of extra groups of plates would increase the efficiency of the units. In many cases, however, this was not proven in practice.

It is generally conceived among air pollution control engineers that particles abide by Stokes' law and the laws of inertial and electrostatic attraction (Table 32-1). Practical application of this law reveals that particles below 3 μ tend to follow the air flow rather than their own path. Thus they resist inertial separation unless:

1. Their size is increased through agglomeration or impaction.

TABLE 32-1

The Frank Chart: Size and Characteristics of Airborne Solids[a]

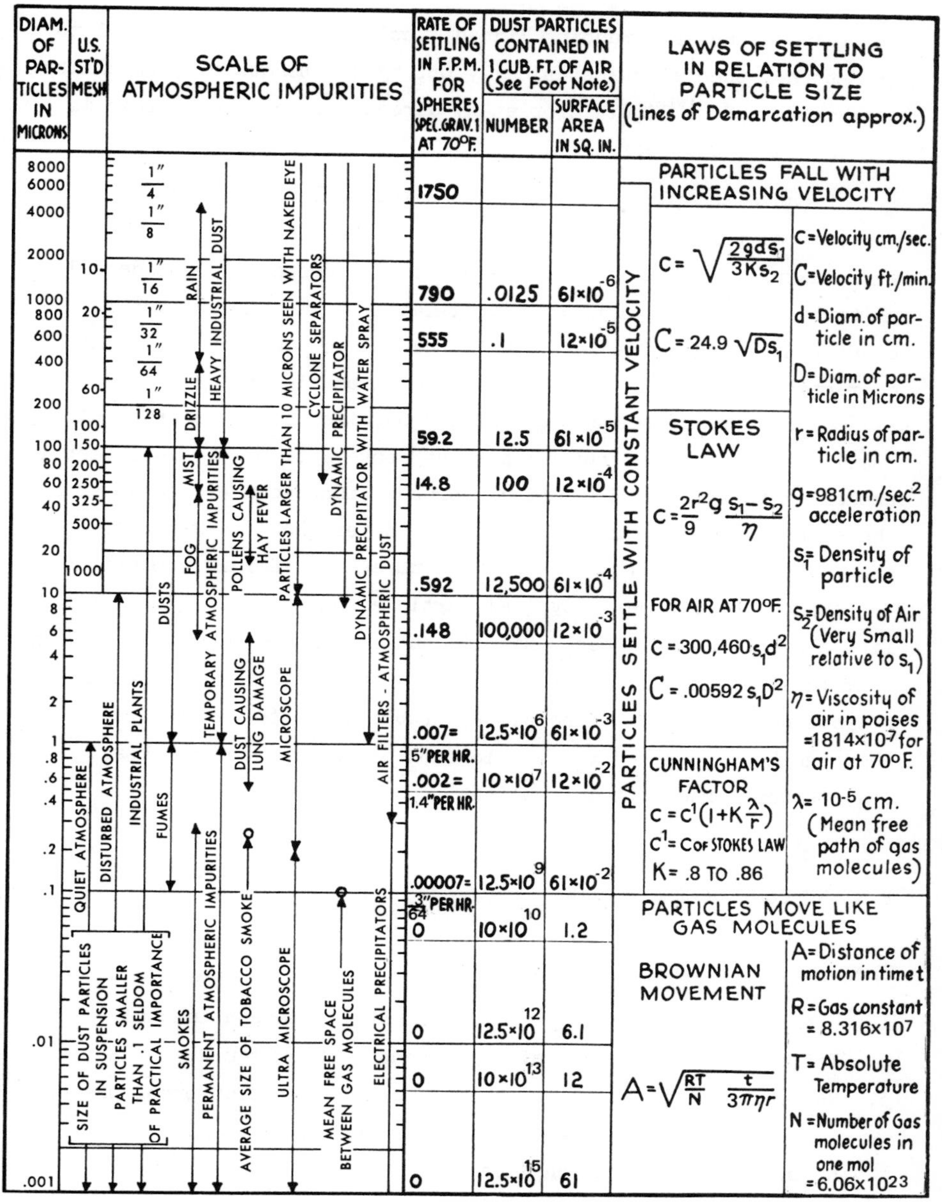

[a]It is assumed that the particles are of uniform spherical shape having specific gravity 1 and that the dust concentration is 0.1 gr/1,000 ft^3 of air, the average of metropolitan districts. (Compiled by W. G. Frank, copyrighted by American Air Filter Co., Inc., Dust Control Products, Louisville, Kentucky.)

2. They encounter scrubbing liquid droplets sufficiently small to attract them electrostatically or, by the tenets of probability, happen upon a liquid droplet by direct interception. This is the "droplet theory."

Because the plate collector uses inertial techniques, the efficiency on particulate above 3 μm is outstanding. Due to the inverse nature of Stokes' law, the efficiency drops off considerably as particle distributions of approximately 3 μm and less are encountered. Studies regarding this phenomenon prompted the Cunningham correction to Stokes' law for these small particles.

Claims of 99+ percent efficiency on lime dust were truthful and justified if the dust was 99+ percent above 2 to 3 μm. This fact is exploited to this day by various scrubber manufacturers to enhance the salability of their products. The buyer should ask for efficiency data on particulates below 1 μm to achieve a fair comparison.

Figure 32-1 shows a typical particle size distribution for a lime sludge kiln scrubber. Significant quantities of dust are below 1 μm in size, including all of the soda fume. This realization, the operational problems of the plate scrubber, tighter economics, and changing laws prompted the application of the venturi scrubber, today's workhorse on this application.

PRESENT CONTROL TECHNIQUES

The venturi scrubber permits:

1. High recycle solids (up to 10 percent solids continuously),
2. Efficiencies in excess of 99 percent all particulate,
3. Removal of spin vanes or other internal corrosion-prone pieces common with the plate-type scrubber, and
4. Recovery of valuable chemicals.

Figure 32-2 shows the basic components of a venturi scrubber. The venturi scrubber uses the induced draft fan static pressure to atomize the scrubbing liquid in a high-velocity venturi throat so

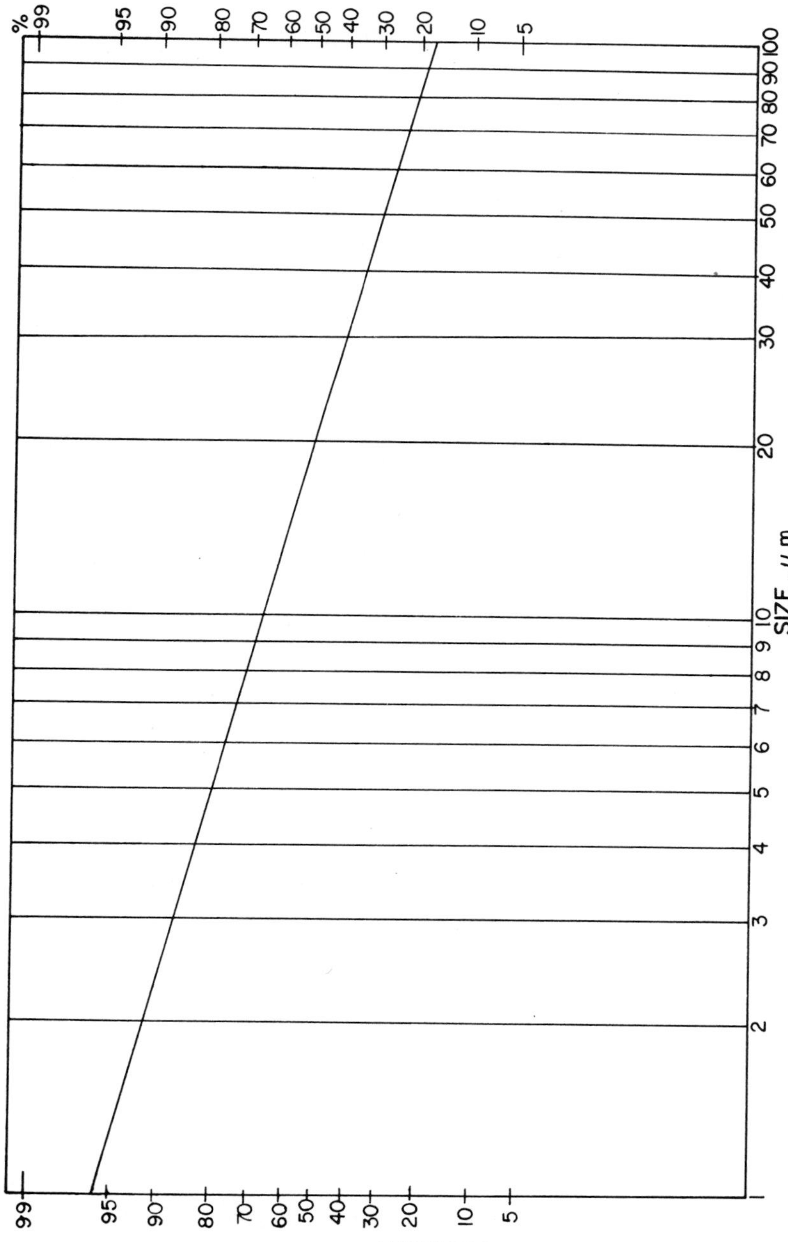

FIG. 32-1. Limestone kiln particle size distribution.

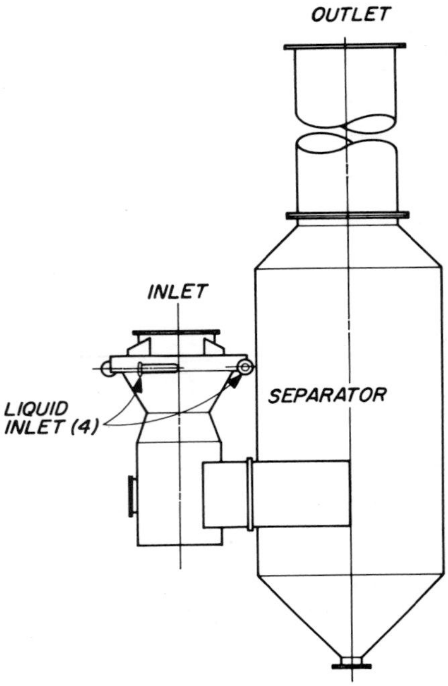

FIG. 32-2. Venturi scrubber components.

that the contaminant will meet a droplet and be removed. Because the lighter dust moves more quickly than the liquid, this differential motion makes the word scrubber an apt expression. Unlike a a scrub brush, the operation is continuous and largely in one direction in many designs. At any given instant, a turbulent mixture of finely divided liquid droplets and contaminant is present in the throat. Contact times of only a few hundredths of a second or less are common. As energy levels are increased, a point at approximately 60 in.wc is reached where the surface tension of the liquid resists further subdivision unless thermal or additional mechanical energy is introduced. Up to that point, however, excellent efficiencies may be obtained not only on dust but on gaseous pollutants as well.

In the late 1950s and throughout the 1960s, venturi scrubbers began replacing plate-type equipment.

In typical business fashion, most companies decided it prudent and economical to comply with the existing legal requirements rather than install equipment that exceeded specified efficiencies. Venturi scrubber efficiency versus pressure loss is expressed as a straight line on log-log paper. Thus, the initial venturi scrubbers were of the low (10 to 15 in.wc) type to achieve 95+ percent on lime and 50 percent on soda fume. Improvements showed a rapid increase in pressure drop with a smaller increase in efficiency.

There are various ways of creating this pressure drop, including:

1. Orifice plate restriction,
2. Simple short-contact venturi throat,
3. Adjustable venturi throat:
    a. Short-contact orifice type and
    b. Long-contact annular type.

Among venturi scrubber manufacturers, the virtues of long contact versus short contact are hotly debated.

Companies not having long-contact scrubbers on their product list contend that the scrubber chemical reactions are much longer than the contact time anyway, so long contact is not necessary. People selling long-contact throats point to the laws of probability, which indicate that the longer one keeps a particle exposed to fine liquid droplets, the more likely it is for removal to occur. They say that this is not important until the problem involves removal of submicron particulate.

The initial lime sludge kiln scrubbers utilized two fan arrangements:

1. Hot fan. The fan is located directly after the feed-end housing and before the scrubber. Handles all particulate at 400 to 600° F, fume thoroughly vaporized, and gases above their dew point. Stack is on scrubber. Fan handles dust.
2. Wet fan. Fan after scrubber. Handles wet vapor, cleaned by scrubber. Separate stack required.

Both arrangements are popular, with each method of fan location having its advantages and limitations (Table 32-2).

TABLE 32-2

Hot fan

Advantages:

1. Cheaper materials of construction. Mild steel, or mild steel housing with stainless steel wheel.
2. No excessive water vapor to cause dust buildup.
3. Draft control between feed-end housing and fan for smooth draft on kiln.
4. Stack may be put on scrubber to save space.

Limitations:

1. Requires heat slingers, special construction for temperature resistance.
2. Handles full dust load at slightly higher horsepower.
3. Requires larger fan, with cooled bearings.

Wet fan

Advantages:

1. Smaller fan is needed.
2. Handles only "cleaned" air flow.
3. Requires lower horsepower (in many, but not all cases).
4. Uses typical roller bearings without cooling.

Limitations:

1. More exotic materials of construction are required (stainless steel throughout).
2. May create liquid-level control problems in separator.
3. Requires separate stack.
4. More difficult draft control because scrubber may introduce fluctuations.

The fan position is not the only consideration in the proper application of a venturi scrubber. The venturi throat itself is the most important and controversial design aspect of the system. We touched on the difference in philosophy, long contact versus short contact, in previous paragraphs. Equally important is how the pressure drop is adjusted.

Nearly all lime sludge kiln venturi scrubbers being installed today are of the adjustable type. That is, their venturi throat velocities can be adjusted to provide optimum kiln draft, production, and clean air under widely varying kiln operating conditions. The need for this adjustability is important because the kiln feed, rotation, feed-end temperature, percent solids in feed, and ambient conditions all affect the volume to the scrubber.

Throat adjustment mechanisms may be categorized into three types:

1. Damper type--The throat contains a damper blade or blades by which the cross-sectional area of the throat may be adjusted.
2. Flooded-disk type--Adjustable orifice-type venturi in which a wetted disk acts as an adjustable plug, actuated externally.
3. Annular type--A central wetted aerodynamic core is used to provide a variable annular gap to vary air velocity.

Table 32-3 lists some of the major advantages and limitations of each type of throat adjustment mechanism.

TABLE 32-3

Damper type

Advantages:

1. Simple, economical construction.
2. Accessible from outside.
3. Falls open but with low turndown (ratio of maximum volume to minimum volume).
4. Easily adjusted.

Limitations:

1. Sixty to eighty percent of gas is directed to one side of throat, creating higher wear on one side than the other.
2. Has two seals.

TABLE 32-3 (Continued)

3. Has low turndown.
4. Restricted throat width (14 to 16 in. maximum) for even distribution of scrubbing liquid.
5. Acts as extremely short-contact orifice when fully closed.
6. Lower theoretical aerodynamic and mechanical efficiency.

Flooded disk type

Advantages:

1. Small annular gap conserves scrubbing liquid energy.
2. Thoroughly wetted venturi approach.
3. Simple external adjustment.
4. High turndown for safe operation.

Limitations:

1. Falls closed unless specifically designed otherwise.
2. Has center liquid circuit more complex than other designs.
3. More expensive to build.
4. High-wear orifice design.
5. Short-contact throat.

Annular type

Advantages:

1. High turndown for safe operation.
2. Long-contact turbulent throat.
3. Easy external adjustment.
4. Totally wetted venturi approach.
5. No sharp edges for erosion and wear.

Limitations:

1. More expensive to build.
2. Requires one additional piping connection.
3. Requires remote throat positioner.

Of the types compared, the damper type is by far the most frequently installed. Its lower cost makes it more attractive in the competitive marketplace. However, within the last five years, the need to offer more sophisticated designs of higher mechanical efficiency and better aerodynamics has placed the annular and flooded-disk types in a more prominent position. Just as hydraulics texts indicate higher, more restrictive flow coefficients for an orifice plate over a venturi, the annular venturi's less drastic direction changes offer more effective mechanical characteristics.

Tests and studies have been made throughout the evolution of the lime sludge kiln scrubber which provide sometimes conflicting yet helpful data. The data today indicate that longer-contact venturis offer power savings through more thorough mixing of the scrubbing liquid with the hard-to-remove submicron fume. (In one application, an annular scrubber operating at 22 in. pressure drop provided 99.7 percent efficiency when another type of scrubber had been quoted at 26 to 28 in. pressure drop for the same efficiency.)

The decision as to which type of scrubber to use is made easier by analyzing the evolution of the techniques used to control this important application. Thought and time are being expended daily to improve the reliability, the performance, and the value of air pollution control equipment operating on a lime sludge kiln.

Many improvements have been made in each of the air pollution control equipment manufacturer's product lines to reflect the lessons learned. The buyer today is able to purchase the latest in this improving technology.

Chapter 33

PARTICULATE AND ODOR CONTROL IN THE ORGANIC FERTILIZER INDUSTRY: A CASE HISTORY

Ken Schifftner

Environmental Elements Company
Division of Koppers Inc.
Ramsey, New Jersey

The growing shortage of fertilizer on a worldwide basis combined with the concern of ecology groups over the type of fertilizers used is causing a major impact on millions of lives. Shortages of fertilizer are threatening the health and lives of millions of people, particularly those in the developing nations. This has heightened interest in the use of organic fertilizers by more and more users.

One company, the Walker-Gordon Laboratory in Plainsboro, New Jersey, formerly the world's largest certified milk farm with 2,400 acres of farmland but now in a beef-feeding operation with some 1,800 feeder steers, is currently drying and processing cow manure for the fertilizer market under the trade name Bovung. They use a process which involves the collection of manure which has been mixed with a bedding material of dried peanut and coconut shells, drying it in a rotating dryer, and packaging and selling the end product. The company processes 9,500 lb/hr of wet manure using this method.

A major problem related to this type of fertilizer production is air pollution control, including the control of particulate matter from the manure dryer product-recovery cyclone and the odors which result from the drying process.

Walker-Gordon decided in May 1969 to try to resolve this problem by engaging a firm to perform stack tests on the dryer to obtain

TABLE 33-1

Stack Test Results

Temperature	272° F
Volume	13,200 scfm
Dust	1.03 lb/hr coarse
	0.77 lb/hr fine
Ammonia	Trace (less than 25 ppm)

specific data on amounts of pollutants in the air. The results are shown in Table 33-1.

These tests showed that the dryer's product-recovery cyclone was performing very well, but that an odor problem existed. A later test conducted by the State of New Jersey in September 1969 confirmed that the particulate emissions were well below state requirements. But the odors remained a problem.

Walker-Gordon purchased a special wet scrubbing system in 1972 to eliminate the odor problem. This system contains an adjustable side damper venturi scrubber, a special disengagement section designed to separate scrubbing liquids, and a packed adsorption section for odor control through chemical oxidation and absorption.

DESIGN CONCEPTS

Since current odor control techniques involve ozonation, incineration, chemical oxidation, or absorption, one of these had to be selected to control the odors emitted from the drying section.

Incineration and ozonation were not considered because of the excess fuel costs and complexity of the processes. Adsorption using activated carbon was judged to be too costly because of the excessive temperature and humidity of the effluent gases. It was decided that proven techniques of wet collection, using potassium permanganate as an oxidant, would not only prove effective, but would provide the most economical solution to the odor problem.

Venturi Design

Because permanganate reacts on a molar basis, it is imperative to remove as much of the particulate matter and odor as possible before this final treatment in order to keep operating costs reasonable. A proven adjustable venturi scrubber capable of operating in high-temperature applications was therefore selected. The adjustable feature of the venturi permitted Walker-Gordon to vary the scrubber pressure drop, thereby varying the scrubber efficiency and permitting the addition of a second dryer in line. Once set, the throat adjustment could be left unattended.

Separator Section

A special cyclone separator section was added to remove liquid droplets that might contain particulate matter. Removal of these particulates greatly reduces costs and chemical usage during operation. A special disengagement sump permits closed-circuit recycling of the particulate-laden scrubbing water, lessening the sewer load.

Adsorption Oxidation Section

It has been determined that manure drying odors are similar to rendering plant odors in that they contain hydrogen sulfide, indole, skatole, diketones, and mercaptans as well as some organic acids. Since manure is acid in water solution, a caustic addition system was incorporated into the venturi water circuit (Fig. 33-1), thus providing a neutral dust-free gas flow to the absorption bed.

Since the scrubbed gases contain CO_2 and other acid-forming substances, the permanganate system incorporates a borax buffer to maintain a pH of 8.5 to 9, the established optimum range for permanganate oxidation without excessive chemical demand. The MnO_2, a flocculant, helps contain stray particulates as well as separating

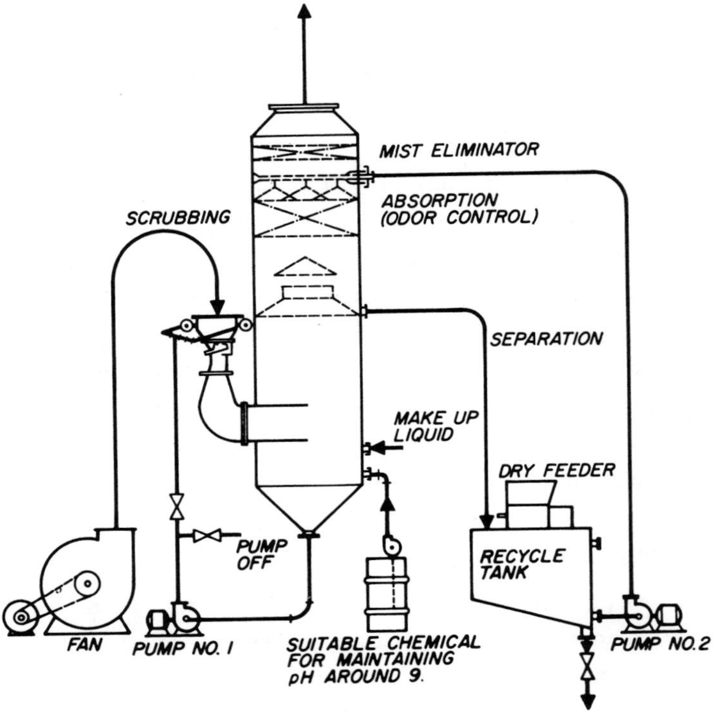

FIG. 33-1. Caustic addition system for venturi water circuit.

the liquid, forming a sludge. A separate recycling tank then confines this insoluble sludge for subsequent disposal. This entire reaction permits continuous operation of the scrubber.

A proportional dry solids feeder is incorporated to meter the permanganate into the tank where it is educted into the packed polypropylene bed circuit in the tower. Two independently operating pumps are used for the system, one for the packed bed and one for the closed-circuit particulate removal section.

RESULTS

In January 1973, stack emission tests were conducted on the Walker-Gordon system, with the following results: The maximum total emission rate (condensible hydrocarbons plus particulates) was well within the allowable 30 lb/hr (based on Walker-Gordon's process rate of 9,500 lb/hr) as listed in Chapter 7, New Jersey Department of Environmental Protection Air Pollution Control Codes.

The worst case of total emission during three test runs showed a system efficiency of 99.96 percent with odors described as trace and no ammonia detected.

Walker-Gordon Farms, a pioneer in production methods for over 70 years, is now producing valuable organic fertilizer at an unprecedented rate, while keeping both particulate matter and odor emissions well within both the New Jersey and federal air pollution codes.

Chapter 34

CONTROL OPTIONS FOR METALLURGICAL OPERATIONS

Jeffrey W. Goller
Nassau Recycle Corporation
Staten Island, New York

Paul N. Cheremisinoff
New Jersey Institute of Technology
Newark, New Jersey

Melting operations in metallurgical industries and refining result in exhaust effluents which must be controlled because of their toxic nature. Processors of metals, such as cadmium and beryllium, have the added problem of eliminating even very small concentrations because of the toxicity of these metals.

A typical small melting operation, using an induction-type furnace with an exhaust fan, may emit dust concentrations of 0.05 to 0.5 gr/scf. Because many melts are alloys of metals, their emissions can contain oxides of metals as well as toxic substances.

The exhaust gas temperature from a melting operation must be in the operating range of the collection equipment. This can be accomplished in some instances by using ductwork from the hood to the collection device as a heat exchanger. Adding atmospheric air to the hot exhaust stream can drop the temperature of the gas to below 100° F when properly designed. Spray chambers may also be used to reduce the gas temperature. However, care must be taken to prevent condensation in the gas stream when dry collectors such as fabric filters are used.

TOXICITY

Beryllium can cause acute irritation of the respiratory tract, including chemical pneumonitis. It can also cause berylliosis, a chronic disease, and skin reactions including dermatitis, ulcers, and tumors. Acute pneumonitis has been produced by the inhalation of beryllium oxide as well as the pure metal. When concentrations exceed 1 mg/m^3, cases of this disease are produced among almost all personnel. The disease runs its course in a few weeks, and unless death occurs during this period, complete recovery may be expected.

The dose that causes berylliosis is even lower than the dose that causes acute disease. In one community a total of 16 cases, including 5 fatalities, has been reported. These cases occurred within three-fourths of a mile of a plant producing beryllium compounds. The concentration of beryllium in the air was between 0.01 and 0.1 mg/m^3. Cases of this disease have been known to occur years after cessation of exposure. In some fatal cases less than 0.1 mg of beryllium was found to be present in a lung. Beryllium oxide exposure has been specifically involved in most of the reported cases. Beryllium compounds have been given a high toxic hazard rating for inhalation--and there is no specific antidote [32].

Cadmium and cadmium compounds are also extremely toxic. Brief exposure to high concentrations may result in pulmonary edema or death. Inhalation of cadmium dust or fumes may cause a range of problems from dryness of throat and nausea to chest pains and fatality. Cadmium oxide fumes from metallurgical burning can cause metal fume fever.

In melting operations where beryllium or cadmium compound emissions are possible, high-efficiency collection equipment must be used to control the fumes. The threshold limit values, as set by the American Conference of Governmental Industrial Hygienists [26], are beryllium, 0.002 mg/m^3; cadmium, 0.1 mg/m^3. Even though these values are low, it is not known positively that these limits will ensure complete freedom from response in all cases. Since hypersusceptibility to beryllium may be possible, supplementary air standards for the community air in the vicinity of these plants have been established. The average emission over 1 month may be only 0.01 mg/m^3 of beryllium with no more than 25 mg/m^3 at any one time. Emissions were further restricted to 10 g in a 24-hr day in the Federal Register (36 F.R. 23239) of December 7, 1971.

The strict regulations are a direct result of the well-established toxicity of these two extremely hazardous metals, beryllium and cadmium. To accomplish such low emissions, collection equipment with efficiencies of 99+ percent is required.

PARTICLE ANALYSIS

A principal factor in choosing collection equipment is particle size of the pollutant. Particulates from metallurgical melting operations range in size from 0.001 to 100 μm [27]. It has been found that cadmium oxide particles as small as 0.01 μm are emitted from high-temperature melting operations [40].

Since efficiency is extremely important in handling toxic metals, the definition of efficiency must be examined. Assume that a volume of gas contains 100 particles of size 1 μm and 100 particles of size 100 μm. The efficiency of separation is 90 percent for 1-μm particles and 99 percent for 100-μm particles. On a weight basis, if a 1-μm particle has unit mass, a 100-μm particle has 10^6 mass units, and the weight efficiency is

$$\frac{90 + 99 \times 10^6}{100 + 100 \times 10^6} = 0.98999$$

which is practically 99 percent. However, on a particle basis, 90 1-μm particles and 99 100-μm particles will be removed out of a total of 200 particles. This is an efficiency of only 94.5 percent.

It is therefore important to consider the size spectrum of the particles when determining the efficiency. An efficiency of 99 percent for dusts with particles down to 1 μm, as claimed by an equipment vendor, may be only 50 percent efficient for dusts with a mean diameter of 1 μm. Before a collection system is designed, it must be determined whether the problem is to remove small particles or to reduce the weight of a dust containing mainly large particles. After defining the efficiency and after determining the particle size distribution of the dust to be treated, the required efficiency is determined as dictated by the final emission concentration requirements.

The smaller the particle, the more concern it is from a health hazard standpoint, since these particles remain suspended in air the longest.

EQUIPMENT SELECTION

Although there are many different types of dust collectors, most of them can be eliminated on the basis of particle size collection efficiency. Spray towers and wet cyclones will remove particles in the 5- to 10-μm range. Packed towers are normally limited to handling 10-μm size particles and larger [38]. Equipment which is capable of handling particle sizes less than 1 μm with an efficiency greater than 90 percent include fabric filters, high-efficiency venturi scrubbers, and electrostatic precipitators. Table 34-1 summarizes collector types and the minimum particle range for efficiencies greater than 90 percent.

TABLE 34-1

Collector Efficiencies [36]

Collector type	Minimum micron range for >90% efficiency
Baffled settling chamber	>44
Simple cyclone	>44
Packed tower [38]	>10
Wet cyclone	5-10
Spray tower	5-10
Electrostatic precipitator	0-5
Fabric filter	0-5
Venturi scrubber	0-5

Precipitators

In general, electrostatic precipitators can handle particles of less than 1 μm with an efficiency of about 99 percent. The effective migration velocity (emv) of particles increases with diameter from 4 to 20 μm. The emv also increases as the diameter decreases below 0.5 μm. Dust between 1 and 4 μm in diameter has the lowest emv, and is the most difficult to precipitate electrostatically. Therefore, air containing large amounts of particles below 5 μm, as in the fumes under consideration, is more difficult to precipitate than an exhaust containing more of the larger particles. This has the effect of lessening the overall efficiency of the precipitator.

Another factor in determining the suitability of an electrostatic precipitator is the resistivity of the particles to be collected. Dusts with a resistivity between 10^4 and 2×10^{10} ohm-cm are the most suitable for electrostatic precipitation [39]. Tests show that there is a fourfold drop in migration velocity from 16 to 4 cm/sec when the dust resistivity is increased from 10^{10} to 10^{11} ohm-cm. If the resistivity of beryllium oxide between the temperatures of 1,000 and 2,100° C, as given by Darwin and Buddery (Table 34-2), are plotted on semilog paper, a straight line is formed. When this is extrapolated to the operating temperature of an electrostatic precipitator, one finds that the limit in resistivity (2×10^{10} ohm-cm) is surpassed.

Basically, the electrostatic precipitator is a poor choice for controlling the particulates from beryllium and cadmium melting operations. The only time that it may be considered is when very high gas flow rates are required. Efficiency is, however, difficult to keep at a high level because of operational difficulties, such as poor electrical settings, full hoppers, excessive dust concentrations, poor electrode alignment, vibrating corona wires, dust buildup on electrodes, etc.

The choice for the most efficient and economical collector is really between the fabric filter and the venturi scrubber. The final decision must be based on a number of factors, including dust

TABLE 34-2

Resistivity of Beryllium Oxide [6]

Temperature (° C)	P/cm^3
1,000	80,000,000
1,100	16,000,000
1,200	4,000,000
1,300	800,000
1,400	250,000
1,500	80,000
1,600	35,000
1,700	15,000
1,800	6,500
1,900	3,500
2,000	1,600
2,100	800

concentrations, available area, capital and operating costs, and gas characteristics.

Fabric Filter

The fabric filter, or baghouse, consists of tubular bags made of woven synthetic fabric or fiberglass. Dust-laden gas is made to pass through the fabric. Particles are trapped and collected by the filtering action of the fabric. The dust retained on the bags is periodically shaken off and falls into a collecting hopper for removal. This type of filter can provide high collection efficiencies, over 99 percent, with pressure drops from 4 to 6 in.H_2O. The maximum operating temperature for a baghouse is 550° F using fiberglass bags. The inlet dust loadings range from 0.1 to 10 gr/ft^3 of gas. If higher concentrations are encountered, a precleaning device, such as a low-efficiency cyclone, can be used to reduce the incoming load. The cyclone can remove particles of 10 μm and larger; however, smaller particles still require higher filter efficiency. Among the items limiting the use of baghouses are their large space requirement, low operating temperature, and high maintenance cost for bag replacement.

The primary consideration in the design of the fabric filter is the choice of the filter material. For beryllium and cadmium melting operations, the design parameters must consider high temperature, small particle size, and high efficiency. There is a variety of materials available, and the choice will depend on the specific characteristics of the operation (gas flow, concentrations, temperature, etc.) and the cost of the material to be used.

Glass fiber can be used up to 550°F with an initial penetration of 0.005 percent and a final penetration of less than 0.0001 percent for particles less than 2 μm [43]. (Penetration figures are given for a flow of 1.5 cm/sec.) Other materials are available which will operate at even higher temperatures, but the higher cost must be taken into consideration. For most toxic metallurgical melting operations, cooling of the gas to the operating temperature of the glass fiber is most practical.

Other factors to be considered are sufficient strength in the material to withstand fabrication, installation, adequate face velocity, and sufficiently low pressure drops. The material should be able to be housed in a filter unit sufficiently small to make the installation practical but with dust loading capacity sufficiently high so that replacement need not be too frequent.

After deciding on the process, a flow sheet and mass balance should be developed. The quantities of air should be questioned, and a decision made as to the use of dilution to reduce the filter performance required. The control of the dust at its source should be examined to reduce the quantity and to limit the amount carried into the ductwork to the filter system.

The filtration of fine particles depends on impaction, interception, and diffusion. When dust particles are smaller than 0.5 μm in diameter, impaction and interception usually become negligible and Brownian diffusion becomes the main factor involved.

The collection efficiency [39] for a single fiber is

$$E = \frac{2Bby}{r_f(1 - B)}$$

where

B = packing density = $\frac{\text{fiber volume}}{\text{filter volume}}$

b = depth of mat

y = index of filtration for mat

r_f = radius of fiber

This efficiency is based on the ratio of volume of air filtered to the volume passing an area equivalent to that of the fiber projection. The index of filtration of the mat is related to the quantity $(N_o - N)/N_o$, where N_o is the concentration of particles entering the filter and N is the concentration of particles penetrating the filter.

Particles ranging from 0.01 μm to coarse screen sizes may be filtered at efficiencies above 99 percent by weight of the material entering the collector [28]. An important advantage in cloth filtration is that collection efficiency is reasonably constant under variable operating conditions once a suitable filtration bed has been established by the deposition of particles on the cloth surface.

Filter resistance is a factor which will influence the performance of the collector. The initial resistance R_i (in.H_2O) is equal to the product of the filtration rate V (cfm/ft^2 or simply filtration velocity in fpm) and the permeability constant K_o of the filter medium (in.$H_2O/cfm/ft^2$ or fpm):

$$R_i = K_oV$$

The permeability of the cloth is determined by laboratory or field experiment and is corrected for variations in gas viscosity and density. The maximum resistance R_f at the end of any given operating period is the sum of the initial resistance and the increase of resistance with time R:

$$R_f = R_i + R$$

The factors which increase the resistance are

1. Particle size,
2. Particle shape,

3. Specific gravity of particles,
4. Particle surface characteristics,
5. Packing characteristics,
6. Amount of particulate matter on filter,
7. Density of gas, and
8. Viscosity of gas.

The increase in resistance with time [28] may be calculated with the following equation:

$$R = \frac{K_1 LTV^2}{7{,}000}$$

where

K_1 = specific resistance or resistance factor of the deposit, inches of water gauge per pound of dust per square foot of cloth area per foot per minute filtering velocity

T = time in minutes for filtering resistance to increase b inches water gauge (cycle time)

V = filtering velocity, fpm

L = dust loading in air coming to filter, gr/ft^3

This equation can be used to determine the cloth area required for the collector design. Assume, for example, that the operating conditions to filter beryllium oxide dust from a melting operation are as follows: (1) the temperature of the dust-laden air is 480°F at the filter inlet, (2) the maximum filter resistance is to be 5.0 in.wg, (3) the dust load is 0.2 gr/ft^3, (4) the cycle time of the collector is 60 min, and (5) the volume of air is 10,000 cfm.

For these operating conditions, a glass fiber filter is chosen, and the laboratory results on this fiber are as follows:

Temperature	64°F
Dust load, L	2 gr/ft^3
Velocity, V	2.8 ft/min
Time, T	60 min
R_i	1.18 wg
R_f	3.40 wg
R, R_f, R_i	2.22 wg
Dust weight on filter per ft^2, W	20 g = 0.044 lb

Calculations on test conditions:

$$K_o = \frac{R_i}{V} = \frac{1.18}{2.8} = 0.42$$

$$K_1 = \frac{R}{W \times V} = 18.0$$

Calculations on operating conditions:

Viscosity correction of air due to temperature:

$$\frac{278\ (\text{at } 480°\text{F})}{182.7\ (\text{at } 64°\text{F})} = 1.52$$

Density correction of air due to temperature:

$$\text{Density} = \frac{0.001293}{10.00367T}$$

$$\therefore \text{ Correction} = \frac{10.00367T_1}{10.00367T_2} = \frac{2.14}{2.99} = 0.716$$

$$\text{Kinematic viscosity ratio} = \frac{1.52}{0.716} = 2.12$$

$$K_1 = 18.0 \times 2.12 = 38.16$$

$$K_o = 0.42 \times 2.12 = 0.89$$

$$R_f = K_oV + \frac{K_1LTV^2}{7,000}$$

$$5.0 = 0.89V + \frac{38.16 \times 0.2 \times 60 \times V^2}{7,000}$$

$$V = 4.31 \text{ ft/min}$$

Therefore, in filtering this particular beryllium oxide dust at a filtration velocity of 4.31 fpm and a concentration of 0.2 gr/ft^3, the anticipated initial resistance will be 3.84 in.wg, and the maximum resistance will be 5.0 in.wg after 60 min at 480°F. These results are valid only if the same cloth as tested is used for the operating unit. It should be noted that the residual resistance, that is, the resistance after cleaning, is the initial resistance, not that of the clean cloth at startup.

The required cloth area can now be determined using the result above:

$$\text{Cloth area} = \frac{Q}{V} = \frac{10{,}000}{4.31} = 2{,}320 \text{ ft}^2$$

This is an air-to-cloth ratio of 4.32.

Operating fabric filters will have a pressure drop of 4 to 6 in.H_2O with dust loading from 0.1 to 10 gr/ft^3 of gas. The permeability in cfm/ft^2 of cloth at 1/2 in.wg resistance is equal to 1/2 K_O. It is a function of filter cloth material as well as the operating conditions.

During the cleaning cycle it is important that complete removal of the dust from the filter cloth is avoided. This buildup may take only seconds to coat the cloth during startup, but it adds to the efficiency of the equipment. By leaving this initial resistance, R_i, on the cloth after the cleaning step, the efficiency of the collector can be maintained.

There are three methods of cleaning the bags: (1) mechanical shaking at the top of the tubes, (2) mechanical rapping of panel support frames, and (3) reverse air cleaning in which air is forced through from the clean air side of the cloth by a traveling ring or hood.

Mechanical shaking imparts rapid horizontal and/or vertical motions to the tops of the tubes, thereby causing loosely adhering dust to fall down into storage hoppers below. The main force is in the flexing of the cloth, and in the rapping type the loose material is freed by the rapping or vibrating of the rigid frames. In both methods the filtering resistance is lowered to a stable residual value, R_i. To prevent shutdown, continuous units are employed in which one section of tubes is shaking while the others are operating. The cloth tube shaking methods are shown in Fig. 34-1.

Reverse air cleaning continuously cleans large-diameter tubes by a traversing jet ring or hood while the unit remains in operation (Fig. 34-2). This results in a relatively constant resistance, although it is possible for the freed dust to redeposit on another

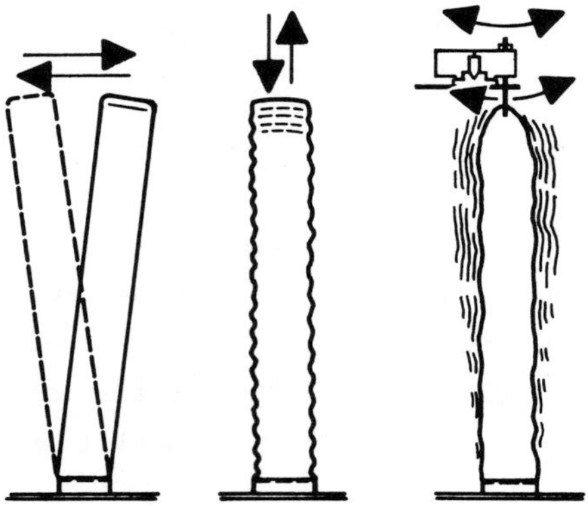

FIG. 34-1. Cloth tube shaking methods [28].

section, thereby increasing the resistance at that point. Filters of this type can operate at filtering velocities in the range of 10 to 30 fpm as compared to 0.5 to 10 fpm for units with mechanical cleaners. Reverse air-type collectors have been successfully used in handling particulate matter averaging 0.3 μm in diameter.

Another type of reverse air cleaning is the introduction of a pulse of high-pressure air into the top of the bag through a specially shaped venturi (Fig. 34-3). The air jet produces a reverse flow capable of cleaning the whole of one filter bag in a few seconds. One of these units accumulated more than 4,800 hr of operating time

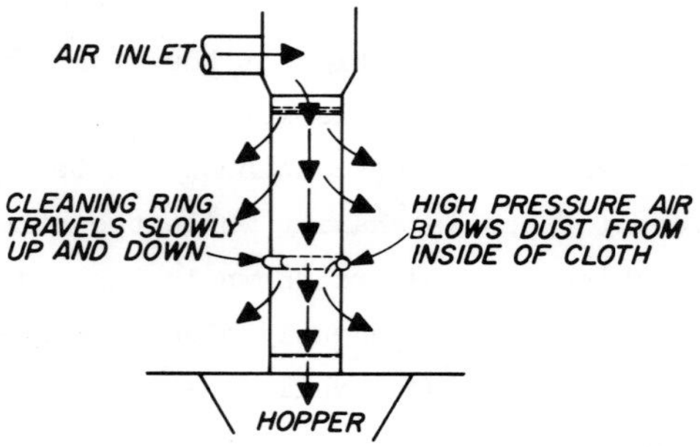

FIG. 34-2. Reverse air cleaning method [28].

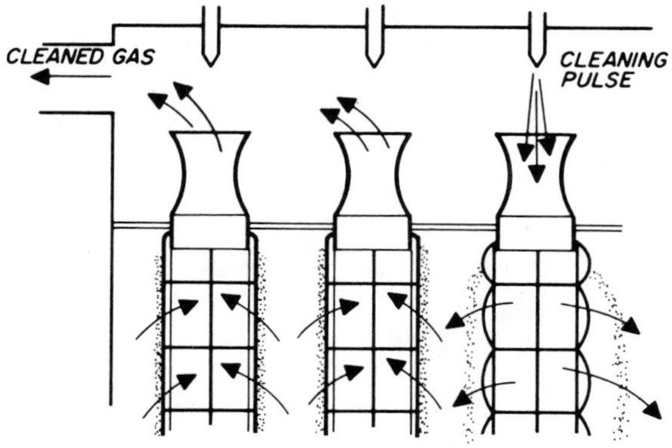

FIG. 34-3. Reverse jet with air pulse [39].

without a shutdown for maintenance, as compared to the average bag life of 240 hr for the mechanical shakers [39].

Filtration of up to 550°F can be accomplished with fabric filters made from woven glass fibers which possess high tensile strength to weight ratios, endurance to thermal shock, and good dimensional stability. Since the glass is able to operate at temperatures above the dew point of the gas, it eliminates condensation and the resulting corrosion. Since the glass fibers are brittle and easily damaged, particularly by shear, the glass fibers cannot withstand severe mechanical shaking. Cleaning is accomplished by the reverse air flow method. The life of the glass bags may be up to 3 years, with a 1- to 2-year life more prevalent in practice.

Venturi Scrubber

The venturi scrubber is essentially a constriction in the gas flow line with an arrangement for injecting water into the gas at or near the constriction. The water is injected at high pressures (between 5 and 25 $lb/in.^2$), resulting in velocities between 200 and 300 fps. This atomizes the water with a high velocity relative to the dust particles, leading to high impaction efficiencies with large areas of liquid exposed for vapor-phase contact. The dust particles are

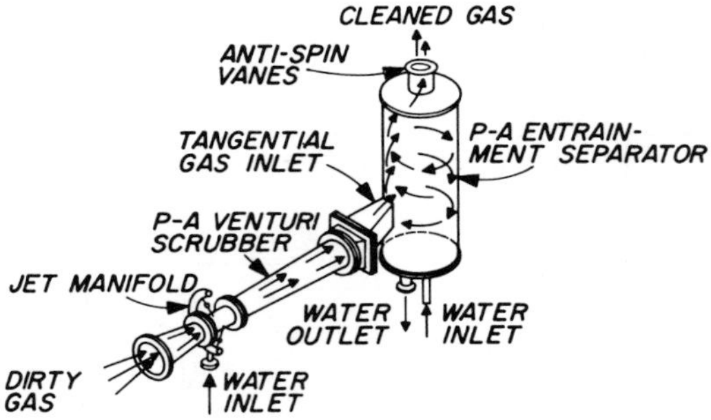

FIG. 34-4. Pease-Anthony venturi scrubber.

enlarged and can be collected by a cyclone or other spray collector (Fig. 34-4).

The liquid rate ranges between 2 and 9 gal/1,000 ft^3 of gas. Even the simplest type of venturi scrubber has the disadvantage of very high energy use, with pressure drops in the range of 8 to 30 in.H_2O across the unit [27]. A graph of pressure drop as a function of operating conditions is given in Fig. 34-5.

For a given efficiency the pressure drop across the venturi increases as the particle size decreases, as can be seen in Fig. 34-6. To remove the extremely small particles under consideration, the operating requirements must be more stringent than for the usual venturi application. For 99 percent removal efficiency of particles less than 1 μm, the pressure drop increases at a rate which may make use of a venturi scrubber uneconomical.

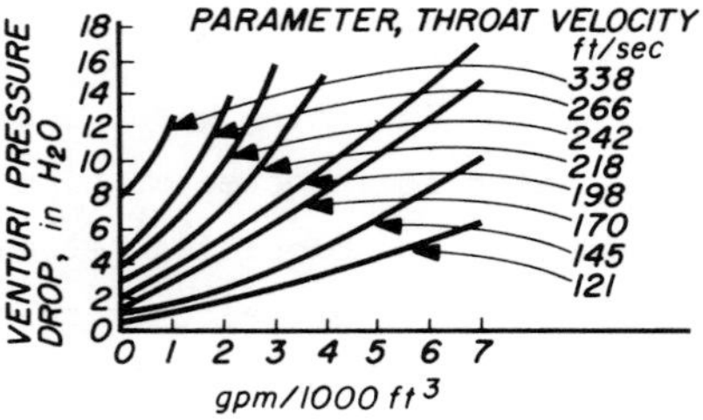

FIG. 34-5. Venturi pressure drop versus water rate [27].

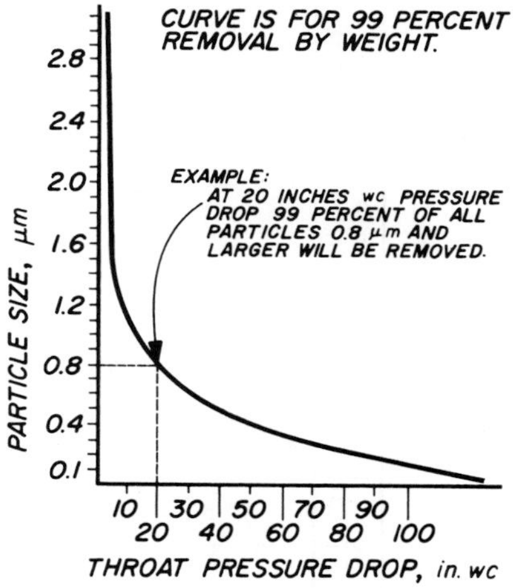

FIG. 34-6. Particle size versus pressure drop. (Courtesy of Canning Associates, Inc.)

A grade-efficiency curve for the venturi scrubber has been given by Smith and White [43] which indicates that for particles larger than 2 μm the efficiency approaches 100 percent (Fig. 34-7). It has also been claimed that an efficiency of 99.98 percent has been accomplished for the removal of boric oxide fume particles of less than 1 μm in diameter. As seen on line 2 of the lower part of Fig. 34-7, removal of iron oxide fumes with particles in the 0.02- to 0.50-μm range has been accomplished with 99 percent efficiency. Although Fig. 34-8 deals with scrubbing gases, it demonstrates the effect of water rate on the efficiency. Most of the claims deal with the venturi discharging into a cyclonic spray scrubber, the spray nozzles of which can be operated or not. In most cases, when the spray nozzles were operated, an increase in efficiency was noted.

It must be remembered that the collecting efficiency by weight does not tell the absolute quantity of pollutant escaping. For meaningful performance data it is necessary to get more complete information on the small particles.

Attempts have been made to reduce the pressure drop and to increase the efficiency of these scrubbers. Finer water particles can

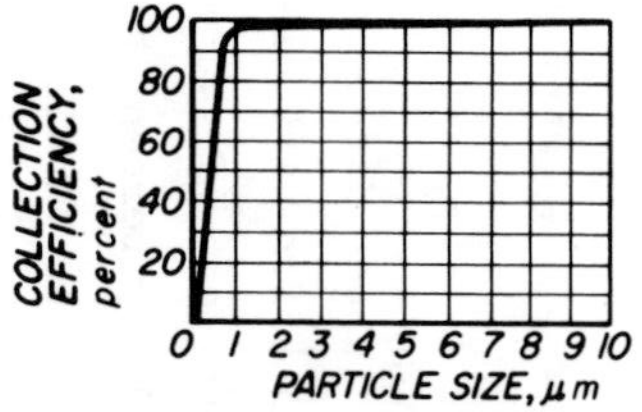

LINE NO.	SOURCE OF GAS	DUST	PARTICLE* SIZE, μm	DUST LOADING, GRAINS/ft³		AVERAGE REMOVAL EFFICIENCY, PERCENT
				INLET	EXIT	
1	BLACK LIQUOR RECOVERY FURNACE	CHEMICAL FUME	0.1 - 1.5	1 - 2.5	0.1 - 0.25	90+
2	STEEL OPEN HEARTH (oxygen lanced)	IRON OXIDE	0.02 - 0.50	1 - 6	0.01 - 0.07	99
3	STEEL OPEN HEARTH (scrap furnace)	IRON AND ZINC OXIDE	0.08 - 1.0	0.5 - 1.5	0.03 - 0.06	95
4	BLAST FURNACE (iron)	IRON ORE AND COKE DUST	0.5 - 20.0	3 - 24	0.008 - 0.05	99+
5	BLAST FURNACE (sec. lead)	LEAD COMPOUNDS	0.1 - 1.0	2 - 6	0.1 - 0.3	95
6	REVERBERATORY LEAD FURNACE	LEAD AND TIN COMPOUNDS	0.1 - 0.8	1 - 3	0.12	91
7	ROTARY DRYER	ACTIVATED CARBON	0.1 - 3.0	1.9	0.04	98

*Small end of particle size spectrum as seen on electron-microphotographs.

FIG. 34-7. Grade-efficiency curve for venturi scrubber [43].

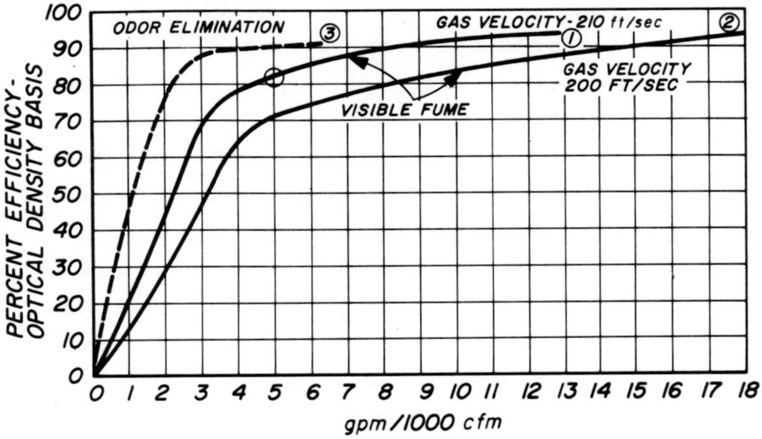

FIG. 34-8. Efficiency versus water rate [18].

be produced at the throat with higher relative velocity between the dust and the liquid by introducing the water spray just ahead of the duct constriction. If the gas is saturated before it enters the venturi, condensation on the particles can take place within the convergent section. This mist can then be collected by a further spray. Efficiencies of 99.9 percent for removal of dust with a mean particle size of 0.3 μm have been claimed for venturi scrubbers of this type. The efficiency may also be improved by using a fine, preatomized spray for the injection into the venturi. However, unless the water particles are considerably larger than the dust, removal of the particle-dust agglomerates may be difficult.

The collection of fine dusts with the use of a venturi scrubber is effected by four mechanisms. The first is impaction or impingement of the high-velocity dust particle on the liquid droplet. This is effective for particles down to 0.25 μm. The second mechanism is collection by diffusion, molecular, or Brownian movement, which can stop particles down to 0.1 μm. Condensation and electrostatic charges probably add to the collection efficiency. As the dirty gas collides with the curtain of liquid, the flow is violently accelerated and disrupted. The gas decelerates in the diverging section, the fine particulate matter in the gas is wetted by the finely atomized droplets, and then coalescence occurs downstream. This results in drops of liquid large enough to be separated from the gas by centrifugal force.

A higher liquid rate is usually more advantageous to use than a higher gas velocity. A high velocity and a low liquid rate could create an unwetted void through the middle of the venturi. Therefore, it is preferable to lean toward a higher liquid rate to ascertain a proper liquid-to-gas impaction level. At the other extreme, a problem can occur wherein a low gas velocity exists with too high a liquid rate. Under these conditions liquid shattering may be reduced to such a point that the scrubber starts to act like an ejector, which means that poor collection efficiencies are achieved.

The example problem given in the last section was for a gas containing 0.2 gr/ft^3 of gas at a rate of 10,000 cfm. The limiting factor is the beryllium oxide portion of the dust. Assume that allowable emissions require a 99.5 percent reduction in the beryllium oxide particles. This would not be the required efficiency of the scrubber, but that of the removal efficiency of the toxic pollutant.

When dealing with such high efficiencies, experimental results must be obtained before the unit is specified. Assume that the chosen scrubber has a throat diameter of 1 ft. The throat area would then be 0.785 ft^2, and the velocity would be 212.3 fps. If the scrubber produced a 99.5 percent efficiency at a water flow of 40 gpm, then the pressure drop across the unit is 12 in.H_2O, using Fig. 34-6. It has been found that for the removal of submicron particles as obtained in the melting operation, the water rate must be in the range of 2 to 6 gal/1,000 ft^3 of gas, and the overall pressure loss is 12 to 20 in.H_2O for efficiencies of 95 to 99 percent.

ECONOMIC COMPARISON

Now that the applicability of the fabric filter and venturi scrubber have been established, which one should be chosen? Each has its own merits under different circumstances.

If, in addition to the particulates, a mist or fume were present and unwanted, then the only choice would be the venturi scrubber since the filter has no effect on mists or gases. In most melting operations, however, the temperature of the gas is well above its dew point and no mist would be present. High-temperature gases created by the furnace may have to be cooled if a fabric filter is employed, but may be in the operating range of the venturi scrubber.

Maintenance of the venturi requires no special skills beyond general fan and pump considerations. To start up a venturi scrubber in a plant requires only fan and water pump attention. The venturi is also easily embodied into existing flow lines. This becomes significant when space is a problem, and it can reduce the installation costs for piping and ductwork. A disadvantage of this type of scrubber is the cooled and wet stack plume which can reach the ground over a relatively small area close to the stack. Also, loss of efficiency may result from power failure or problems with the spray flow. Control of the gas and water flow rates must be maintained if the unit is to work properly.

The baghouse offers operation which can handle a change in gas flow rate and still maintain its efficiency. The maximum allowable concentrations of beryllium and cadmium dusts are given by White and Smith [43], Fig. 34-9, along with the equipment capable of such collection. High-efficiency filters are applicable, but venturi scrubbers are not included in the chart. White and Smith do state later in the same work that fabric filters are used in the metallurgical industry for beryllium and cadmium to meet the acceptable levels for these dusts.

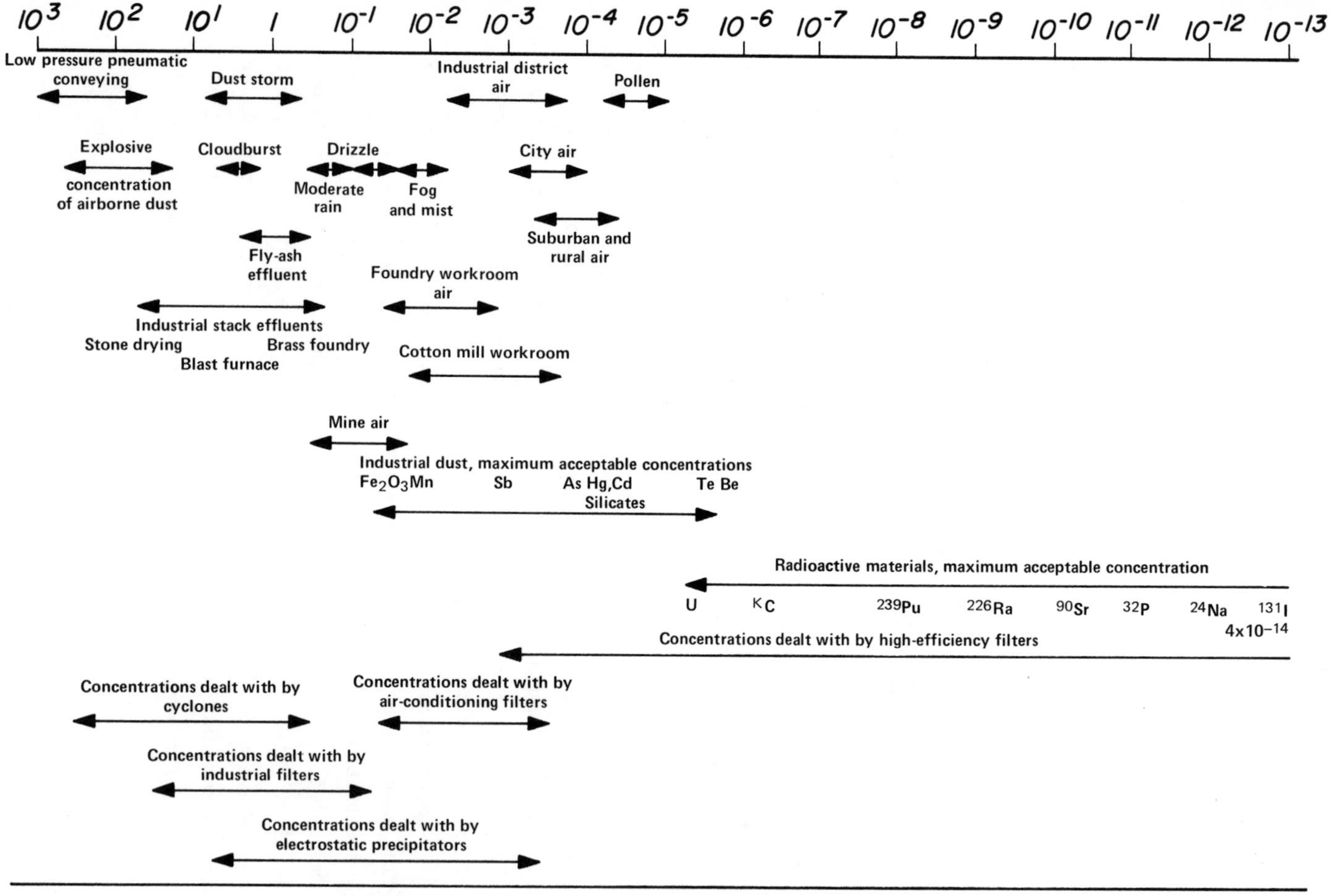

FIG. 34-9. Allowable dust concentrations (g/m^3) [43].

The available space may also be a factor in determining the unit to be used. The floor space requirement for a baghouse using the blow-ring filter is from 5 to 12 ft^2/1,000 cfm, while the requirement for the venturi scrubber is only 3 to 6 ft^2/1,000 cfm [39].

Whether the filter or scrubber is used, the toxic dust which is removed from the air must still be disposed of. After the dust is shaken or blown from the fabric filter, it drops into hoppers which are used as storage for intermittent dust removal, or simply as chutes when the installation includes facilities for continuous dust removal. Careful consideration must be given to the handling of the dust after it is removed from the fabric arrester hoppers. Unloading and transportation facilities such as closed conveyors or covered carts must be used to ensure dustless conditions. Depending on the nature of the melting operation, the metal content may be recoverable.

The dust discharged from the scrubber is in a wet condition which makes it transportable without recontaminating the air. In many cases the scrubbing water can be recirculated (with reduced water requirements) and the concentrations built up for cheaper dewatering of solids. Due to the extremely small particle size, settling may not be a useful method for removing the toxic particles and a filtration system must be employed. This will add to the floor space required and to the capital and operating costs.

The capital cost for a venturi scrubber to clean 100,000 cfm is approximately $50,000, with an installation cost of nearly that sum again. The average cost of a blow-ring bag filter is generally 1.7 times that of the scrubber.

Typical initial costs of fabric filters are as follows [12]:

ft^2	Inside length (ft)	No. of bags	Cost/ft^2
940	4	96	1.36
3,760	16	384	0.87
7,520	32	768	0.80
11,750	50	1,200	0.76

Usual capital costs for the venturi scrubber are as follows:

cfm	Mild steel cost/cfm	Stainless steel cost/cfm
500	2.0	3
1,000	1.5	2
10,000	0.75	1.5
$>$25,000	0.5	1

Although capital costs for the venturi are low in comparison, operating costs are high. Power and water requirements are much greater than for the baghouse. The power requirement for a venturi scrubber is usually 3 to 5 hp/1,000 cfm. As previously stated, pressure drops in the scrubber are 12 to 20 in.H_2O as compared to 2 to 8 in. for the baghouse. Water requirements are again high for the scrubber, 2 to 9 gal/1,000 ft^3 of gas, and the baghouse has no water requirement. The power consumption for the venturi scrubber ranges from $2.25/cfm for 500 cfm units constructed of mild steel to $0.60/cfm for units greater than 25,000 cfm. Stainless steel units have operating costs from $3.35 to $2.25/cfm for the same units [27], but stainless steel would generally not be required for the melting operation unless a corrosive gas was involved. Annual maintenance costs would run about $11,000 for a 100,000 cfm unit with power costs as much as $75,000. Maintenance costs for a baghouse to treat the same gas flow would be approximately $15,000 for the reverse jet cleaning unit with only $20,000 for power. The following is an example of the operating needs for a baghouse treating cadmium smoke [12]:

Recovered tons in 6 months	No. of bags replaced	No. of bags in baghouse	cfm per bag	Operating cost per ton of dust recovered
583	0	135	67	10.37
878	2	288	220	2.52

With all these factors to consider in selecting the correct unit, there is still another to be considered. This is the concentration of dust in the unclean air. As a general rule, if the concentration is less than 10 gr/1,000 ft^3, then the fabric filter is probably more efficient; and if the concentration is greater than 1 gr/ft^3, then the scrubber is usually more practical. If the concentration lies between these two figures, then experimental tests may be required before a choice can be made based on efficiency.

Any air control system which requires high efficiency demands standards of design, construction, testing, operation, and maintenance higher than those generally accepted of a more conventional nature. The selection cannot be done merely on the basis of collection efficiencies reported for other pollutants, as toxic dusts deserve careful consideration. The more that is known as to dust loading and particle size distribution of the material in the gas to be cleaned, the better are the chances of making the right selection. The cost of equipment of this nature, added to the fact that so many installations are unsatisfactory, suggests that more emphasis should be placed on making pilot installations before large sums are invested in equipment.

Considering the toxic properties of beryllium and cadmium, and the small particle size generated by melting operations, only two types of equipment indicate the efficiency required. These are fabric filters and venturi scrubbers. The final choice depends on many factors, and careful analysis of the given situation should be made before the choice is finalized. Proper furnace hooding is a factor for efficient ventilation into ductwork, which is an integral part of the dust collection system. Economics, of course, is paramount. Whichever system is chosen, operational control must be geared to meet the high demands of emission requirements.

REFERENCES

1. A. B. Adams: Filtering Apparatus for Dust-Laden Air Streams, Potteries Ventilating & Heating Co., Brit. 1,059,301, Feb. 15, 1967.

2. E. Anderson: Some Factors and Principles Involved in the Separation of Dust, Mist and Fumes from Gases, American Institute of Chemical Engineers, 16:69 (1925).

3. C. S. Blacktin: "Dust," Chapman & Hall, London, 1934.
4. A. D. Brant: "Industrial Health Engineering," John Wiley & Sons, New York, 1947.
5. T. D. Cremer: "Chemical Engineering Practice, Vol. 3, Solid Systems," Academic Press, New York, 1957.
6. G. Darwin and J. Buddery: "Beryllium," Butterworths, London, 1960.
7. P. Drinken and T. Hatch: "Industrial Dusts," McGraw-Hill Book Co., New York, 1954.
8. Anon., Future Bright for Fabric Filters, Environmental Science and Technology, 8:6 (1974).
9. W. C. Faith: "Air Pollution Control," John Wiley & Sons, New York, 1959.
10. W. L. Faith and A. A. Atkisson: "Air Pollution," Wiley-Interscience, New York, 1972.
11. C. R. Flodin: Specification and Selection of Dust and Fume Collection Equipment, Consulting Engineer, May 1961.
12. Friedlander, Silverman, Drinker, and First: "Handbook on Air Cleaning-Particulate Removal," U.S. Department of Industrial Hygiene, Washington, D.C., 1952.
13. W. E. Gibbs: "The Dust Hazard in Industry," Ernest Benn, London, 1925.
14. Alan Gilpin: "Control of Air Pollution," Butterworths, London, 1963.
15. H. L. Green and W. R. Lane: "Particulate Clouds, Dusts, Smokes, and Mists," Sopon, London, 1957.
16. W. Heneon: Gas Cleaning Efficiency Requirements for Different Pollutants, Journal of the Air Pollution Control Association, 12:105 (1962).
17. H. R. Jones: "Fine Dust and Particulate Removal," Noyes Data Corporation, N. J., 1972.
18. W. P. Jones and A. W. Anthony: Pease-Anthony Venturi Scrubbers, in "Air Pollution," McGraw-Hill, New York, 1952.
19. W. P. Jones and A. W. Anthony: Pease-Anthony Venturi Scrubbers, Pease-Anthony Equipment Company, Newtonville, Mass.
20. C. E. Lapple: Dust and Mist Collection, in "Air Pollution Abatement Manual," Manufacturing Chemists Association, Washington, D.C., 1951, chap. 9.
21. H. F. Lund: "Industrial Pollution Control Handbook," McGraw-Hill, New York, 1971.
22. Magill, Holden, and Ackley: "Air Pollution Handbook," McGraw-Hill, New York, 1956.

23. W. J. Mead: "Chemical Process Equipment," Reinhold, New York, 1964.

24. National Air Pollution Control Administration, Control Techniques for Particulate Air Pollutants, NAPCA Publication No. AP-51, 1969.

25. G. Nonhebel: "Processes for Air Pollution Control," CRC Press, Cleveland, Ohio, 1964.

26. F. A. Patty: "Industrial Hygiene and Toxicology," Interscience, New York, 1958, pp. 1002-1016.

27. John H. Perry: "Chemical Engineers' Handbook," 4th ed., McGraw-Hill, New York, 1963, pp. 18-55-56, 20-75-82.

28. Robert T. Pring: Bag-Type Cloth Dust and Fume Collectors, American Wheelabrator & Equipment Corporation, Mishawaka, Ind.

29. R. D. Ross: "Air Pollution and Industry," Van Nostrand Reinhold, New York, 1972.

30. R. D. Ross: "Industrial Waste Disposal," Reinhold, New York, 1968.

31. G. D. Sargent: Gas/Solid Separations, Chemical Engineering, Feb. 15, 1971, pp. 11-22.

32. I. N. Sax: "Dangerous Properties of Industrial Materials," Reinhold, New York, 1963, pp. 501-504, 558-561.

33. K. T. Semrau: Dust Scrubber Design, Journal of the Air Pollution Control Association, 13, 1963.

34. Sheeby, Achinger, and Simon: "Handbook of Air Pollution," U.S. Department of Health, Education, and Welfare, National Center for Air Pollution Control, Durham, N.C., Publication No. 999-AP-44.

35. B. J. Squires: Fabric Filter Dust Collectors, Chemical and Process Engineering, 156, 1962.

36. A. C. Stern: "Sources of Air Pollution and Their Control," Academic Press, New York, 1968.

37. I. R. Tabershaw: Chemically Active Air Pollutants, Liberty Mutual Insurance Co., New York.

38. A. J. Teller: Air Pollution Control, Chemical Engineering, May 8, 1972, pp. 93-98.

39. G. Underwood: Removal of Sub-Micron Particles from Industrial Gases, Department of Fuel Technology and Chemical Engineering, The University of Sheffield, Dec. 28, 1961.

40. Vandegrift, Shannon, and Gorman: Controlling Fine Particles, Chemical Engineering, June 18, 1973, pp. 107-114.

41. D. W. White, Jr., and J. E. Burke: "The Metal Beryllium," The American Society for Metals, Cleveland, Ohio, 1955.

42. Harry White: "Industrial Electrostatic Precipitation," Addison-Wesley, Reading, Mass., 1963.

43. P. A. White and S. E. Smith: "High Efficiency Air Filtration," Butterworths, London, 1964.

44. R. Whytlaw-Gray: "Smoke," Edward Arnold, London, 1932.

Chapter 35

COMMERCIAL SO_2 REMOVAL USING LIMESTONE IN A HIGH-VELOCITY SPRAY TOWER

Howard P. Willett
Peabody Process Systems, Inc.
Stamford, Connecticut

Peabody Engineered Systems, in cooperation with Detroit Edison Company, undertook the development of a limestone slurry process in 1971. A 1-MW pilot plant was installed at the utility's River Rouge Station in Detroit. The pilot plant consisted of a Peabody-Lurgi venturi Scrubber followed by an upflow Peabody Tray Absorber, along with associated recycle pumps and tanks. Early experience was rather discouraging and revealed scaling and plugging problems typical of the limestone slurry process. In late 1972 the pilot plant was drastically modified to overcome the scaling and plugging problems. The major modifications centered around replacing the tray-type absorber with a high-velocity countercurrent spray tower, larger recycle tanks and pumps, automatic pH control system, slurry density control, and wash tray. The modified pilot plant was started up in February 1973 and operated round the clock for 21 days. The results were most encouraging; no plugging or scaling was experienced. The entire operation was smooth and easy to control. Additional runs were performed to develop design data.

The green light for the full-scale system was given in February 1974. Detroit Edison's St. Clair Station was chosen for installation of the full-scale system. The scrubbing system is installed on Unit No. 6 and is designed to handle 175 MW of the unit capacity. The full-scale unit has been operated continuously for 30- to 60-day periods on a hands-off basis. The fully automated system, operating under closed loop conditions, removed better than 90% of the sulfur dioxide at all times. Most importantly, no scaling or plugging has been experienced.

CHEMISTRY AND SYSTEM DEVELOPMENT

Sulfur dioxide removal from stack gases by scrubbing requires the use of an aqueous solution or suspension of an alkaline material. Since limestone is the cheapest and an abundantly available natural alkaline rock, it has found wide acceptance for stack gas desulfurization. However, its use in scrubbers is fraught with dangers of severe equipment scaling. Therefore, it is imperative that certain fundamental facts about chemistry are understood and accounted for in developing a limestone slurry scrubbing system.

The end products of sulfur dioxide reaction with limestone or lime are calcium sulfite and calcium sulfate, both of which are sparingly soluble in water. The primary product is calcium sulfite, which upon further reaction with dissolved oxygen is converted to calcium sulfate. Because of their low solubilities, both calcium sulfite and calcium sulfate in the aqueous phase very rapidly become supersaturated. Unless adequate safeguards are taken, rapid, uncontrolled crystallization will ensue with attendant plugging. Controlled crystallization is the key to developing a trouble-free scrubbing system. The following factors must be considered:

1. Relatively high liquid-to-gas ratio: Because of the low solubilities involved, production of the end products per unit volume of liquid must be kept low. Some guidance in this respect can be derived from the British work [1] of the 1930s. They found it necessary to use liquid-to-gas ratios in excess of 100 gal/1,000 ft^3 of gas when burning 1.75 percent sulfur coal.

2. Isomorphic surfaces: The crystallization of dissolved salts is increased by the presence of isomorphic surfaces, i.e., the seed crystals of the salts being crystallized. Thus, scrubbing liquor must contain a high suspension of both calcium sulfate and calcium sulfite.

3. Delay time: Since the process of crystallization occurs at a finite rate, the liquor leaving the scrubber is generally still supersaturated; hence it must be delayed outside the scrubber before being contacted with the flue gas. This delay time is also important for alkali dissolution, pH recovery, and overall stoichiometry.

4. Control of pH: pH control of the recirculating slurry is essential to maintain high efficiency of sulfur dioxide absorption while preserving low stoichiometry. With limestone the optimum pH is in the range of 5.8 to 6, which is slightly acidic. With lime a wider pH control can be exercised. However, operation on the slightly acidic side gives most economical use of either alkali.

5. Low liquid holdup: When dealing with scale-forming liquors and slurries, it is essential to keep the liquid holdup low within the scrubber. The longer the liquid remains within the scrubber, the more likely it is to cause scaling. Fast movement of slurry, hence its low residence time within the scrubber, avoids scale formation even under relatively supersaturated conditions.

A spray-type scrubber is well suited for this use because it handles a large liquid-to-gas ratio, has high suspension of slurries, has low liquid holdup and a minimum of internal surfaces, and a very low pressure drop. Further, it has infinite turndown capability, with little possibility of plugging because of its simple, open design.

SYSTEM DESCRIPTION

The scrubbing system for Detroit Edison's St. Clair Station is installed on Unit No. 6, which is a coal-fired boiler. The scrubbing system is designed for 175 MW capacity. It consists of two identical scrubbing trains with a common recycle tank, induced draft fan, and oil-fired reheater. Each train consists of a Peabody-Lurgi Radial Flow Venturi followed by a high-velocity spray tower. A wash tray, irrigated with fresh makeup water, is placed above the slurry spray section to act as an interface between the heavy slurry zone and the mist eliminator. The clean gas, after passing through the radial vane mist eliminator, passes through the I.D. fan and is reheated to 250° F prior to its discharge to the stack.

System Design

The following design parameters were used for the St. Clair scrubbing system:

Total system capacity = 175 MW, 500,000 acfm

Fuel analysis (%)

Hydrogen	= 4.12
Carbon	= 64.1
Nitrogen	= 1.07
Oxygen	= 4.00
Sulfur	= 4.00
Ash	= 16.00
Moisture	= 5.9

Equipment performance

Inlet SO_2	10,000 lb/hr	(3,000 ppm)
Outlet SO_2	1,060 lb/hr	(300 ppm)
SO_2 removal efficiency	90%	
Inlet particulate	18,048 lb/hr	
Outlet particulate	48 lb/hr	

Venturi Scrubber

A schematic diagram of the Peabody-Lurgi Venturi is shown in Fig. 35-1. Hot flue gas is introduced through a conical wetted-wall quench section. This wetted approach is important to eliminate the deposits associated with wet-dry lime. The slurry from the recycle tank is used for irrigation, part being sprayed through a bull nozzle and part being introduced tangentially through a distributing shelf. The quench gas, along with the slurry, then passes radially through the adjustable orifice. The orifice consists of two opposing, replaceable rings. The lower ring, placed in a fixed cup, is automatically adjusted to maintain the desired pressure drop.

Because of the ring's light weight and its thin edge moving against the gas, the throat adjustment requires comparatively little force when compared with conventional venturis. The gas makes two 90° turns, for maximum entrainment separation. The venturi not only removes particulate matter but also some sulfur dioxide. The sulfur

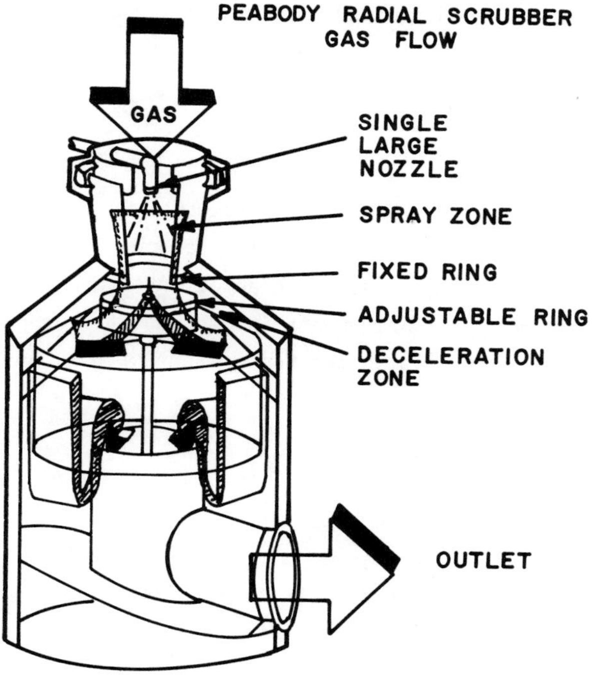

FIG. 35-1. Schematic diagram of the Peabody-Lurgi Venturi.

dioxide removal efficiency of the venturi is expected to be in the range of 35 to 50 percent when using limestone. The quench section and the orifice of the venturi are made of 316L stainless steel, and the balance is rubber-lined carbon steel.

Absorber

Gas leaving the venturi passes upward through a high-velocity spray tower absorber. The slurry from the recirculation tank is sprayed through six spray banks; each pair of spray banks is fed with a single rubber-lined pump. Thus, there are three operating pumps for each absorber. Rubber-lined piping is used for slurry service. The slurry is sprayed through large hollow-cone silicon carbide nozzles, which are nonclogging and virtually erosionproof for this service. The absorber shell is 316L stainless steel. Above the slurry spray section an impingement wash tray is installed. Above the tray a radial vane-type mist eliminator is used. Gas leaving the two absorbers is fed into a common duct, which leads it to the I.D. fan.

Reheater

The clean gas leaving the I.D. fan is reheated by a direct oil-fired reheater. The combustion chamber is located outside the main duct. The hot combustion gases are then injected into the main gas stream through a diffuser. The oil burner is designed to burn No. 6 fuel oil and is capable of raising the temperature of the flue gas from 120 to 250° F.

Recycle Tank

A single recycle tank is used for both scrubbing trains. The tank is constructed from carbon steel with a plastic coating. A set of four agitators is placed inside the tank to keep the solids in suspension. The tank is 48 ft in diameter and 38 ft high. It is sized to provide a retention time of 10 min. Waste slurry is discharged from the tank by an overflow nozzle, collected in a sump, and then pumped to the disposal pond.

SYSTEM CONTROL

The control system is designed for minimum operator attention. No control valves are used for the main recycle lines. Instead, turndown is achieved by simply shutting off the pumps stepwise. This eliminates the use of control valves in abrasive slurry service. The system operation is not sensitive to gas flow, hence no control dampers are used for controlling the gas flow through each module with load fluctuations. In duct design, allowance is made for proper gas distribution through each module. The system can accommodate 0 to 100 percent load variations, while maintaining high SO_2 removal and tight closed-loop operation.

Overall flue gas system flows are controlled by the boiler draft and combustion control system without complications introduced by the absorbing system.

Basically, system control is achieved by the addition of proper amounts of fresh limestone or lime slurry to react with sulfur dioxide and discharge the reaction products while maintaining the required concentration of suspended solids in the circulating slurry. This control is maintained through two process variables, pH and density.

The addition of alkali and discharge of waste products are automatically achieved by monitoring the pH. Similarly, the suspended solids concentration is maintained at the proper level by a density monitor which automatically regulates the amount of supernatant to the recycle tank. This control principle makes the system control independent of boiler operation, thus making the system control simple and reliable.

Each scrubbing train is supplied with a pH and density controller. Fresh alkali is continuously pumped in a piping loop which is tapped. Alkali is added to the recycle tank by a control valve operated by the pH control signal. Waste slurry in the recycle tank is allowed to overflow by gravity into the waste slurry sump and be pumped to the disposal pond. It is important that liquid velocity in the pipe from the sump to the pond be maintained constant under all load conditions to prevent the solids from settling out under reduced load operation. This is achieved by maintaining a constant-flow loop to and from the pond. The supernatant return from the pond is tapped into the recycle tank by the density control; the balance is put into the waste slurry sump. Since the tank also overflows into the sump, the net flow going to the thickener is always constant.

Normal operation of the system is automatic. The heart of the control system is pH and density control. The particular instruments used have been proven in pilot plant operation for reliability, accuracy, and compatibility with actual process requirements.

The system is equipped with an SO_2 analyzer on the inlet and outlet of the system. No control adjustments are required to accommodate load variations. Increased SO_2 absorption will result if recycle pumping rates for full load are maintained for reduced load conditions. The SO_2 analyzer may be used to determine if pumping rates on the absorber can be reduced.

REFERENCE

1. J. L. Pearson, G. Nonhebel, and P. H. N. Ulander: The Removal of Smoke and Acid Constituents from Flue Gases by a Non-Effluent Water Process, Journal of the Institute of Fuel, 8(39):119-156 (Feb. 1935).

Chapter 36

THE CITREX PROCESS FOR SO_2 REMOVAL IN COAL-FIRED BOILERS

Srini Vasan

Peabody Process Systems, Inc.
Stamford, Connecticut

Sulfur dioxide from flue gases is a pressing ecological problem, especially since the energy crisis has accelerated the use of high-sulfur fuels. The currently available flue-gas desulfurization process, which is proven practicable, is the throwaway system with its attendant sludge-disposal problem.

SO_2 removal using throwaway systems such as limestone or lime scrubbing have been well demonstrated and adopted at many installations where availability of land and limestone have helped its choice.

A regenerable process that yields sulfur as by-product gives high flexibility to the user. One such regenerable process is the citrate process, on which pioneering work was done by the U.S. Bureau of Mines at Salt Lake City, Utah. This process involves a new regenerative process for SO_2 removal that yields high-quality sulfur as by-product [1]. In this process a buffered citrate solution is used to effectively scrub SO_2 from a gas (at 110 to 140°F and after particulate removal) in a countercurrent absorber. Since absorption in aqueous solution is self-limiting, the use of a buffering agent such as citric acid is to shift the equilibrium to the right as shown below:

$$SO_2 + H_2O = HSO_3^- + H^+$$
$$Cit^{2-} + H^+ = H\ Cit^{2-}$$
$$H\ Cit^{2-} + H^+ = H\ Cit^-$$

The SO_2-laden liquor from the absorber is contacted in a reactor at atmospheric pressure and 150° F with H_2S, to convert all the SO_2 completely to sulfur and water.

The chemistry of this conversion is complex; but the overall reaction can be shown to be the Claus reaction, which goes to completion in the liquid phase according to

$$2H_2S + SO_2 \rightarrow 3S + 2H_2O$$

In practice, SO_2 actually forms complex thionates with thiosulfate originally generated in situ in the system by reaction of H_2S and SO_2 according to

$$4HSO_3^- + 2HS^- \rightarrow 3S_2O_3^{2-} + 3H_2O$$

In the reaction, hydrogen sulfide reacts with the thiosulfate and polythionates to form sulfur, as shown below:

$$S_2O_3^{2-} + 2H_2S + 2H^+ \rightarrow 4S + 3H_2O$$

The sulfur formed in the reaction is separated by a flotation technique to yield a sulfur froth, and the clear liquor is recycled to the absorber. The sulfur froth is melted and decanted into a sulfur product. An H_2S generator supplies the H_2S for the reactor, wherein a part of the by-product sulfur is reacted with natural gas, propane, or LPG to generate the H_2S-containing gas.

Peabody Engineered Systems adapted its prior experience in another sulfur-recovery process (the Holmes-Stretford process) to increase the effectiveness of sulfur recovery. Their experience with their own citrate pilot unit, operated by Pfizer at its Terre Haute, Indiana, plant and the introduction of proprietary techniques in absorption, sulfur separation, etc., qualify the naming of the process as the Citrex® process.

SPECIAL ROLE OF CITRIC ACID

The unique aspect of a buffered process like the citrate process is illustrated by the following: SO_2 solubility in water at 50° C is only 0.17 g/liter (with 1,000 ppm SO_2 in the gas phase), while a solution buffered with citrate has a solubility of 8.7 g/liter (at pH = 4.5), which is a 50-fold increase.

Among the many potential buffering agents, citrate seems to be superior, as seen from the comparative buffering capacities of four buffering agents shown in Table 36-1. Large values of buffer index show strong resistance to pH change. Hence citric acid, with a high buffer index, appears to be eminently suitable for the absorption of SO_2 in this type of buffered system.

As operated in the pilot plant at Terre Haute, the process consistently removed SO_2 down to the 25- to 50-ppm level. During the entire run, the Peabody absorber trays operated without any malfunction or lowered efficiencies, and operated successfully even with 1 percent sulfur slurry being returned to the absorber.

TABLE 36-1

Comparison of Buffer Indices at 0.5 M Concentration

Buffer	pH range		
	3.5	4.0	4.5
Citric acid	0.39	0.39	0.32
Glycolic acid	0.25	0.28	0.18
Phosphoric acid	0.05	0.02	0.01
Orthoboric acid	8×10^{-4}	2×10^{-4}	6×10^{-5}

Where buffer index = $\Delta C_A/\Delta pH$, and C_A = concentration of acid in moles per liter.

A distinct feature of the Citrex process is its low SO_2 oxidation rates. The U.S. Bureau of Mines has reported [3] that their test unit operations showed the oxidation rate to be about 1.3 percent, even with 20 percent O_2 in the feed gas. Test data from Pfizer's laboratories (Table 36-2) illustrate the effect of sodium thiosulfate (produced in situ in the process) in inhibiting oxidation to sulfate, even at the higher pH of 5.0. Another buffering agent has higher oxidation rates, which are not reduced by chelating agents such as EDTA.

Chemical consumption in this process is also lower with lower oxidation, and alkali consumption is but a fraction of that of a process using sodium scrubbing. Citrate loss was reported by the Bureau at its tests to be 7.5 lb/long ton of recovered sulfur [3]. Sulfur produced at the pilot plant at Terre Haute analyzed 99.96 percent sulfur content with 0.03 percent carbon, while the Kellogg unit of the Bureau has reported sulfur purity of 99.5 percent with 0.3 percent carbon content (partly due to the kerosene additive used for flotation in the Bureau's process).

In the citrate process, any liquid effluent (from spills or leaks) can be easily handled since citric acid is a biodegradable

TABLE 36-2

Laboratory Data on Oxidation of SO_2 During Absorption [2]

Acid	Conc. (M)	pH	Additive	Percent oxidation
Citric	0.5	4.0		0.4-0.5
Citric	0.5	5.0		0.8-1.0
Other	0.5	5.0		5.6-6.0
Other	0.5	5.0	EDTA	5.6-6.0
Citric	0.5	5.0	Thiosulfate 0.3 M	0.15

Temperature = 50° C.
Inlet gas = 2,000 ppm SO_2, 3.5% O_2, remainder N_2.

and ecologically acceptable component. With the Peabody tray design, high-efficiency (95 to 97 percent) SO_2 removal is achieved with L/G of 7 (gpm/1,000 scfm), as contrasted with the L/G of 10 in the packed tower absorber at Kellogg and the L/G of 100 used in limestone-scrubbing processes for SO_2 removal. This leads to lower power requirements for the flue-gas blowers in the Citrex process. About 40 percent lower overall power requirement is indicated for the Citrex process as compared with a Peabody limestone scrubber process.

The flywheel effect of the thiosulfate generated in the processing step allows the citrate process a great degree of flexibility. Tests at Terre Haute showed that the circulating liquor has a high capacity for short-term overloads of SO_2 or H_2S, which eliminates the need for precise, instantaneous control of H_2S feed rate to exactly match variations of SO_2 input from the flue gas. The role of citrate as an efficient buffer (with a buffer index of 0.38 at pH = 4.5) is also helpful in this flywheel effect. Since there is no need for precise stoichiometry, operational control of the citrate unit is also simplified. At Terre Haute, the plant operator performed routine checks for pH and thiosulfate levels only for operational monitoring purposes.

An efficient absorber and a simple stirred-tank regenerator, followed by a flotation unit, constitutes the major SO_2 removal and sulfur-separation steps. The sulfur recovery and H_2S generation system is also made up of small compact units in the Citrex version of the citrate process. Thus the Citrex process is a compact, low-cost process composed of proven components for H_2S generation along with a well-tested absorber, regenerator, and sulfur-separation units.

OPERATIONAL EXPERIENCE

The U.S. Bureau of Mines tested the citrate process first in a pilot plant in Arizona with 300 scfm smelter gas. Later another pilot plant was installed by the Bureau at Kellogg, Idaho, to treat 1,000 scfm of gas containing 5,000 ppm SO_2.

TABLE 36-3

Operational Experience with the Citrate Process

	Pfizer Terre Haute unit	U.S. Bureau of Mines Kellogg unit
Stack gas volume	2,000 scfm	1,000 scfm
Inlet SO_2 concentration	1,000-2,000 ppm	5,000 ppm
Exit SO_2	25-50 ppm	300 ppm
SO_2 removal	95-97%	95%
Sulfur production	400-500 lb/day	600 lb/day
Absorber	Peabody tray 2.5 x 10 ft	Packed tower 2.5 x 30 ft
Absorber, L/G	7.0	10
Reactor	Stirred reactor	Stirred reactor
Flotation	Air flotation	Flotation with kerosene additive
Sulfur purity	99.96% (0.03% carbon)	99.5% (0.3% carbon)
Total operating hours, cumulative	2,300 hr	1,300 hr
Longest sustained run	180 hr	160 hr

A pilot plant, built by Peabody, was installed at Pfizer's Terre Haute plant for treating 2,000 scfm gas from a coal-fired boiler. The diverted flue gas slip stream (equivalent to 1 MW) contained 1,000-2,000 ppm SO_2.

The comparative data on the Terre Haute unit and the Kellogg unit are shown in Table 36-3. The key differences between these two pilot operations are (1) use of the Peabody tray absorber with higher efficiencies of SO_2 removal and lower L/G, and (2) the use of air flotation at Terre Haute as compared to flotation with kerosene additive at the Kellogg unit. These two factors have made the Citrex process more economical and easier to operate.

It is interesting to note that the purity of sulfur is very high (of the order of 99.96 percent with less than 300 ppm carbon content), even though there was the possibility of chelated impurities originating from coal ash.

Another significant result from both the pilot plant operations is the low citrate consumption encountered; even on the pilot plant scale, where leaks and spills are proportionately larger than in a commercial unit. The citrate loss is of the order of 6 to 7 lb of citrate per ton of sulfur recovered. This loss can be further reduced in the process modifications introduced into the Citrex process by the experience gained in an analogous sulfur-recovery process (viz., the Holmes-Stretford process).

COMPARISON OF CITRATE VERSUS LIMESTONE SCRUBBING

Peabody Engineered Systems, in cooperation with Detroit Edison, developed a high-velocity spray scrubber for SO_2 removal using limestone after a successful demonstration of the unit in a 1-MW pilot plant using flue gas from a coal-fired boiler. The three key problems encountered in other limestone scrubbers hitherto were scaling, plugging, and sludge settling. Results from the pilot unit showed no plugging or scaling in the Peabody limestone system. Also, high conversion to $CaSO_4$ was achieved in the pilot tests, leading to a fast-settling sludge. Based on the encouraging pilot run of nearly a year, Detroit Edison decided to install the Peabody limestone system on its 175-MW coal-fired boiler.

A paper from TVA [3] compares several limestone scrubbing processes on the basis of investment and operating costs. In their survey of several alternative SO_2 scrubbing processes, the authors concluded that for a coal-fired boiler using 3.5 percent sulfur coal, the limestone process has the lowest annual operating cost as compared to the Wellman-Lord process, the MgO process, etc. Hence, we have selected the limestone process for a comparison with the Citrex process.

Table 36-4 shows the comparative evaluation of a few key process features as well as economic parameters for the Citrex process and

TABLE 36-4

Comparison Between Citrex and Limestone SO_2 Scrubbing Processes
(Basis: 200-MW Coal-Fired Boiler)

	Peabody Citrex process	Peabody Limestone process
By-product (per lb of SO_2)	0.5 lb of sulfur	6-8 lb of sludge (as 50% solids)
SO_2 removal	95%	90%
Exit SO_2	25-50 ppm	200-300 ppm
Absorber, L/G	7.0	100
Absorber pressure drop	8 in.	4 in.
Investment costs (battery limits--Nov. 1974)	\$40-\$50/kW	\$60-\$70/kW
Raw materials required	15,600 scf of natural gas per ton of sulfur recovered Citrate, 7 lb per ton of sulfur recovered	2 lb of $CaCO_3$ per lb of SO_2
Operating requirements:		
Power	7 kW/MW	12 kW/MW
Fuel	4,000 Btu/lb SO_2	
Annualized costs (annual capital and operating costs, mils/kWh)	2-2.5	2.5-3.0
Cost (\$/ton coal burned)	\$7	\$8

the limestone process. It appears that the Citrex process can reach higher SO_2 removal efficiencies (of the order of 95 percent) and lower SO_2 levels (about 25 to 50 ppm) in the treated gas, at a lower investment and operating cost, as compared to a throwaway process like the limestone process. Data shown are for a coal-fired boiler of 200-MW capacity range.

The Citrex process produces 0.5 lb of by-product of elemental sulfur per pound of SO_2 removed, which could be either marketed or easily stored without environmental hazard; whereas the limestone process produces nearly 16 times that amount or about 8 lb of sludge per pound of SO_2 removed, which has to be stored on a long-term basis.

As shown in Table 36-4, one of the key economic factors in favor of the Citrex process is its L/G requirements (7.0 versus 100 for limestone), with an accompanying benefit of lower pressure drop and pumping requirement. These factors lead to the investment costs being about 25 percent lower than for the limestone system. The power requirements are also about 40 percent lower. Hence, annualized operating costs (which is the total of annual capital and operating costs) appear to be about 15 to 25 percent lower than a limestone system for a comparable plant.

The above cost differential can be expressed as the cost per ton of coal burned for the two cases studied. The Citrex process costs about $7 per ton of coal burned, whereas the Peabody limestone process has a cost of about $8 per ton of coal burned with current coal price differentials of the order of $15 to $25 per ton of coal (delivered cost difference between high-sulfur and low-sulfur coal). There is sufficient incentive for the long term for an intensive interest in a regenerable process such as the Citrex process. All the above costs are for a typical one-train unit for a 200-MW coal-fired boiler using coal with 3.5 percent sulfur content.

REFERENCES

1. D. R. George, L. Crocker, and J. B. Rosenbaum: Mining Engineering, 22(1):7577 (1970).
2. W. A. McKinney, W. I. Nissen, D. A. Elkins, and J. B. Rosenbaum: Pilot-Plant Testing of the Citrate Process for SO_2 Emission Control, paper presented to the EPA Flue-Gas Desulfurization Symposium at Atlanta, Ga., November 4, 1974.
3. G. G. McGlamery and R. L. Torstrick: Cost Comparisons of Flue-Gas Desulfurization Systems, paper presented at the EPA Flue-Gas Desulfurization Symposium, Atlanta, Ga., November 7, 1974.

2. [illegible] TRENDS

[illegible]

[illegible] programs [illegible] requirements [illegible] and [illegible] requirements. [illegible]

[illegible] at about [illegible]

[illegible]

REFERENCES

1. [illegible]

2. [illegible]

3. [illegible] the Health and [illegible] the [illegible] November [illegible]

Chapter 37

pH CONTROLS FOR SO_2 SCRUBBERS

F. G. Shinskey

The Foxboro Company
Foxboro, Massachusetts

The most expedient technology to control sulfur dioxide being emitted from smelters, industrial boilers, and central station power plants is the wet scrubber. A variety of scrubbing media have been tried, each with its own advantages and disadvantages.

Where recovery of the sulfur is desired, a regenerable medium such as magnesia or ammonium citrate is required. However, reagent expense is minimized by using limestone, although this process yields the largest amounts of sludge for disposal. The use of a sodium-based reagent reduces the possibility of scaling in the scrubber, but it must be regenerated by lime, complicating the process.

Whatever method is used, absorption of the sulfur dioxide into the scrubbing medium is controlled by the concentration difference between SO_2 in the gas and the liquid. Fortunately, sulfur dioxide is quite soluble in water. But removal down to the parts-per-million level in preference to the preponderance of carbon dioxide present in most flue gases requires that its solubility be enhanced. This is accomplished by removing dissolved SO_2 either to the ionized or solid state by dosage with an appropriate reagent. The shift in equilibrium is controlled by the nature of the scrubbing medium and its pH.

LIME SCRUBBING OF SMELTER GAS

Sulfur dioxide is generated when sulfide-bearing ores are smelted to obtain metals or their oxides. One means of preventing the sulfur dioxide from escaping to the atmosphere is countercurrent scrubbing with lime slurry. The sulfur dioxide is absorbed and reacts with the lime to form calcium sulfite as a precipitate. The calcium sulfite crystals are grown to a filterable size by recirculating through a holding tank and back to the scrubber as shown in Fig. 37-1. Lime addition rate is adjusted to control pH at the scrubber exit. This is the only point where the pH is sensitive to SO_2 absorption and lime addition--the pH of the liquid entering the scrubber approaches 12.5, being saturated with lime.

Experience with the system in Fig. 37-1 has shown that emissions and corrosion rates are both high when the pH at the scrubber exit falls below 5. While a pH above 9 reduces the SO_2 in the stack to 20 ppm or below, sustained operation under these conditions is impossible. Solids accumulating in the area around the flue gas inlet force a shutdown in a few hours, the operating time decreasing as the pH increases.

Equilibria in the Lime-SO_2 System

The relationship between pH and absorption can be inferred after studying the equilibria existing in this system. The principal reaction proceeds as follows:

$$SO_2 + H_2O + Ca(OH)_2 \rightleftarrows CaSO_3 + H^+ + OH^- + HSO_3^- + SO_3^{2-} + Ca^{2+} \quad (1)$$

Ionic equilibrium demands that the total concentration of all positive charges equals that of all the negative charges in the system:

$$[H^+] + 2[Ca^{2+}] = [OH^-] + [HSO_3^-] + 2[SO_3^{2-}] \quad (2)$$

Brackets represent the concentration of the species within, in gram-moles per liter (abbreviated M).

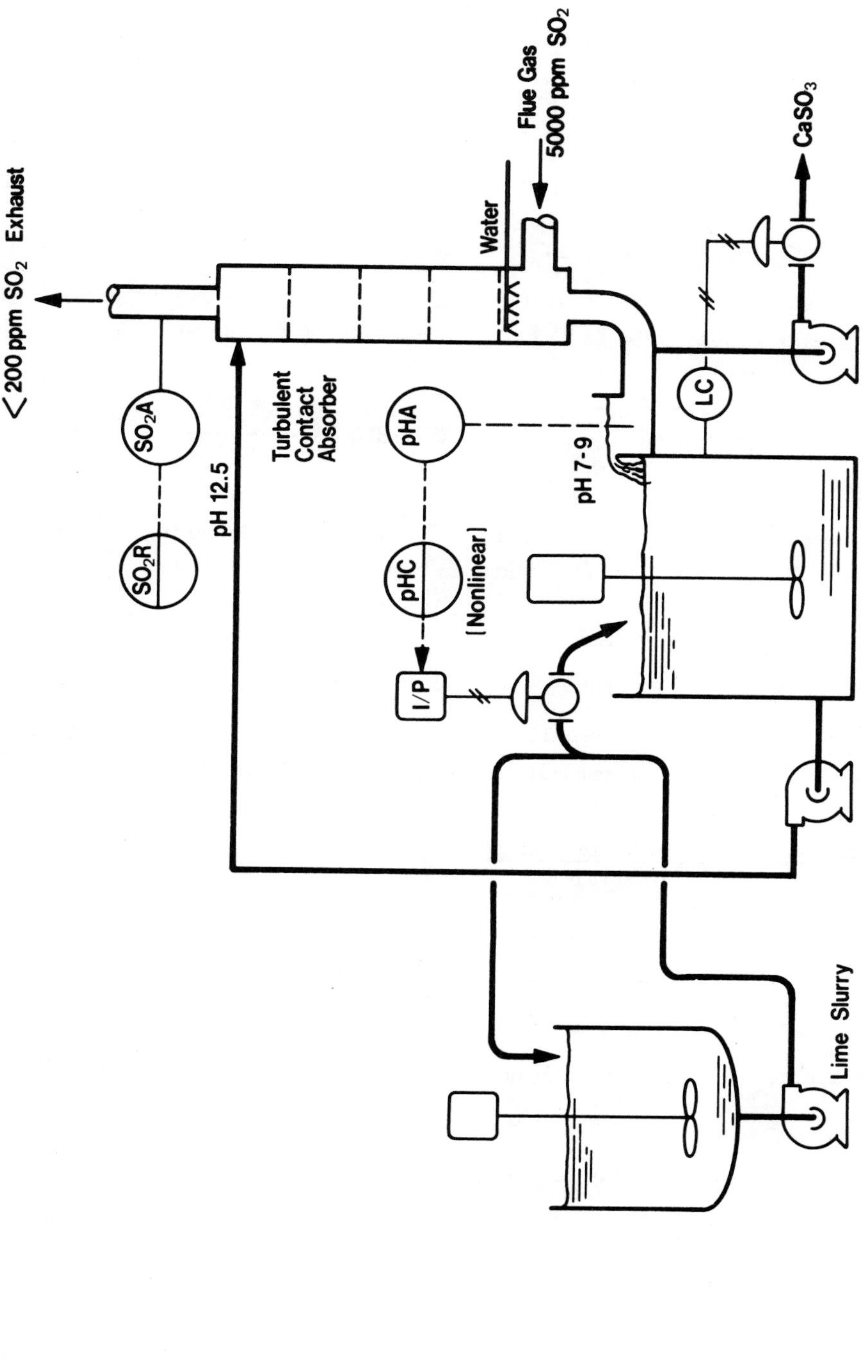

FIG. 37-1. Control of pH is mandatory when scrubbing flue gas with lime.

Four different ionic equilibria exist in this system:

$$[H^+][OH^-] = K_w = 10^{-14} \tag{3}$$

$$\frac{[HSO_3^-][H^+]}{[SO_2]} = K_1 = 10^{-1.8} \tag{4}$$

$$\frac{[SO_3^{2-}][H^+]}{[HSO_3^-]} = K_2 = 10^{-6.8} \tag{5}$$

$$[Ca^{2+}][SO_3^{2-}] = K_s = 10^{-5.7} \tag{6}$$

Substituting equilibria (3), (5), and (6) into the charge balance (2) yields a quadratic of $[HSO_3^-]$ in terms of $[H^+]$ and the equilibrium constants:

$$\left(1 + \frac{2K_2}{[H^+]}\right)[HSO_3^-]^2 + \left(\frac{K_w}{[H^+]} - [H^+]\right)[HSO_3^-] - \frac{2K_s[H^+]}{K_2} = 0 \tag{7}$$

Hydrogen ion concentration may be changed into the more familiar and measurable pH by substituting 10^{-pH} for $[H^+]$. Entering numerical values for the equilibrium constants then gives:

$$(1 + 2 \times 10^{pH-6.8})[HSO_3^-]^2 + (10^{pH-14} - 10^{-pH})[HSO_3^-]$$
$$- 2 \times 10^{1.1-pH} = 0 \tag{8}$$

The above quadratic is readily solved for values of $[HSO_3^-]$ in terms of solution pH. A set of solutions appears in Table 37-1.

Incorporating the Material Balance

Having determined the concentration of bisulfite ions at each pH, the next task is to relate pH to the material balance.

The total sulfur concentration x_S in the system at any time is

$$x_S = [SO_2] + [HSO_3^-] + [SO_3^{2-}] + [CaSO_3] \tag{9}$$

Similarly, the total calcium concentration x_{Ca} is expressed as

$$x_{Ca} = [Ca^{2+}] + [CaSO_3] \tag{10}$$

These two statements may be combined by eliminating the common $CaSO_3$:

$$x_{Ca} - x_S = [Ca^{2+}] - [SO_2] - [HSO_3^-] - [SO_3^{2-}] \tag{11}$$

TABLE 37-1

Concentration Versus pH for the Lime-SO_2 Reaction

pH	$[HSO_3^-]$ (M)	$x_{Ca} - x_S$ (M)	$[SO_2]$ (M)
3	0.159	-0.090	0.010
4	0.050	-0.0254	3.15×10^{-4}
5	0.0156	-7.81×10^{-3}	9.84×10^{-6}
6	4.36×10^{-3}	-2.18×10^{-3}	2.75×10^{-7}
7	7.7×10^{-4}	-3.85×10^{-4}	4.86×10^{-9}
8	8.77×10^{-5}	-4.34×10^{-5}	
9	9.0×10^{-6}	$+5.0 \times 10^{-7}$	
10	9.0×10^{-7}	$+4.96 \times 10^{-5}$	
11	9.0×10^{-8}	$+5.0 \times 10^{-4}$	
12	9.0×10^{-9}	$+5.0 \times 10^{-3}$	

The $[Ca^{2+}]$ and $[SO_3^{2-}]$ terms may be eliminated by combining Eq. (11) with the charge balance (2), yielding:

$$x_{Ca} - x_S = 0.5([OH^-] - [H^+] - [HSO_3^-]) - [SO_2] \quad (12)$$

The $[OH^-]$ term may be converted to $[H^+]$ by Eq. (3), and the $[SO_2]$ term may be converted to $[HSO_3^-]$ using Eq. (4). Then, by combining terms and substituting 10^{-pH} for $[H^+]$ along with numerical values of K_1, K_2, and K_s, we have:

$$x_{Ca} - x_S = 0.5(10^{pH-14} - 10^{-pH}) - [HSO_3^-](0.5 + 10^{1.8-pH}) \quad (13)$$

Values of the concentration difference $x_{Ca} - x_S$ for each pH unit are listed in Table 37-1.

The Titration Curve

Since all the sulfur is introduced as SO_2 in the flue gas, and all the calcium as lime reagent, $x_{Ca} - x_S$ represents the difference between reagent addition and absorber loading. A titration curve taken

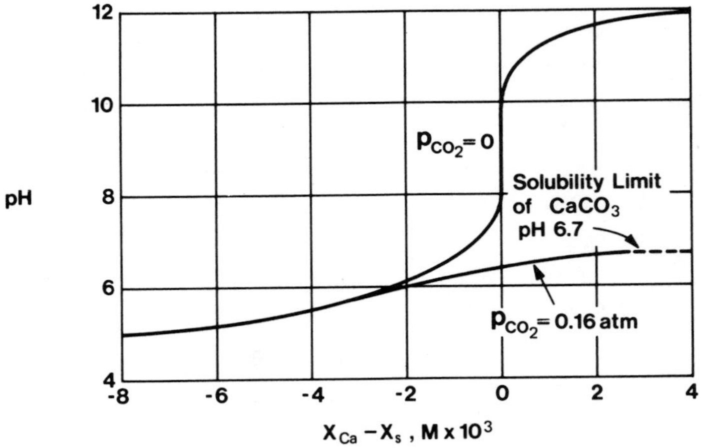

FIG. 37-2. The lime-sulfur dioxide titration curve exhibits a sharp rise in the absence of carbon dioxide.

from the data in Table 37-1 is plotted in Fig. 37-2. The end point or equivalence point is indicated by the sharp change in pH. Equivalency is achieved just below pH 9.

If the pH of the slurry leaving the scrubber is allowed to deviate much below 7, some of the SO_2 will not be removed from the flue gas. To illustrate, the concentration of SO_2 in solution is listed in the last column in Table 37-1. This dissolved SO_2 is in equilibrium with that in the flue gas, and so the concentrations given are indicative of relative concentrations in the gas phase. Note that SO_2 concentration in the liquid changes 50-fold between pH 6 and 7.

While a true equilibrium state cannot be reached in the scrubber due to limitations in mass transfer, the tabulations presented illustrate the relative driving forces present in the system. When the pH is 7 or less, absorption of SO_2 probably takes place across the entire scrubber. Should the pH creep up to 11, most of the absorption seems to focus at the flue-gas inlet. Extended operation at pH levels above 9 result in plugging the scrubber with calcium sulfite solids in that area. Solids do not tend to collect in the body of the scrubber, as its packing is designed to be self-cleaning.

The sheer sensitivity of the titration curve in the desired operating range warns of the difficulty in controlling pH there. Variations in flue gas flow or composition can easily force the pH below 7 or over 11. Furthermore, a conventional controller is prone to cycling the pH constantly through the operating range, because of both the sensitivity of the curve and the delay in pH response to changes in lime addition rate. A controller with special nonlinear characterization to match the titration curve is needed to successfully control pH in this process.

LIME AND LIMESTONE SCRUBBING OF PRODUCTS OF COMBUSTION

Lime scrubbing of the products of combustion differs from scrubbing of smelter offgas due to the dominance of carbon dioxide. Being an acid gas, CO_2 is absorbed along with SO_2, forming carbonate and bicarbonate ions which alter the titration curve. In the extreme, the scrubbing medium may become saturated with calcium carbonate, placing an upper limit on its pH.

Limestone ($CaCO_3$) scrubbing of the products of combustion is characterized, in the steady state, exactly like lime scrubbing. The reason for this identity is the existence of an equilibrium among calcium, carbonate, and bicarbonate ions in both cases. Whether the carbonate and bicarbonate ions originate in the flue gas or in the limestone is of no consequence in the steady state. In the unsteady state, however, the differences are considerable. Lime may be removing some CO_2 from the flue gas, while limestone is releasing it. The limitations in mass transfer between the solid and gaseous states through the liquid result in the discharge of considerable unreacted limestone, compared to relatively little unreacted lime.

Equilibria in the Lime-CO_2 System

In addition to the reaction between lime and SO_2 described earlier, CO_2 reacts in a similar fashion:

$$CO_2 + H_2O + Ca(OH)_2 \rightleftarrows CaCO_3 + H^+ + OH^- + HCO_3^- + CO_3^{2-} + Ca^{2+} \quad (14)$$

To assess the effect of CO_2 on SO_2 absorption, the charge balance of Eq. (2) must be augmented to include the carbonate and bicarbonate ions:

$$[H^+] + 2[Ca^{2+}] = [OH^-] + [HSO_3^-] + 2[SO_3^{2-}] + [HCO_3^-] + 2[CO_3^{2-}] \quad (15)$$

Some additional equilibria appear:

$$\frac{[HCO_3^-][H^+]}{[CO_2]} = K_{10} = 10^{-6.35} \quad (16)$$

$$\frac{[CO_3^{2-}][H^+]}{[HCO_3^-]} = K_{20} = 10^{-10.25} \quad (17)$$

Because of the preponderance of CO_2 (16 percent) in the flue gas compared to SO_2 (5,000 ppm or 0.5 percent), its concentration may be considered constant throughout the scrubber. The scrubbing medium may then contain a constant level of dissolved CO_2 in equilibrium with that in the gas. The concentration c of CO_2 in the liquid can be related to its partial pressure p in the gas through Henry's law(1):

$$c = \frac{p}{H} \quad (18)$$

At 140°F, H = 3,410, with c in mole fraction and p in atmospheres. Selecting 0.16 atm for p yields

$$c = \frac{0.16}{3,410} = 4.69 \times 10^{-5} \text{ mol fr}$$

Conversion of c to moles per liter is achieved by dividing by the molecular weight of water (18 g) and multiplying by the weight of a liter of water (1,000 g):

$$[CO_2] = 4.69 \times 10^{-5} \frac{1,000 \text{ g/liter}}{18 \text{ g/mol}} = 2.6 \times 10^{-3} \text{ M}$$

Then equilibria (16) and (17) may be solved for $[HCO_3^-]$ and $[CO_3^{2-}]$ in terms of $[H^+]$ and the known $[CO_2]$, and substituted into the charge balance. As before, $[Ca^{2+}]$ and $[SO_3^{2-}]$ may be replaced by equivalent

TABLE 37-2

Concentration Versus pH for the Lime-SO_2-CO_2 Reaction

pH	$[HSO_3^-]$ (M)	$[Ca^{2+}]$ (M)	$[Ca^{2+}][CO_3^{2-}]$ (M)	$x_{Ca} - x_S$ (M)
3	0.159	0.079	5.17×10^{-15}	-0.090
4	0.050	0.025	1.64×10^{-13}	-0.0254
5	0.0156	8.07×10^{-3}	5.27×10^{-12}	-7.75×10^{-3}
6	3.96×10^{-3}	3.18×10^{-3}	2.07×10^{-10}	-1.60×10^{-3}
6.7	6.65×10^{-4}	3.78×10^{-3}	6.19×10^{-9}	$+2.58 \times 10^{-3}$
6.8	4.58×10^{-4}	4.35×10^{-3}	1.13×10^{-8}	$+3.44 \times 10^{-3}$

expressions of $[HSO_3^-]$, leaving a quadratic in terms of pH. The first (A) and last (C) terms of the quadratic are identical to Eq. (8), but the middle (B) term is augmented by the presence of CO_2:

$$B = 10^{pH-14} - 10^{-pH} + [CO_2]10^{pH-6.35}(1 + 2 \times 10^{pH-10.25}) \quad (19)$$

Solution of this quadratic for values of pH from 3 to 6.8 is given in Table 37-2. Comparison of the first column in Tables 37-1 and 37-2 shows that the influence of CO_2 is negligible below pH 6.

Carbonate Solubility Limit

The above formulation assumes that the scrubbing medium is not saturated with $CaCO_3$. Saturation occurs when

$$[Ca^{2+}][CO_3^{2-}] = 7.9 \times 10^{-9} \quad (20)$$

To determine whether saturation does occur, the system must be tested by substituting the concentrations of calcium and carbonate ions into Eq. (20) at each pH level. Their individual concentrations may be estimated by combining equilibrium statements in terms of known $[HSO_3^-]$ and $[CO_2]$ concentrations:

$$[Ca^{2+}] = \frac{K_s[H^+]}{K_2[HSO_3^-]} = \frac{10^{1.1-pH}}{[HSO_3^-]} \quad (21)$$

$$[CO_3^{2-}] = [CO_2]\,\frac{K_{10}}{[H^+]}\,\frac{K_{20}}{[H^+]} = [CO_2]10^{2pH-16.6} \tag{22}$$

The carbonate ion increases in concentration by two decades for each unit rise in pH. By contrast, the calcium ion level falls as pH increases due to transformation of HSO_3^- to SO_3^{2-} ions. However, above pH 6, calcium ion concentration rises again; the reason for this reversal is that lime addition is continuing to raise the pH, but most of the sulfur has already been removed from solution. Table 37-2 lists the calcium ion level at each value of pH.

The product of calcium and carbonate ions rises sharply with pH until saturation is reached between pH 6.7 and 6.8, apparently closer to 6.7.

If lime addition is continued above pH 6.7, calcium carbonate will be precipitated. Because of the huge reservoir of CO_2 in the flue gas, the concentration of CO_2, and therefore carbonate ions, in the scrubbing medium will not be significantly changed by this precipitation. As a result, the pH cannot be increased above that point by lime addition--an upper limit is reached.

Again, these relationships apply rigorously only in steady state. An excess of lime may raise the pH somewhat above the limit of 6.7 in proportion to the rate of carbonate precipitation. In the case of limestone addition where the carbonate must be transferred into the gas phase, it is doubtful if a pH as high as 6.7 can ever be reached.

The Titration Curve

In order to generate a titration curve below the solubility limit of $CaCO_3$, the material balance must be solved. The statements given representing total calcium and total sulfur are still valid. Their difference, Eq. (11), must be substituted into the new charge balance (15) to yield $x_{Ca} - x_S$ in terms of pH. The resulting equation is equal to Eq. (13), augmented by

$$\Delta(x_{Ca} - x_S) = [CO_2]10^{pH-6.35}(0.5 + 10^{pH-10.25}) \tag{23}$$

Adding Eq. (23) to Eq. (13) yields the right-hand column in Table 37-2.

A plot of $x_{Ca} - x_S$ for an atmosphere of 16 percent CO_2 is shown in Fig. 37-2 to allow comparison with the CO_2-free curve. In addition to the solubility limit, the presence of CO_2 produces tremendous buffering. In this system, pH is extremely stable and relatively insensitive to reagent addition. For best results, the pH should be controlled near the equivalence point of 6.3. A narrow-span (e.g., pH 5 to 7) instrument should be used, carefully installed and scrupulously maintained if effective results are to be achieved.

If limestone is the reagent in the 16 percent CO_2 atmosphere, the titration curve will be similar, although in the unsteady state some $CaCO_3$ will always remain unreacted even below pH 6.7. If limestone is used in the absence of CO_2, the pH at the $CaCO_3$ solubility limit will increase somewhat, but probably not beyond 7.5.

THE DOUBLE-ALKALI PROCESS FOR FLUE-GAS DESULFURIZATION

In the double-alkali process, SO_2 is absorbed from the flue gas by a solution of sodium sulfite. Absorption lowers the pH of the solution while forming bisulfite ions. The spent solution then proceeds to a regenerating system where lime is added. At this point, calcium sulfite is precipitated, producing in the process the sodium sulfite solution to be used again. Each stage of this process will be investigated to determine the pH corresponding to optimum operating conditions.

Reactions in the Scrubbing System

The reaction between sodium sulfite and the acid gases proceeds as follows:

$$Na_2SO_3 + H_2O + SO_2 + CO_2 \rightleftarrows Na^+ + H^+ + OH^- + HSO_3^- + SO_3^{2-} + HCO_3^- + CO_3^{2-} \quad (24)$$

The charge balance is

$$[Na^+] + [H^+] = [OH^-] + [HSO_3^-] + 2[SO_3^{2-}] + [HCO_3^-] + 2[CO_3^{2-}] \quad (25)$$

The sodium mass balance is simply

$$x_{Na} = [Na^+] \quad (26)$$

while the sulfur mass balance is

$$x_S = [SO_2] + [HSO_3^-] + [SO_3^{2-}] \quad (27)$$

The terms on the right of the sulfur balance can all be converted into $[HSO_3^-]$; the same can be done with the sulfur terms in the charge balance, allowing a substitution of x_S for $[HSO_3^-]$. With appropriate substitution of $[CO_2]$ for the carbonate ions, we have

$$x_{Na} = 10^{pH-14} - 10^{-pH} + x_S \frac{1 + 2 \times 10^{pH-6.8}}{1 + 10^{pH-6.8} + 10^{1.8-pH}} + [CO_2]10^{pH-6.35}(1 + 2 \times 10^{pH-10.25}) \quad (28)$$

In this scrubbing solution, the absolute concentration of the sulfite ion or sodium ion is not fixed by a solubility constraint as in the calcium-based system. Consequently, Eq. (28) cannot be solved without arbitrarily assigning a value for either x_{Na} or x_S. Accordingly, values of 0.1 M and 0.01 M were assigned to x_S in order to generate the titration curves in Fig. 37-3.

Note that the curves cross at the equivalence point of pH 4.4. Increasing solution strength moderates the curve, but the control point remains the same. The only departure from this rule would be changes in the equilibrium constants at high solution concentrations. In actual operation, the pH at equivalence can be located by adjusting the flow of scrubbing solution until the inflection point of the curve is found.

The strength of the scrubbing solution is controlled by the water balance in the system. As water evaporates, it must be replaced or concentrations will increase.

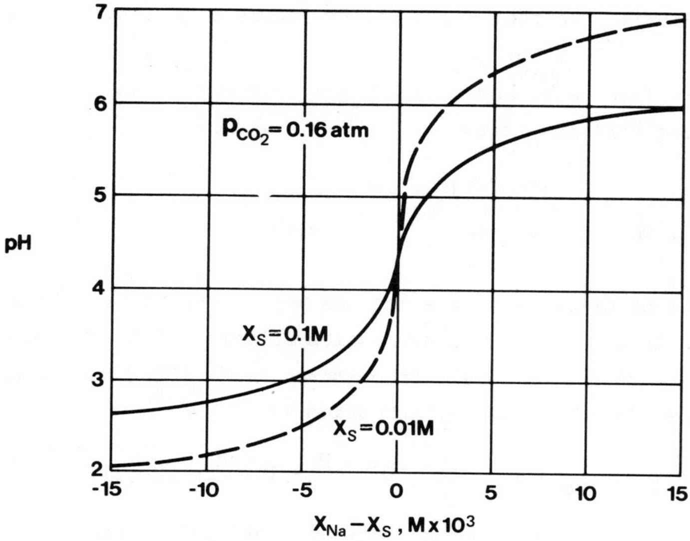

FIG. 37-3. The equivalence point for the sodium-sulfur dioxide titration is independent of solution concentration.

Regeneration of the Sulfite Solution

The spent scrubbing solution leaves the absorber with a certain sulfur concentration x_S and a pH in the range 4 to 6. Presumably, below pH 6, little CO_2 will have been absorbed. The equilibrium in the regeneration tank is essentially that between calcium and sulfite ions:

$$NaHSO_3 + Ca(OH)_2 \rightleftarrows CaSO_3 + H^+ + OH^- + HSO_3^- + SO_3^{2-} + Na^+ + Ca^{2+} \qquad (29)$$

The charge balance is

$$[Na^+] + 2[Ca^{2+}] + [H^+] = [OH^-] + [HSO_3^-] + 2[SO_3^{2-}] \qquad (30)$$

The only difference between this and Eq. (2) is the presence of sodium ions in concentration x_{Na}. Concentrations of calcium and sulfite ions are limited by their solubility product. Substitution of equilibria into the charge balance forms a quadratic. Whereas Eq. (8) was solved for $[HSO_3^-]$, a solution in $[SO_3^{2-}]$ is more appropriate in this case, due to its higher concentration in the pH range of the regeneration system:

$$(2 + 10^{6.8-pH})[SO_3^{2-}]^2 + (10^{pH-14} - 10^{pH} - x_{Na})[SO_3^{2-}] - 2 \times 10^{-5.7} = 0 \quad (31)$$

Next, the material balance equations may be substituted into the charge balance to yield an expression in terms of the known $[SO_3^{2-}]$ and $[H^+]$. At the pH levels encountered here, $[SO_2]$ in Eq. (9) may be dropped. The resulting expression is

$$x_{Ca} + 0.5x_{Na} - x_S = 0.5(10^{14-pH} - 10^{-pH} - [SO_3]10^{6.8-pH}) \quad (32)$$

If the pH of the absorber is controlled at equivalence, $x_S = x_{Na}$, and equivalence in the regenerator then gives $x_{Ca} = (1/2)x_S$. For $x_S = x_{Na} = 0.1$ M, 0.05 M lime is used to precipitate 0.05 M sulfur, leaving 0.05 M sulfur and 0.1 M sodium in solution as Na_2SO_3. Upon return to the scrubber, an additional 0.05 M sulfur is absorbed at equivalence, producing an equimolar solution of sodium and sulfur as $NaHSO_3$.

The titration curve for Eq. (32) is shown in Fig. 37-4 for sodium ion concentrations of 0.1 and 0.01 M. Equivalence is between pH 9 and 10. Control over the regeneration actually should be maintained below pH 9, however, to avoid CO_2 absorption from the atmosphere. Since clarification and filtration of the regenerated solution are open to the atmosphere, CO_2 absorption cannot be avoided, and its rate increases exponentially with pH. At pH levels above 10, calcium carbonate will form, eventually being deposited as scale in the piping and scrubber.

In the scrubber, if the pH is allowed to go too low, emissions will increase. However, in the regeneration system, a low pH presents no overt difficulty, since whatever sulfur is not precipitated will be returned later. Consequently, operating the regenerator as

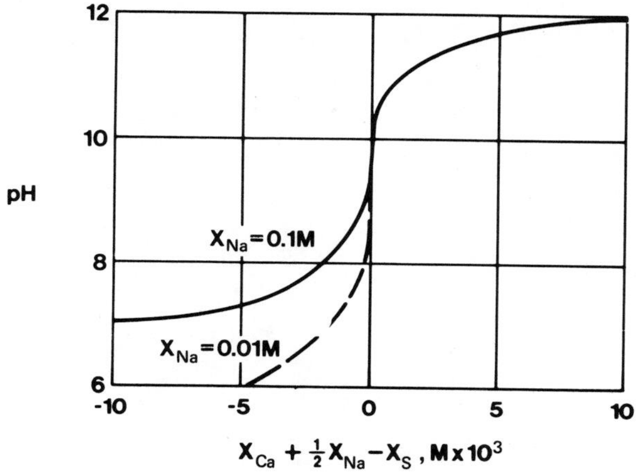

FIG. 37-4. The end point for regenerating the scrubbing solution is about pH 9.

low as pH 8 would seem to be advantageous in restricting carbonate formation.

CONCLUSIONS

Titration curves for all reactions described except the lime or limestone scrubbing of combustion products indicate end points with sharp breaks. Although the curves were derived on a theoretical basis and their true shape and position may differ somewhat, they nonetheless are very helpful guides in establishing operating conditions. If control is enforced at the optimum pH level for each reaction, difficulties in scaling, plugging, and corrosion will be minimized in addition to achieving emission control.

REFERENCE

1. J. H. Perry: "Chemical Engineer's Handbook," 4th ed., McGraw-Hill Book Co., New York.

Chapter 38

GRAVEL BED FILTER

J. W. Simonini
Air Pollution Control Division
Rexnord, Inc.
Louisville, Kentucky

INTRODUCTION

The gravel bed was used for the filtration of gases and liquids as far back as the nineteenth century. Specific patents for this method were granted in 1888 when it was applied for the cleaning of molasses in sugar factories. Around the turn of the century, patents were issued and equipment was constructed with a view toward using a filter for the separation of dust from gases. However, gas cleaning on an industrial scale developed in the direction of intermittent-type equipment, and the whole range of methods involving continuous gas cleaning was neglected.

One of the first examples of industrial gas cleaning was a system applied to the effluent from a carbide furnace in a German chemical plant. Here, the dusty waste gases from the carbide furnace were precleaned in cyclones and then cleaned of fines in a bed of granular coke, which was also the feed material. When the coke bed became saturated with dust, the bed was replaced and the dust-saturated coke was fed into the furnace. A similar method was known to have been applied in the Soviet Union, where waste gases from open hearth furnaces were cleaned through a bed of magnesite granules. Cleaning of the filter bed was reported to take place intermittently by means of water.

Not until January 1957 did Max and Wolfgang Berz, in Germany, succeed in devising a practical solution which permitted a continuous, fully automated cleaning of the filter bed from the dust load within the collector. They used a horizontally arranged filter bed consisting of uniform quartz grains housed in a flex-spring support casing. To clean the filter bed, the casing was vibrated by means of an eccentric motor. This motion separated the dust and moved it down into a dust-collecting chamber. Thus, the granular bed was used for the first time for separaing dust from effluent gases on an industrial scale and on a continuous basis.

There were several problems associated with the operation of the vibrating bed-type filter, principally with the bed-to-shell seals; and a simple and effective backflush method was devised, employing reverse air and agitation of the gravel bed with a stirring arm. Evolved from this basic design are the filters marketed by Rexnord, Inc., for the cleaning of gases from clinker coolers, lime kilns, and sinter machines and other large-scale industrial applications.

GENERAL DESCRIPTION OF THE FILTER AND ITS OPERATION

A typical gravel bed filter installation consists of several filter modules (Fig. 38-1). Figure 38-1 (a) shows the filter module in the filtering or forward flow mode, while Fig. 38-1 (b) shows this module in the backflush mode. The filter modules operate in parallel from a main inlet chamber and the operation of each module is identical.

Referring to Fig. 38-1 (a), the raw dust-laden gas enters the module from the inlet chamber (1) into the cyclone section of the module. The cyclone (2) removes the coarse dust and the gas flow moves up the vortex tube (4) into the filter chamber (5). The removed dust is discharged through the tipping gates (3). The gas flow is then directed through the filter medium (6), where the remaining fraction of dust is removed. Gas then passes through the

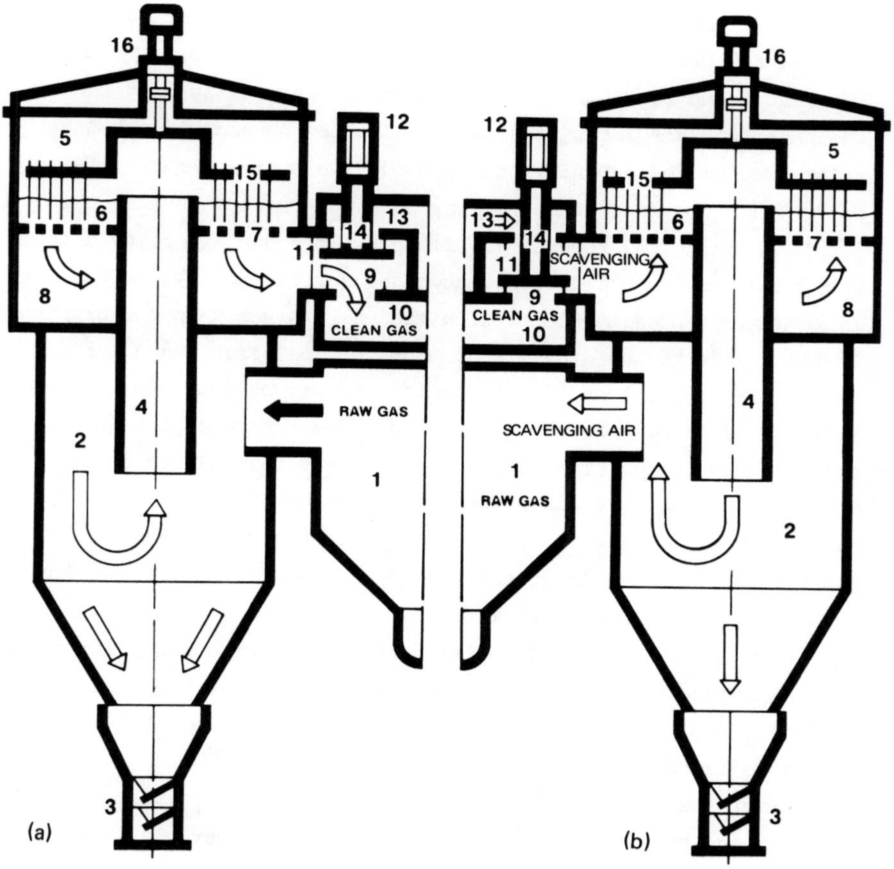

FIG. 38-1. (a) Gravel filter during cleaning phase. (1) Raw gas duct, (2) cyclone, (3) double-tipping valves, (4) vortex tube, (5) filter chamber, (6) gravel bed, (7) screen, (8) clean air chamber, (9) exhaust post, (10) clean air duct, (11) valve disk, (12) cylinder, (13) backflush duct, (14) backflush port, (15) rake mechanism, (16) rake drive. (b) Gravel filter during backflush phase.

medium support screen (7) into the clean gas chamber (8) and through the clean gas outlet duct (9). The process described above continues until the filter bed becomes sufficiently laden with dust to require cleaning. At this point, the filter module is isolated from the clean gas stream by actuation of the backflush control valve (11) powered by the valve cylinder (12). At this point, the module has entered the bed cleaning or backflush mode.

Referring to Fig. 38-1 (b), clean air enters the unit through the backflush air inlet (13) and is directed upward through the filter medium (6). During this period, the stirring rake (15) is activated by means of the stirring rake motor (16). During the combined raking and reverse air flushing, the dust is removed from the filter medium and directed down the vortex tube (4). This dust is in agglomerated particles significantly larger than the original incident dust to the gravel bed. Some of the resuspended dust is removed in the cyclone (2); the remainder passes into the inlet chamber (1) where more of the dust is removed because of the greatly reduced velocities. The backflush gases and the balance of the agglomerated dust are distributed to the other filter modules for removal by the companion cyclones. Any dust not captured by the companion cyclones is reintroduced to a companion gravel bed for another opportunity to agglomerate.

The backflush cycle is often controlled in frequency and duration by adjustable interval timers. It may also be controlled by pressure drop across each filter bed or total headloss across the entire collector.

After each filter is backwashed, the backwash valve changes position and puts the bed back on line while another bed is taken offstream for backflushing. The most common method of backflush valve actuation has been the use of double-acting hydraulic cylinders. Of course, air or electric actuation is entirely possible where their use seems advantageous. The backflush air itself is generally supplied by a small separate fan using preheated ambient air.

The scalping cyclones under each module are relatively low-headloss, low-efficiency collectors that remove the larger particles (and backflush agglomerates) only; the smaller particles that have a strong tendency to agglomerate are introduced to the gravel bed.

The most popular medium currently used in gravel bed installations is a granular quartz material. The granule size used varies with the application but is generally in the range of 1 to 6 mm in diameter. The medium need only be fine enough to begin to form and retain a dust layer, since it is this dust layer which produces the most effective filtering and agglomeration. Quartz aggregate has been the most common medium choice, chiefly due to its low cost, hardness, and durability.

PRINCIPAL ADVANTAGES

One of the most appealing aspects of the gravel bed filter is its inherent simplicity. The only moving parts within the collector itself are the backflush valve and the raking mechanism. Since both of these operate intermittently, the wear on these mechanisms is extremely low. The velocities used within the collector are, in general, quite low, and the effect of even highly abrasive dust, such as clinker dust, is very small. On clinker cooler applications, even the cyclones themselves receive a ceramic liner only in the lower cone section.

The inlet raw gas duct serves not only as a distribution plenum but also as a low-velocity settling chamber where much of the coarser, more destructive dust is dropped out. This, of course, greatly reduces the duct wear in the rest of the system.

It has been shown by experience that the filter medium itself, when consisting of quartz granules, is subject to almost insignificant wear. In addition to the fact that the quartz itself is very hard, the granules remain coated with a fine layer of dust even after backflushing. Thus, even during the brief rake agitation period, the dust contacts only itself. The quartz granules used are those found in natural formations and will easily withstand at least 1,500°F. Thus, the operating temperatures of the gravel bed filter are limited only by the steel itself. Inlet temperatures to 700°F can be accommodated with the standard design. The seals on the backflush valves consist of machined steel surfaces. Since they operate only in the backflush and clean gas streams, the wear on these valve plates is very low. Of course, in addition to low maintenance and simplicity, the gravel bed filter is attractive in that it uses no water.

Depending upon the application, steady-state headloss of a typical gravel bed filter installation is 5 to 15 in.wc. Once on line, the operation is quite steady; and using ordinary control damper techniques, the process control achieved is generally as good as the original process control without the collector. For more economical and flexible control, the use of variable-speed induced-draft fans has proven to be an excellent approach. One of the filter modules is always in the backflush mode; thus, the number of modules in the forward-flow mode remains constant. Of course, the system is insensitive to dust resistivity characteristics and, in general, is extremely tolerant of process upsets. Lower-than-designed gas volumes generally enhance filtering capabilities of the gravel bed, while higher-than-designed gas volumes will increase the efficiency of the primary cyclone collectors.

The collector's modular design permits easy increase in collector capacity for future expansion. The modular design also allows one to isolate individual filter modules for service, inspection, or even close tailoring of filter bed area to process gas flow. This feature has also made it possible to accept gases from multiple sources, since the collector can be serviced on a continuous basis and modules can be blanked off when gas flow is reduced during process downtime. This filter module isolation is easily accomplished manually or automatically and can be done for short- or long-term isolation.

DESIGN CRITERIA

Shown in Fig. 38-2 are the approximate number of modules required for a range of gas flows. Figure 38-3 shows approximate space requirements for handling the designated gas flows.

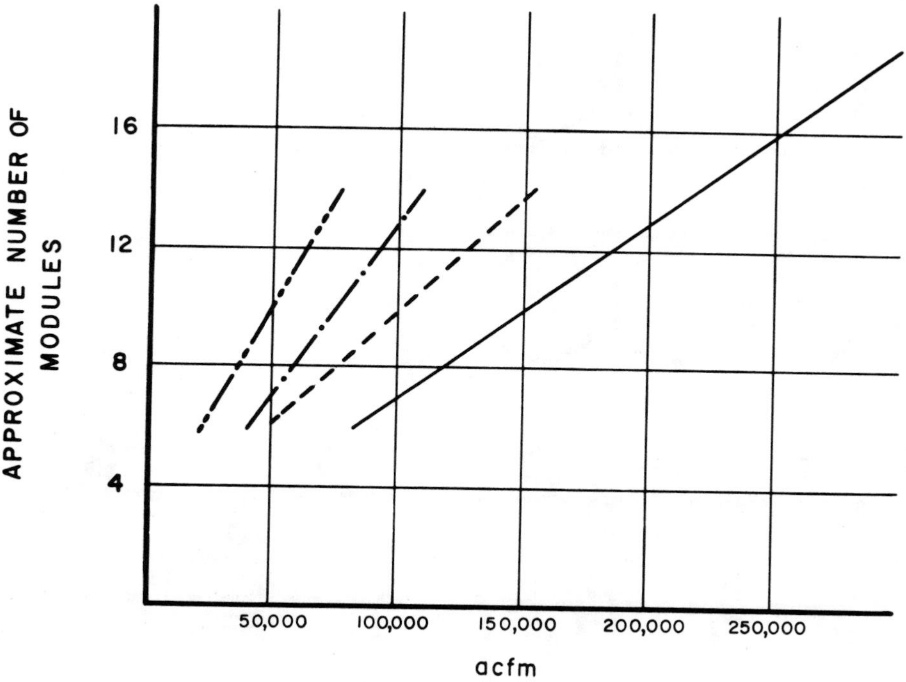

FIG. 38-2. The approximate number of modules needed for a gravel bed installation can be determined using the above graph. ——— double bed 110, ----- double bed 90, —·— single bed 110 and double bed 80, —--— single bed 90.

The size and design details, of course, vary with the application. Particle size distribution of the inlet dust is often an important factor, especially in light of the fact that three different collection methods are actually employed within the gravel bed filter. The low-velocity inlet duct acts as a settling chamber for the coarser fraction, while conventional cyclonic action removes a second fraction; and the finest fraction is then filtered through the granular medium. Calculated values for the settling duct cut size are of the order of 40 to 50 μm, while the cyclone cut size is of the order of 10 to 20 μm. (This is assuming a particle specific gravity of 2 and nominal flow values.)

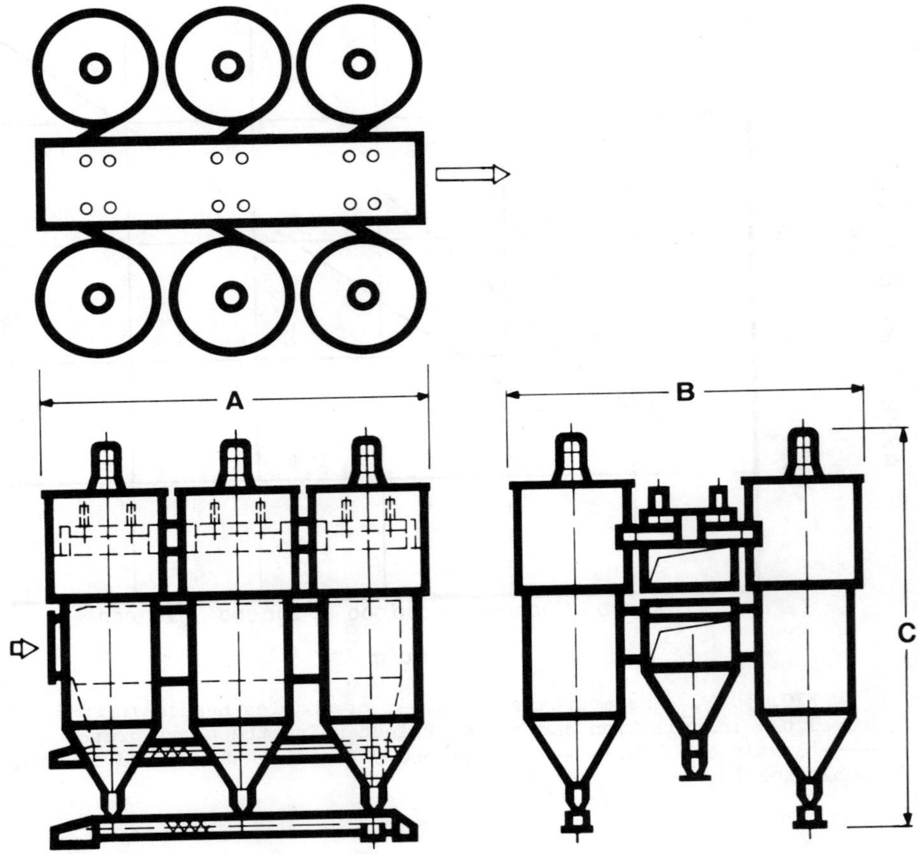

Gas flow ACFM	A	B	C
60,000	40′	28′	32′
120,000	44′	28′	32′
180,000	52′	34′	42′

FIG. 38-3. Typical space requirements needed by a gravel filter. The requirements given in the table are for approximation purposes only. Actual space requirements are dependent on complete specifications.

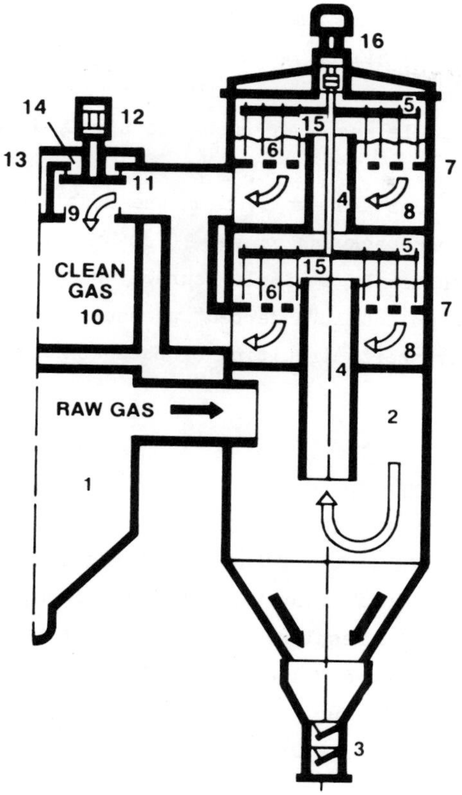

FIG. 38-4. For higher-volume systems, economy and space savings are possible by using two filter chambers per module. This reduces the relative number of rake arm drives, counterweight gates, length of screw conveyors, etc. The operating and backflush operations are the same as those already described. Nearly double gas volumes may be handled with the same diameter modules. Shown is a diagram of the filter with two superimposed gravel beds. (1) Dusty gas duct, (2) primary collector, (3) counterweighted valve, (4) vortex tube, (5) filter chamber, (6) gravel bed, (7) screen, (8) clean air chamber, (9) exhaust port, (10) clean air duct, (11) valve disk, (12) cylinder, (13) backflush duct, (14) backflush port, (15) rake mechanism, (16) rake drive.

Increasing the number of filter modules results in a smoother operation, and this fact is considered where steady-state process control is critical. On the other hand, where space requirements must be kept to a minimum, the filter area can be effectively doubled by using a double-bed filter module, i.e., two gravel beds concentrically located over one common cyclone precleaner (Fig. 38-4). This double-bed module has been used extensively in United States applications.

Two backwash fans are employed on the double-bed module, and each bed has its own backwash valve. (For hydraulic actuators, these valves can be actuated by a single four-way solenoid.) This ensures that each bed has an equal "underbed" pressure during the backwash cycle and prevents short-circuiting of backwash air through the cleaner bed.

INSTALLATIONS AND PERFORMANCE

To illustrate the viability of the gravel bed filter as an air-cleaning device, two different U.S. installations will be described in some detail and the results of performance tests given.

The first gravel bed filter in the United States was installed at the Paul Lime Plant, Inc., Douglas, Arizona, on the feed end of a rotary lime kiln. This kiln was originally equipped with a cyclone arrangement, the performance of which was not good enough to meet the newly adopted particulate emission requirements of the state of Arizona.

In late 1971, after considering all available systems, Paul Lime decided to install a gravel bed filter to meet the emission limitations of a process weight table which, for instance, limits particulate emissions from a lime kiln, having a feed rate of 15 tons/hr, to 25.2 lb/hr.

The following design criteria were set forth as inlet conditions to the collector:

Kiln production	220 tons/day
Gas volume	55,000 acfm

Gas temperature	570° F
Inlet concentration	1 gr/scf
Outlet concentration	0.05 gr/scf
Resulting stack emissions	less than 11.9 lb/hr
Allowable emissions	less than 25 lb/hr

For these conditions, an arrangement was proposed consisting of two rows of four gravel bed filter modules arranged after the existing multiclones. The unit was ordered in November 1971 and placed in operation in July 1972. During engineering, several changes in the actual filter design had to be made, having to do with converting the German design to U.S. specifications and manufacturing standards.

As can easily be appreciated, there were some initial problems associated with the design and operation of a piece of equipment new in this country. The problems were relatively minor and could be addressed in a straightforward manner.

The problems generally were in the following areas: (1) dust conveying system, (2) high draft loss, and (3) condensation problems.

Problems (1) and (3) were related in that the collector was inadvertently allowed to fill with dust while it was shut down due to leakage in the isolation gate. The slow-moving (leakage) gases caused condensation. Subsequent solidification of the dust took place. During subsequent cleanout, the high dust load and solid chunks resulted in conveyor problems, and efforts were begun to eliminate the leakage in the isolation gate.

Problem (2) was caused largely by conservative design regarding cyclone performance and slight undersizing of the backwash valves. Rexnord, at a later date, gave Paul Lime two more filter modules which were installed. This resulted in a proportionately lower headloss and a better effluent.

There have been several series of stack emission tests conducted by independent laboratories, all of which indicated compliance with the Arizona regulations. The individual results are as follows:

	July 1972		
	No. 1	No. 2	No. 3
Kiln feed (tph)	15	15	15
Gas volume (cfm)	61,300	60,700	58,700
Gas temperature (°F)	350	340	370
Gas moisture (%)	7.9	6.9	7.5
Dust concentration (gr/cf)	0.02	0.014	0.016
Actual emissions (lb/hr)	10.5	7.3	8.1
Allowable emissions (lb/hr)	25	25	25
Test duration (min)	60	60	120

	September 1972		October 1972	
	No. 4	No. 5	No. 6	No. 7
Kiln production (tph)	15	15	15	15
Gas volume (cfm)	51,300	51,500	53,500	54,800
Gas temperature (°F)	360	360	375	380
Gas moisture (%)	5.6	7.3	6.0	7.2
Dust concentration (gr/cf)	0.033	0.033	0.026	0.026
Actual emissions (lb/hr)	15	15	12	12
Allowable emissions (lb/hr)	25	25	25	25
Test duration (min)	120	60	120	60

Since the first installation, Paul Lime has ordered another gravel bed filter for cleaning the gases from their No. 5 kiln, a substantially larger operation, as shown in Fig. 38-5. The design of this unit took into account all design changes made in the interim, and represented what was installed in several large industrial plants in early 1973.

To illustrate the application of the gravel bed filter on clinker coolers, the installations at General Portland, Inc., Tampa, Florida, will be discussed. Three separate units were installed at this plant on coolers No. 4, 5, and 6. (See Figs. 38-6 and 38-7.) All kilns are of the wet process. The clinker coolers for kilns No. 4 and 5 are of identical size and are No. 744 Fuller inclined-grate coolers. The No. 6 clinker cooler is of an F. L. Smidth

FIG. 38-5. Gravel bed filter dust collection system installed at Paul Lime Plant, Inc., Douglas, Arizona, Kiln No. 5.

"Folax" design, with the first section from the kiln discharge on an incline discharging into a clinker crusher, which reduces the clinker size for further cooling in the second straight section.

For cleaning the gases from No. 4 and 5 clinker coolers, two separate gravel bed filters were employed. Each gravel bed filter consists of seven filter modules of the double-bed design. Each module has a diameter of 78 in. Because of limited space requirements, both filters had to be installed on a common support structure, well above the kiln, with difficult ductwork runs. Again, because of space limitations, both filters were arranged on a common structure, so that actually two rows of seven filter modules are fitted together but with separate gas flows; each system has its own induced-draft fan. Each filter is designed for a capacity of 60,000 acfm and will permit the easy installation of additional filter modules if future production increases would so dictate.

FIG. 38-6. General Portland, Inc., Tampa, Florida, uses gravel filter dust collector system on clinker coolers No. 4 and No. 5.

The No. 6 clinker cooler dust collection system consists of two rows of six filter modules of the double-bed design. The module diameter is 110 in. The filter is arranged between the kiln burner building and an adjacent finish mill building by utilizing existing building columns as actual filter supports. The gases are vented through an induced-draft fan, which is equipped with an electromagnetic coupling for varying the speed of the fan. The filter design can treat a maximum gas volume of 180,000 acfm at 700° F under upset conditions, while the actual normal operating condition is in the neighborhood of 130,000 to 140,000 acfm at 400° F. The fan capacity is 200,000 acfm at an equivalent temperature. It should be noted that the original cyclones and induced-draft fan system were removed during the tie-in of the gravel bed filter with the clinker cooler with a minimum amount of ductwork. In the case of No. 4 and 5 clinker coolers, the existing multiclone arrangement was retained as a standby, but new and larger vents were connected to the clinker coolers from which the inlet ductwork to the gravel bed filters

FIG. 38-7. These are the 12 filter modules of the double-bed design used in the No. 6 clinker cooler dust collection system of General Portland, Inc., Tampa, Florida.

takes off in a scroll-type configuration. The combination of a large and, hence, low-velocity outlet from the cooler with a scroll exit to a smaller ductwork with a substantial amount of horizontal runs has caused no problems so far; and it would appear that this arrangement, designed for maintenance-free ductwork, is indeed satisfactory. The philosophy behind this arrangement was to create a treacherous path ahead of the actual horizontal run, so that particles which would escape the path would be fully carried in the substantial amount of horizontal ductwork.

The actual pressure drop across the gravel bed filter for the No. 6 kiln is at about 8 in.wc at a fan speed of 800 rpm, with a maximum fan speed of 1,200 rpm. This is lower than originally anticipated, which was a design uncertainty as U.S. emission standards are substantially more stringent than in Europe and Canada, where prior experience was available.

All three gravel bed filters have satisfactorily complied with prevailing air pollution control regulations, including the Federal Standards of Performance for new and substantially modified Portland cement plants. For example, tests run in December 1973 on cooler No. 4 produced the following results:

	Test No. 1	Test No. 2	Test No. 3
Legal maximum emission (lb/hr)	22.63	22.63	22.63
Measured emissions:			
gr/acf	0.0039	0.0035	0.0034
gr/scf	0.0047	0.0043	0.0041
lb/hr	2.06	1.84	1.75

From the foregoing discussion, it is apparent that the gravel bed filter is a viable industrial-scale dust collector, especially suited for hot gases laden with abrasive dusts. Ongoing research has shown that many dusts having the tendency to agglomerate can be collected in gravel bed filters; and with appropriate media choices, certain adsorptive or absorptive collection of gaseous contaminants may also be feasible and practical.

Chapter 39

DEVELOPMENTS AND TRENDS FOR SO_2 AND PARTICULATE SCRUBBERS

Robert W. McIlvaine

The McIlvaine Company
Northbrook, Illinois

Scrubber technology is by no means static. Historically, however, progress has been made on a rather unsteady basis. Considerable scrubber work was done about the turn of the century and then again in the 1940s. In the last ten years, there has been a new surge of research and development; consequently, we are now in a phase of rapid development.

New developments can be classified in three areas. One is new hardware; the second is new applications for present hardware; and a third is evaluation techniques.

NEW HARDWARE DEVELOPMENTS

Progress has been made recently in the development of scrubbers which are considerably more efficient for the energy consumed. Scrubbers can very efficiently remove above-micron-sized particles with only modest energy consumption. A pressure drop of 5 or 6 in.wg will give 95+ percent removal of all particles larger than 1 or 2 μm. For particles which are between 0.5 and 2 μm, substantial energy is required for removal. Particles in the size range 0.25 to 0.5 μm are very difficult to remove with scrubbers. There are some applications such as prilling towers in the fertilizer industry and ferrosilicon furnaces where even with several hundred inches of pressure drop across high-energy scrubbers results have been unsatisfactory.

Three new principles have shown a great deal of promise in reducing the energy consumption for many applications. One of these involves the use of the energy in the waste gas. Aronetics, now a division of Chemico, utilizes this principle in its two-phase jet scrubber (Fig. 39-1). In the Aronetics system, heat is transferred to the scrubbing liquid in a heat exchanger which precedes the scrubbing mechanism. Without the use of external energy, the water is heated to somewhere between 300 and 400° F. It is then pumped to a two-phase jet nozzle scrubbing chamber. As the water is released, it expands and partial flashing occurs. The remaining liquid is atomized; thus, a two-phase mixture of steam and small droplets leaves the nozzle at high velocity. Gases are entrained by this high-velocity mixture, and mixing and cleaning take place. The nozzle provides the motive force to move the gas through the system.

This type of scrubber has now been applied to ferroalloy furnaces and to coke oven emissions in the steel industry. Results to

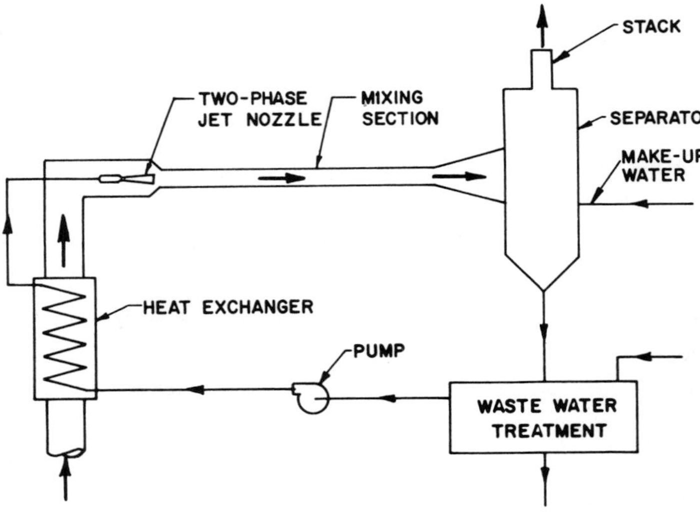

FIG. 39-1. Schematic diagram of basic Aronetics system.

date have been quite promising. The obvious drawback of this system is the need for high-temperature inlet gas to preheat the scrubbing water. There is also concern about the ability of heat exchangers to remain clean for long periods in many types of applications. In addition to Aronetics, Lone Star Steel Corp. also makes a scrubber using these principles.

Another promising potential is the use of condensation in scrubbers. Small-scale tests by Stanford Research, APT, and others have shown that the efficiency of scrubbers can be improved greatly by the use of condensation principles. Condensation scrubbing relies on particle growth followed by impaction of the particles as with conventional scrubbers. The efficiency depends greatly on the amount of water vapor that is condensed per unit mass of dry gas. It looks economically attractive where there is a hot gas to form the water vapor. Work by Seymour Calvert has shown that these scrubbers are capable of relatively high collection efficiency on fine particles. In order to maximize the condensation and particle growth effect, it is necessary to use multiple-stage or continuous-contact towers to allow for the growth of particles. At this point, there are no commercial uses of this equipment.

A third area which shows, perhaps, the greatest promise of any new hardware concept is the foam scrubbing area. As an indication of the interest in an activity in this field, there have been at least three companies which have received patents on scrubbers using foam within the last year. The advantage that foam scrubbers have is that they require very little energy. Brownian movement of the small particles is sufficient to impinge them on bubble surfaces. High efficiency is achieved by allowing the gas encapsulated in bubbles to be retained in this condition long enough for impingement to take place. Since Brownian movement of the particles increases as the diameter decreases, the smaller the particle the more likely it will be to collide with a bubble surface. This principle has long been known, but the problem has been that the surface area of liquid required for impingement by this principle is extremely large and the time requirement substantial. This means that the equipment is necessarily larger than conventional equipment.

Monsanto Research has a contract with EPA to develop a prototype of a 1,000-cfm model. Companies which have recently received patents in this area are Alfa Laval and Environeering. The foam scrubber marketed by Rexnord is shown in Fig. 39-2.

Another new development is the TRW charged droplet scrubber which uses an electrostatic spray technique for the capture of fine particles. A spray of electrically charged liquid droplets is produced and passes through an electrostatic field between the spray tube and the collector plate. As the contaminated gases pass through this field, they impinge upon the droplet by direct collision or indirect charging encounters (Fig. 39-3).

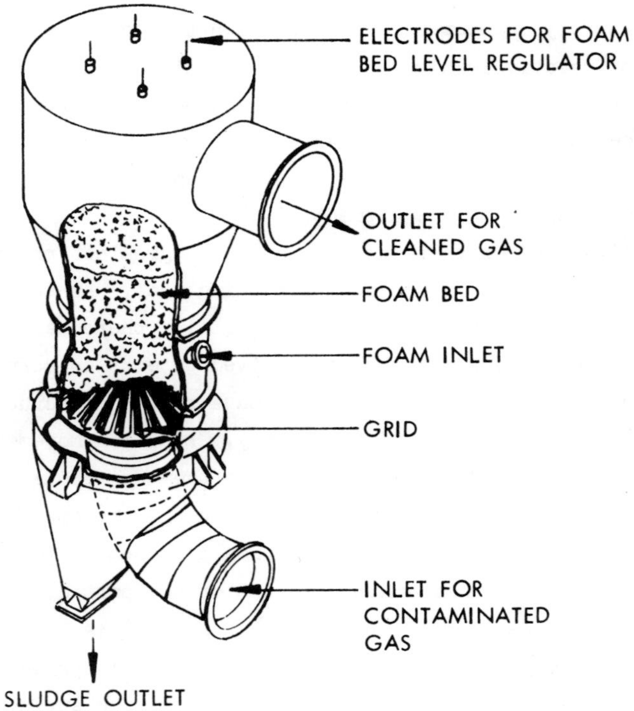

FIG. 39-2. The centrifoam scrubber.

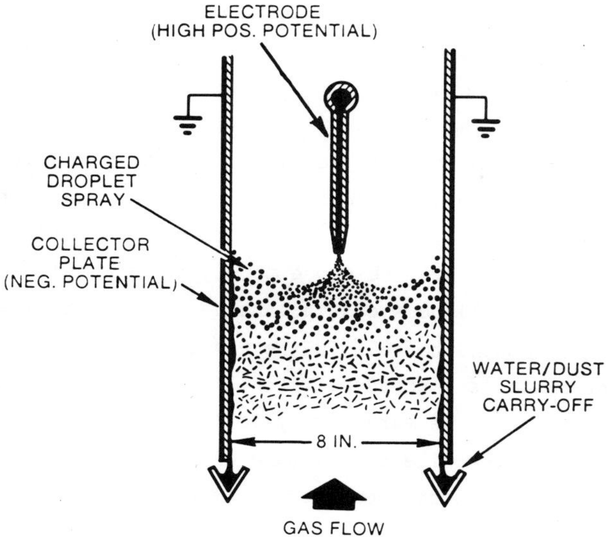

FIG. 39-3.

Tests run on the unit have shown relatively high efficiencies on particulate in the 0.1 to 1-μm range. Liquid requirements are less than a gallon per minute per 1,000 cfm. Installations are limited to inlet loadings with less than about 2 gr/scf. Where loadings exceed this figure, a precleaner or other collection device may be required. This unit is limited to 350° F, but temperature reductions to this level can be effected by spray nozzles in the inlet. A full-scale unit of 10,000 cfm is operating in Japan on a hot grinding operation in a steel mill. In the United States, a 30,000-cfm unit is scheduled to be put on stream exhausting coke oven emissions at the Kaiser Fontana plant. Another unit is being installed downstream of an SO_2 scrubber for acid mist removal in an unspecified plant.

A fourth hardware area that offers a potential breakthrough is in a high-speed moisture eliminator. American Air Filter has, in its Type R Rotoclone, designed a high-speed eliminator and achieved as a result a very compact scrubber. An even more compact scrubber has been introduced by Fisher-Klosterman. The velocity through the scrubber and the eliminator is 2,000 fpm. It consists of a tangential venturi which enters the wet cyclone at the top rather than at the bottom. As the gases pass downward, liquid is separated in a rotary spinner vane prior to the gas discharge. This unit can handle about three times the cfm that can be handled by a conventional unit of similar cubic volume. In other words, the cfm per cubic foot of scrubber internal volume will be close to 90, as opposed to the conventional 30. There will be a cost advantage also. Carbon steel scrubbers are anticipated to sell in the 23¢ to 25¢ per cfm range. If Fisher-Klostermann is successful with this high-speed eliminator, scrubber companies would be very likely to introduce high-speed designs of their own. The size of moisture eliminators has been the one factor that has prevented scrubber manufacturers from reducing the size of their equipment in the past.

APPLICATIONS FOR SCRUBBERS

There have been a number of new applications in which scrubbers have been used. The most notable is the utility SO_2 removal field. Ten years ago, practically no one, including the scrubber manufacturers themselves, considered that scrubbers would ever be used for removal of SO_2 from coal-fired boilers. Erosion problems were known to be extremely high. It was felt that utilities would never spend the money for SO_2 removal. In one short decade that situation has changed dramatically. Hundreds of millions of dollars worth of scrubbers have been purchased.

Government forecasts show a predicted purchase of nearly $1 billion per year of SO_2 systems by utilities for the next seven or eight years. Some of the problems which have been overcome in this period are as follows:

1. Corrosion. A polyester lining with glass flakes has been one of the most successful materials for corrosion prevention. 316L stainless steel has also proved effective in certain situations,

but has not proved acceptable for low-pH, high-chloride concentrations. Other successful anticorrosion techniques include the use of rubber linings in pumps, valves, and piping, and certain system design techniques such as the placement of the fan in the dry gas either prior to the scrubber or subsequent to the scrubber but after the reheater.

2. Long-term reliability. Both manufacturers and utility operators have quite a bit to learn about operating scrubber systems, but a great deal of progress has been made in this area. Louisville Gas & Electric has operated a scrubber system with reliability comparable to that of the boiler itself. Arizona Public Service has operated a scrubber system with availability and efficiency in excess of 90 percent for a period of one year.

In meeting the National Academy of Engineering's criteria for adequate demonstration of scrubber technology, the situation at Commonwealth Edison Will County Station has been of particular interest. In 1972 and 1973, the very poor availability was cited as an example by opponents of scrubbers. However, through the summer of 1974 and since, the availability was virtually 100 percent.

Other utility operating systems have shown steadily increasing availability also. Experience with high-sulfur coals has not yet been as extensive as with lower-sulfur coals, but there is no reason to suspect that the long-term reliability of higher-sulfur coal systems cannot be as good as for lower-sulfur coal systems.

Regenerable flue gas desulfurization systems dissociate the SO_2 from the additive used originally to capture it and offer the opportunity to recover the sulfur in a usable form. Sulfur can be recovered in a variety of ways; end products range from commercial sulfuric acid to elemental sulfur. These systems are not as advanced as the nonregenerable systems, although there have been successful full-scale prototype installations.

The first such installation was a magnesium oxide recovery system at Boston Edison. With this process, magnesium oxide is used to absorb the SO_2. The magnesium sulfite is then calcined at high temperature to drive off SO_2 and regenerate the MgO. The SO_2 stream is then processed in a sulfuric acid plant to make 98 percent acid. Another regenerable system which is being applied commercially is the Wellman-Lord process of Davy Powergas. The end product at Northern Indiana Public Service, where Wellman-Lord is installing a 115-MW system, will be elemental sulfur. A larger system of this type is being engineered by Wellman-Lord for Public Service of New Mexico.

A third regenerable process that has had commercial application is the Cat-Ox process by Monsanto. This process uses catalytic oxidation of the SO_2 to H_2SO_4, which is recovered as a weak sulfuric acid. The initial installation at Illinois Power of Monsanto's Cat-Ox system has been less than fully successful. Direct oil-fired reheat fouled the catalyst and has required extensive shutdowns to modify the reheat system. Subsequent attempts at operation have met with other mechanical-type problems. Unlike sulfuric acid plants designed for chemical applications that can run on a steady-state basis, acid plants for use in this application must be designed to run intermittently. It would appear that a number of the problems have been associated with this intermittent operation.

Regenerable systems that will be used in the future include the Chiyoda scrubbing process which has been widely applied in Japan and produces commercial gypsum. Since the United States is now importing millions of tons of gypsum per year, it is conceivable that there is a place for this process in the United States. In the Chiyoda process, a weak sulfuric acid is used as the absorbing solution. A portion of this recirculating acid is reacted with lime to form a high-purity gypsum. Other promising regenerable or recovery processes are shown in Table 39-1.

The advantages of recovering sulfur as elemental sulfur are primarily in the ability to store and transport the material. Elemental sulfur can be stored in open piles without undue concern about environmental contamination. In contrast, sulfuric acid does not lend itself to storage and is comparatively expensive to transport

TABLE 39-1

New Recovery Processes

Process	Product	Developer
Dry systems		
Sodium, potassium, lithium carbonates (molten)	S	Atomics International
Manganese oxide	$(NH_4)_2SO_4$	Mitsubishi Heavy Industries (Japan)
Copper oxide	SO_2	Shell, Showa, Yokkaichi Sekiyu (Japan), U.O.P.
Sodium aluminate	SO_2	U.S. Bureau of Mines, Central Electric Generating Board (U.K.)
Wet systems		
Magnesia (MgO)	SO_2	Chemico, Grillo, United Engineers
Caustic soda/soda ash	SO_2	Davy Powergas, Stone & Webster-Ionics, U.O.P., Showa Denko (Japan)
Ammonia	S, SO_2, H_2SO_4, $(NH_4)_2SO_4$	Cominco, IFP (France), Monsanto, TVA
Sodium citrate	S	Bureau of Mines, Morrison-Knudsen, Pfizer-Peabody-McKee
Potassium formate	S	Consolidation Coal
Carbon adsorption-dry contacting	S, SO_2, H_2SO_4	Chemibau, Lurgi, Bergbau-Forschung, Westvaco, Hitachi & Sumitomo (Japan), Rust Eng.
Wet contacting	Dilute H_2SO_4	Hitachi
Phosphate	S	Stauffer-Chemico

any distance. On the other hand, high-strength sulfuric acid can be used directly, whereas elemental sulfur in most cases must be converted to sulfuric acid for use.

It is generally agreed that the ultimate idealized situation is one in which combinations of regenerable and nonregenerable systems are used depending on the local market demands for sulfur and sulfuric acid. It has been shown that only a few utilities in some areas can generate sulfuric acid or elemental sulfur without materially affecting the market price.

Assuming that there will be a need by 1980 for scrubbers on 83,000 MW which would require 260 million tons of coal, there would be a potential production capability of 28 million tons of 100 percent sulfuric acid. This would satisfy an estimated 50 percent or more of the commercial demand for sulfuric acid and obviously would have substantial impact on the price. However, the traditional suppliers of elemental sulfur are taking an active role in the future of flue gas desulfurization source supply. Both Texas Gulf Sulfur and Stouffer Chemical have aligned themselves with companies having complementary capabilities so that consortiums are being formed that would handle the entire activity from removal of the SO_2 from the stack gases through the marketing of the end-product sulfur or sulfuric acid. Utilities need not invest in the capital equipment but can elect to pay only an operating charge per ton of product produced or of SO_2 removed.

One possible effect of this approach is the development of new markets for sulfur products. Availability of low-cost sulfur or sulfuric acid is likely to extend the market potential. This is important since the utilities are capable of producing more sulfur or sulfuric acid than is used presently by all sources. Therefore, most experts believe that the eventual pattern will be of a mix of regenerable and nonregenerable systems.

Nonregenerable, nonrecovery systems are systems that produce a disposable material which can be used generally as landfill. Limestone systems are by far the most prevalent of the systems being used for SO_2 removal. Lime systems and double-alkali systems using sodium and calcium as the two alkalis comprise other nonrecovery systems. In comparing these three processes, the end products are quite similar. Each has a calcium sulfite/sulfate sludge. The consistency of the sludge differs in each of these processes primarily because of the amount of the oxidation from sulfite to sulfate which takes place. In addition, in the double-alkali system, there are some sodium losses which end up with the sludge.

The limestone system involves the least costly additive, but requires greater initial investment. There are two reasons for this. First, more material has to be handled than with the calcined lime, so that the feeding system has to be larger. Second, the efficiency of limestone scrubbing is less than lime scrubbing, therefore the size of the scrubbers has to be larger. In certain circumstances, a single-stage lime scrubber may do what a two-stage limestone scrubber may do. For larger installations, the economics tend to favor limestone scrubbing; but for smaller installations, lime scrubbing is favored. This is because the capital costs for equipment do not increase proportionately with the size of the utility. The double-alkali system offers the advantage of lack of scaling in the scrubber circuit. This was a particularly important advantage in the early stages of lime and limestone scrubbing, when the reliability factor was quite low. Part of this advantage, however, has now been lost in that lime and limestone scrubbers are operating at relatively high availability. Therefore, for double-alkali systems to compete with limestone systems, the overall equipment costs will have to be nearly as economical as the reliability advantage narrows.

Part of the advantage of lime systems over limestone systems is the higher scrubbing efficiency that can be obtained. Lime scrubbing systems can easily achieve 90 to 95 percent SO_2 removal in one- and two-stage scrubbers. It may take a five-stage limestone scrubbing system to achieve equal performance. However, one new development is the use of magnesium in limestone scrubbing systems to increase efficiency. The magnesium stays in the recirculated scrubbing liquor and increases the effectiveness of the limestone without being consumed.

The addition of magnesium can also improve lime scrubbing systems, as witnessed by the performance of the Duquesne Light scrubber system using thiosorbic lime. The single-stage venturi scrubber achieved over 90 percent efficiency with this additive.

Costs

In a study for EPA, TVA has analyzed the comparative costs of various SO_2 removal systems. Information is shown on a 500-MW system in Table 39-2. It was concluded that costs for a 200-MW system would be approximately 30 percent higher in initial capital investment and that costs for a 1,000-MW installation would be about 25 percent lower per dollar invested per kilowatt. Comparison shows that magnesia and Cat-ox systems could be competitive with lime and limestone scrubbing systems with by-product credit of $8/ton for 100 percent sulfuric acid and $6/ton for the more dilute acid produced with the Cat-ox system.

The utility scrubber situation can be summarized by stating that reliability and sludge disposal are no longer major issues. Issues which are still being hotly debated are the very substantial costs, the need and the timing. There is a question whether SO_2 is as dangerous a contaminant as previously thought. The greatest debate now centers on the capability of scrubber manufacturers and the utilities to install systems. Therefore, it would appear that the installations of these $7 billion worth of systems will be staggered through 1985.

Another area where scrubbers are being successfully applied is in odor removal. The shortage of natural gas and its cost have made the scrubber an attractive alternative to incineration for odor control. The operating costs of earlier scrubber systems were high because they used expensive additives like potassium permanganate exclusively. Newer scrubber designs are more sophisticated and often use combination additive systems. One system is a three-stage scrubbing system in which the chlorine gas is injected ahead of the first scrubbing stage. A 5 to 10 percent solution of sulfuric acid is

TABLE 39-2

Annual Operating Costs
500-MW Plant--New

Process	Cap. costs $ per kW	Total $ x 10^6	Without by-product credit: $ per ton coal	Without by-product credit: Mills per kWh	Without by-product credit: $ per ton S removed	With by-product credit: By-product CR $ x 10^6	With by-product credit: $ per ton coal	With by-product credit: Mills per kWh	With by-product credit: $ per ton S removed
Limestone	50.3	7.702	5.87	2.20	215	–	–	–	–
Lime	44.8	8.102	6.17	2.31	226	–	–	–	–
Magnesia	52.8	9.211	7.02	2.63	255	0.880[a]	6.34	2.38	230
Wellman-Lord	61.0	11.602	8.84	3.31	323	0.818[b]	8.20	3.07	300
Cat-ox	85.0	8.873	6.76	2.54	247	0.660[c]	6.25	2.35	229

[a] 111,000 tons of 100 percent H_2SO_4 at \$8 per ton = \$880,000.
[b] 32,700 tons of elemental sulfur at \$25 per ton = \$817,500.
[c] 141,000 tons of 60-deg-Bé H_2SO_4 = 110,000 tons of 100 percent H_2SO_4 at \$6 per ton = \$660,000.

used in the first scrubbing stage, followed by an entrainment separator. The second scrubbing stage uses a caustic soda solution. Using this combination system on rendering plants, removal efficiencies of 96 to 99 percent have been achieved. The odor intensity was reduced from as high as 32,000 odor units to 300 units. Combination systems have also been used where potassium permanganate is used in the final stage to maintain the highest possible efficiency with the least consumption of permanganate.

Another new application for scrubbers is in coke oven emission control. Fumes from coke ovens have been very difficult to eliminate. During the pushing operation, smoke billows out in large quantities for a relatively few seconds, but then there is no further emission until another oven is pushed, so the emissions are quite sporadic. They consist of combinations of large and small particles. To date, a number of system designs have been used. One uses a shed-type hood which encloses the entire oven area; other designs use traveling hoods that control only the pushing emissions. Venturi scrubbers with 20- to 50-in. pressure drop have proved effective in the elimination of the visible plume. The high-temperature two-phase jet scrubber previously discussed has also been successfully applied to this application.

Scrubbers have a bad image in asphalt batching processes because most of the early scrubbers used were low-energy scrubbers and had emissions as high as 0.3 gr/scf. In fact, for a long while the standard air pollution equipment was a 5- or 6-in. pressure drop scrubber on an aggregate dryer. Consequently, when control agencies demanded higher efficiency, they generally favored the installation of baghouses which would have relatively high efficiencies. It was often overlooked that venturi or other high-efficiency scrubbers would also give good results. Whereas the baghouse would provide uniformly high efficiency, the venturi scrubber would perform at higher or lower efficiency depending on the aggregate mix. It is this necessity of more complete analysis of the problem to determine just how much pressure drop is required that has probably been a stumbling block in the use of scrubbers on asphalt plants.

SCRUBBER IMPROVEMENTS IN THE DEVELOPMENT OF NEW PREDICTION TECHNIQUES

There are too many scrubber systems installed to remove particulate that do not achieve the emissions that were expected or guaranteed. The crux of the problem is that particle-size analysis techniques are unreliable, and even if they were reliable, they cannot be used to predict the operation of scrubbers. To circumvent this problem, several scrubber manufacturers have designed small pilot units. Universal Oil Products and Environeering are two such manufacturers. Environeering introduced the Dust Difficulty Determinator in 1968. It is an 80-cfm unit and has been used in hundreds of applications to predict the performance of large-scale scrubbers. Even though it is an 80-cfm pilot unit and an orifice scrubber, it has been used very successfully to predict the performance of large-scale venturi-rod scrubber systems. This is possible because virtually any scrubber will perform equally if the pressure drop is the same and the energy is used efficiently. Universal Oil Products has used a 100-cfm venturi scrubber in a similar manner.

The most recent development in this field is the use of a 1-cfm pilot unit called the McIlvaine Mini-Scrubber. Early work with this unit shows that it can be used to predict the performance of large-scale scrubbers. The advantage of this device is that it is small enough to be used in a standard sampling train, for it eliminates the large expense of pilot units. It could prove to be as important to scrubbers as horsepower is to motors. The method of use is to draw a sample from the exhaust gas. The rate of flow can be varied to obtain isokinetic conditions. The gas passes through a dropout chamber and then into the Mini-Scrubber where the particulate is removed. The exhaust passes through an absolute filter where residual dust is captured. By comparing the residual dust leaving with the pressure loss across the orifice of the Mini-Scrubber at a number of different pressure levels, a curve can be drawn relating pressure drop to residual emissions. Then, based on the other information available for the process in question, an optimum pressure drop selection can be made.

The purchaser would normally approach the supplier with specifications that the supplier's scrubbing device must equal in efficiency the Mini-Scrubber operating at "X" pressure drop. The vendor, knowing that his device has approximately the same capability at the same pressure drop as the McIlvaine Mini-Scrubber, should select a pressure drop slightly higher than that in the specifications relating to the Mini-Scrubber to allow a safety factor. This factor would compensate for some mist carryover and other minor problems associated with commercial-sized units. After installation, the purchaser would then require that a demonstration of efficiency performance be made by operation of the Mini-Scrubber in parallel with the production unit. If the production unit shows a lower residual emission than the pilot Mini-Scrubber, the guarantee is not met.

The concern over this approach is that the purchaser does not have a guarantee to meet the code, but this is the way that most equipment is purchased. One does not receive a sweeping guarantee in purchasing a motor that it will move a given fan. One reserves only a guarantee that it will develop 500 horsepower. It is incumbent upon the purchaser to determine in advance what size motor he wants before he talks to a motor manufacturer. It is easy to visualize the additional cost of motors to cover guarantees if they were sold to perform certain operations instead of by horsepower rating.

The McIlvaine Mini-Scrubber furnishes the purchaser with the same measuring device as does the horsepower system in motors. Two motors guaranteed for 500 horsepower are assured to be equivalent. Two scrubbers guaranteed to be equivalent in performance to the Mini-Scrubber can be assumed to be equivalent. But two scrubbers guaranteed to meet a code might be vastly different in performance.

The customer's process holds all the variables which ultimately determine the pressure requirement. These variables can be controlled to a greater or lesser degree only by the customer. If the customer asks the vendor to bear risks involving variables over which he has no control, this vendor is forced to add contingency pricing to his quotations to cover his risks. It is therefore more economical for the customer to assume the responsibility of determining how efficient a scrubber is needed for the process. With the Mini-Scrubber, he has the necessary tool.

Chapter 40

SPECIAL CONSTRUCTION MATERIALS FOR SCRUBBERS

Murray Borenstein

Environmental Elements Company
Division of Koppers Inc.
Ramsey, New Jersey

The air pollution engineer is frequently faced with the challenging problem of selecting materials of construction for wet scrubber systems. Processes that are severe polluters are also processes that are usually very corrosive and at very high temperatures. For example, at high temperature chlorides are the most volatile of the inorganic chemicals, and are also one of the most corrosive to metals when wetted. The problem is further compounded by the presence of severe abrasion that one would expect to encounter in scrubbers. Thus, even mildly corrosive liquids become severely corrosive when the factor of abrasion is added. The various parameters which the engineer then must take into account when selecting equipment include chemical attack, abrasion, and temperature. All the while he must bear in mind keeping costs reasonable, especially for systems which usually do not yield a return.

Let us take a hypothetical case in which the air pollution control engineer must design a scrubber system incorporating the conditions mentioned above. Let us assume that the exhaust gas evolves from a rotary kiln processing an ore which contains chloride. The kiln is fired by No. 6 fuel oil containing 1-1/2 percent sulfur, and the temperature of the exhaust gas is 1,500°F. The user has specified that the exhaust gas will contain some HCl resulting from the decomposition of the metallic chlorides, iron oxide dust, nonferrous metallic chloride fume, and SO_2 from fuel oil combustion. The nonferrous fume has some valuable components such as copper which the user would like to recover. He also must meet the local code for particulate of not over 0.05 gr/scf and not more than 500 ppm of SO_2.

The engineer must now use all his resources to design a system to handle this severe duty. He immediately decides that the presence of HCl and other chlorides rules out the use of any of the stainless steels. While there are metals that possibly can take this service such as the Hastelloys, titanium, or high-nickel alloys, they are too expensive to be used as the basic material of the scrubber. The next obvious choice would be the nonmetallic materials such as rubber-lined steel, polyester fiberglass, or fiberglass- or other organic-coated steels. However, the 1,500°F temperature sets off a warning bell against use of these materials.

The engineer might then consider cementatious materials such as lumnite cement or other acid-resistant castables. These cement linings, although acid resistant, unfortunately cannot handle extremely low pH, as will be experienced in this system. Any slight crack in the lining will result in corrosion, with the acid attacking the metal with disastrous results. Thermal shock or differences in coefficient of expansion could cause such cracking. Furthermore, somewhere in the system there will be an interface at which the 1,500°F gas meets the wetted walls of the scrubber. At this point, any castable or cementatious lining would be subject to severe spalling; any organic coating would be subject to heat attack; and any metal would be subject to corrosion as well as thermal shock.

A scrubbing system has been designed to handle the problem as described above. First, the gas is cooled down to the saturation temperature, which will generally be in the temperature range of 160 to 180°F for high-temperature gas streams. Once the gas is cooled, it can be handled with conventional materials of construction, such as rubber lining or polyester fiberglass. The key to this system is then the cooling device, which can be referred to as the quench tower. This is a relatively large vessel in which the gas enters a nozzle at the bottom and flows upward countercurrent to falling sprays of water from spray nozzles located at the top of the tower. The vessel is sized to permit about 1-1/2 sec retention time for the gas. Thus for a 20,000-acfm gas stream, the volume of the vessel is 20,000 divided by 60 times 1.5, which equals 500 ft^3 with the proportion of length to diameter being about 2:1. The quench tower is constructed of carbon steel lined with rubber, covered with acid-proof brick. The brick is grouted against the rubber wall by acid-resistant cements, such as epoxy or Furan-based resins.

The purpose of the brick is to protect the rubber from heat, while the rubber membrane protects the steel from acid that will seep behind the brick wall. The sprays at the top of the tower inject massive amounts of water down through the rising gas stream, far in excess of the theoretical amount required to cool the gas by evaporative cooling. This assures that the brick walls of the vessel are always wetted (except at the inlet nozzle) and that the saturation temperature will be reached. The quench water is then recycled (Fig. 40-1).

The inlet gas nozzle is the critical part of the quench tower since this part is subject to the hot gas stream (Fig. 40-2). The construction of this nozzle is steel, rubber lining, a layer of foam glass or rockwool insulation, then acid-proof brick. The insulation provides a heat barrier to protect the rubber. The engineer should calculate the temperature at the face of the rubber using the heat-transfer coefficients of the brick and the insulation; then use a thickness of insulation such that the limitations at the surface of the rubber will not be exceeded. It is not critical if the temperature at the surface of the rubber exceeds the limitation during the operations. Minor charring at the inside surface of the rubber will not be a problem, since this area is always dry and therefore not subject to corrosion.

The gas stream traveling through the nozzle at a velocity of upwards of 100 fps assures that the water will not back up into the

FIG. 40-1. Top of quench tower. Two independent spray headers are shown: one for recycle water, one for emergency cooling water.

nozzle, at least not while the fan is on. The nozzle is tilted down at an angle of about 5° so that when the fan is shut down, water will not travel down the hot brick to cause spalling. Acid-proof brick has good resistance against spalling; but some minor spalling may occur at the face of the brick next to the inside wall if there are frequent fan shutdowns while the gas is still hot. In this case, repairs can be made by using an acid-resistant castable cement. Even this condition may be prevented if an air-bled damper is used to bleed in cooling air just prior to fan shutdown. Acid-proof brick has excellent resistance to all inorganic acids except HF, and to mild alkali conditions.

The brick lining extends up the vertical wall in the quench tower, but not on the roof since the bricks would have to hang upside down. At this point the gas is already quenched to its saturation temperature; therefore, brick is not needed at the top. The

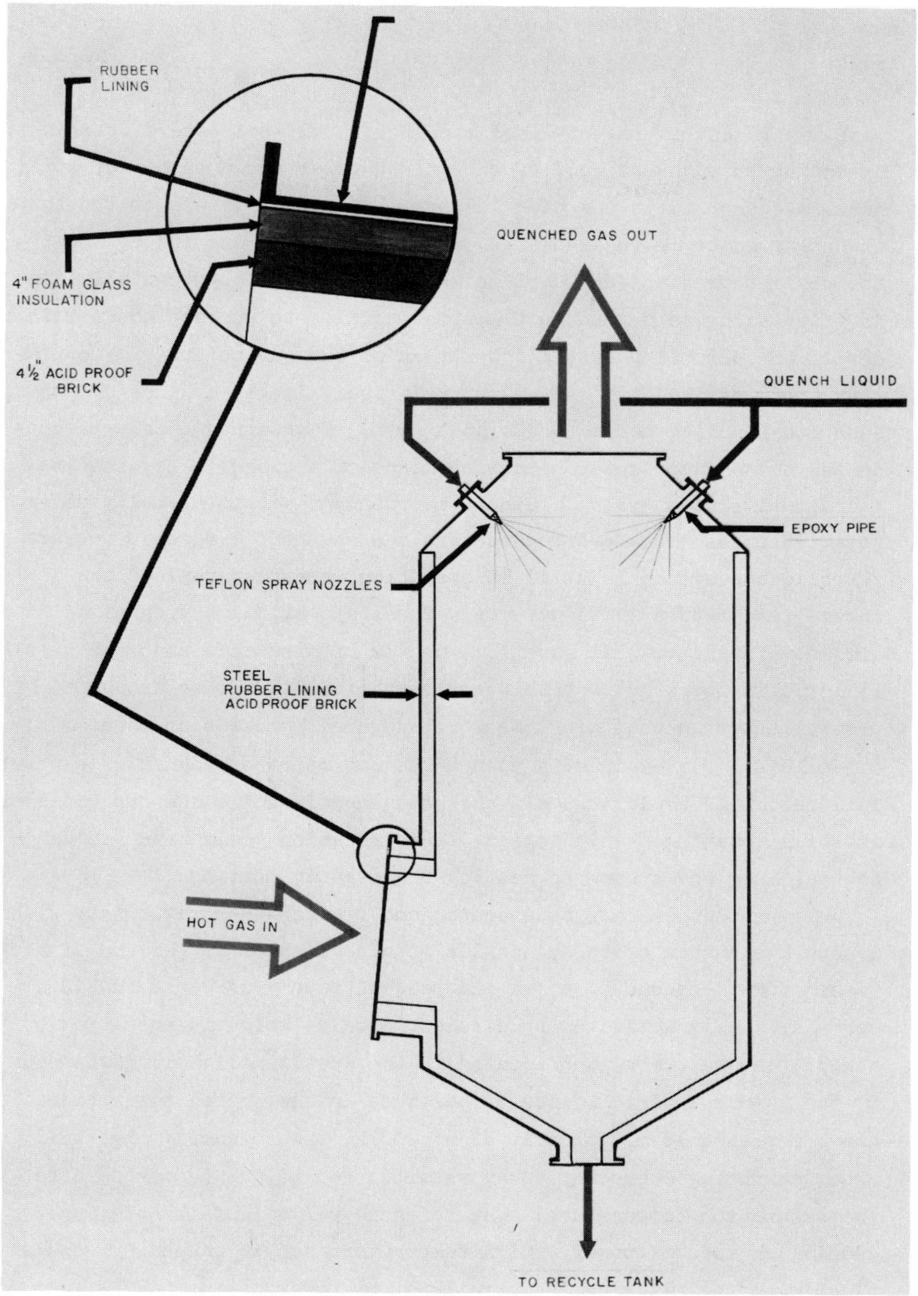

FIG. 40-2. Cross section of quench tower. Special construction features of the inlet nozzle are shown.

roof may be either rubber-lined steel or an all-polyester fiberglass. Extending through the roof of the quench tower are a number of spray nozzles connected to the pipe header. Previously it was stated that the spray water is recirculated, which might make one suspect that the nozzles are subject to plugging. This could be potentially disastrous, since we depend on the spray nozzles to protect the remainder of the scrubbing system downstream of the quench tower. To prevent this possibility, a few, but relatively large, nozzles are used, each one handling about 50 gpm at 15 psi. Each spray nozzle is connected to the main spray header by means of a rubber hose with an individual shutoff valve. Since the quench tower is normally under negative pressure, a maintenance man can shut off a nozzle, retract it while the system is still in operation, clean or replace the nozzle, and return it to service. Thus, a continuous program of nozzle maintenance will give some degree of assurance that the nozzles will remain operable; but this, of course, would not be enough to protect a system that might be worth several hundreds of thousands of dollars. For example, a pump might not operate because of a power failure. This would shut off the water supply while the fan impeller is still rotating at high speed. This rotation could bring enough hot gas into the system to destroy the organic coatings.

To protect against this contingency, a second spray header with a separate system of spray nozzles is also located at the top of the quench tower, connected to an independent supply of water. This spray system is activated by a solenoid valve which is energized by a high-temperature switch located in the duct immediately downstream of the quench tower. If the temperature in the quench tower rises above a preset level, usually about 200°F, the solenoid valve will open, admitting emergency spray water to reduce the temperature to the saturation temperature. The solenoid valve is fail-safe type, opening on loss of power. High temperature can be caused by nozzle plugging, pipe plugging, pump failure, or power failure. The emergency water will stay on until the situation is corrected, at which point it will shut off (Fig. 40-3). An operator at the control panel will be alerted to this situation by an alarm. At the panel board,

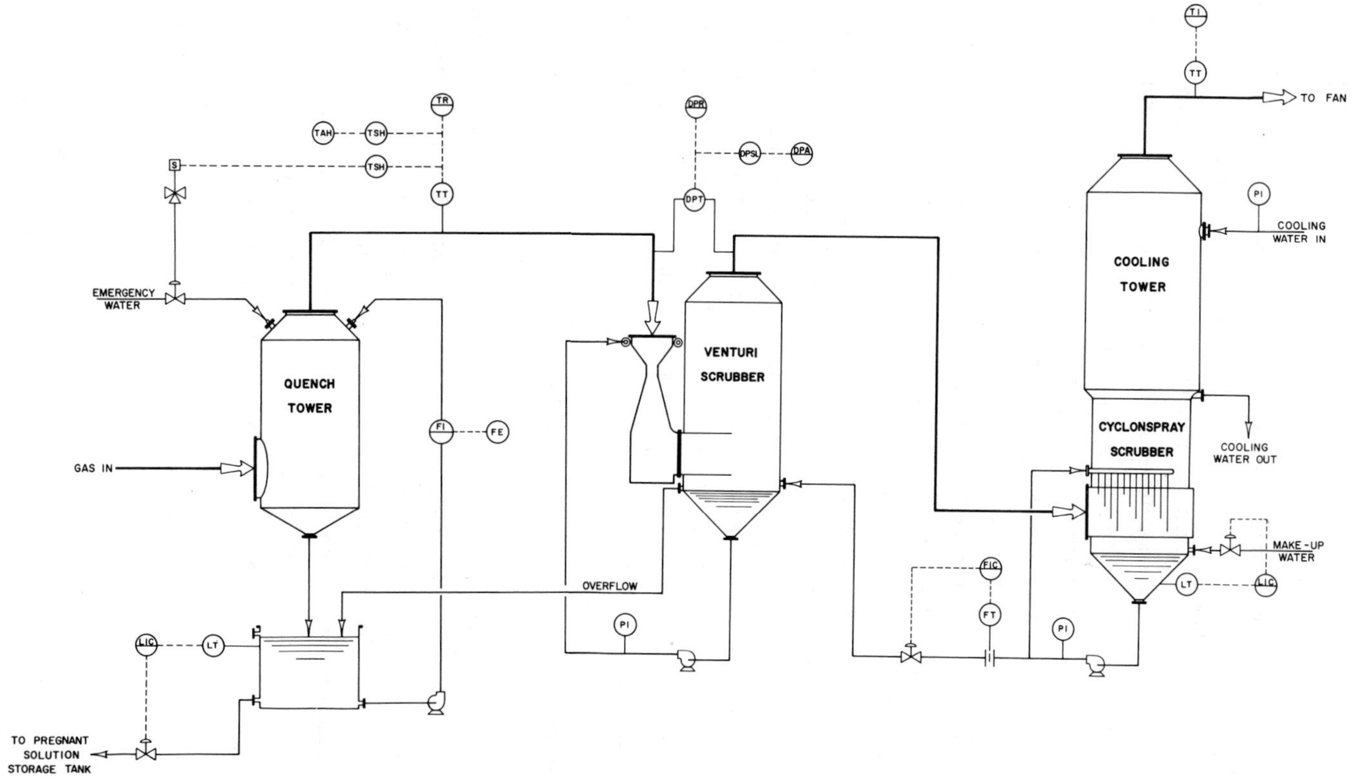

FIG. 40-3. Instrumentation associated with a scrubbing system that may be required to protect the materials of construction.

the flow rate to the quench tower and the temperature should be recorded so that if there is gradual plugging of the spray nozzle, it will be indicated by a gradual dropoff in the flow rate. The operator can then notify the maintenance department that the nozzles need cleaning, thus averting an emergency shutdown.

In this scrubbing system, the quench tower serves another purpose in addition to cooling the gas stream. Since the exhaust gas of the kiln contains vaporized solids, this vapor should be condensed into particulates before entering a high-energy scrubber. Condensed particulate (fume) is present in the submicron particles, requiring relatively high pressure drop for efficient cleaning. The quench tower, while it may be considered a scrubber, will do little to remove these particulates. Therefore, the gas enters a high-energy scrubber, which in this case is a rubber-lined venturi scrubber. This scrubber operates at a pressure drop of 40 in.wg.

Selection of the type of rubber for both the quench tower and the venturi scrubber is an extremely important decision. Soft rubber is flexible, exhibits good resistance against cracking, and has excellent resistance against abrasion. It also has very good bonding characteristics. Its weakness is that it is easily punctured, and has low resistance against gaseous diffusion. This last characteristic is significant in scrubbers where corrosive vapors such as HCl and SO_2 are frequently encountered. Hard rubber has excellent resistance against diffusion, but it is somewhat brittle and has relatively poor resistance against abrasion.

A combination of the desirable properties of rubber has been achieved by laminating three layers into a single sheet, with two soft layers on the outer surface and a hard layer on the inside. The outside layer against the steel provides good adhesion, the internal layer provides a good barrier against gaseous diffusion, and the inside layer gives the desired qualities for abrasion resistance. After the vessel is lined with rubber, it should be steam cured. If the vessel is too large to fit into an autoclave, it can be steam cured by covering all the openings and injecting steam into it for a period of about 24 hr.

While natural rubber is resistant to most inorganic chemicals, it will be attacked by traces of certain organics. Neoprene can be used as a lining and will resist most of the organics where natural rubber will fail; but it cannot resist strong halogen acids. When selecting a rubber lining, it is best to contact a representative of the rubber manufacturer for recommendations.

One of the advantages of a rubber lining over other organic coatings is that it provides a good, thick membrane, usually 3/16 to 1/4 in. thick. One of the disadvantages of rubber is its low resistance to heat. A saturation temperature of a gas stream of 185°F is too hot for a rubber lining. It would be better to consider other organic coatings such as glass flake-reinforced polyester resin, polyurethane coatings, or baked-on epoxy or phenolic coatings.

The advantages of a polyester-reinforced lining over rubber lining is that it can take temperatures of up to 220°F while displaying excellent chemical resistance. These linings are formed with highly resistant polyester resin reinforced with flakes of glass. They are applied by spray or troweled on in two layers, giving a thickness of about 80 mils. The glass flakes are only a few microns thick and about 1/8 in. in diameter. Millions of these flakes present an intricate barrier which, in combination with the cross-linked polyester resin, gives a hard and impervious membrane. These coatings are easy to apply, easy to repair, and cure quickly.

Certain areas of scrubbers have liquids in extreme turbulence with severely abrasive conditions, for example, the throats and elbows of a venturi scrubber. Scrubbers with polyester-reinforced coatings wear out in these critical areas after even a few months of extreme service. When the design engineer faces a problem where he cannot use any other material because of high temperatures or corrosive conditions, he can use organic coatings but should cover the critical areas of the scrubber with acid-proof brick. This combination of materials should stand up well to the most corrosive and abrasive conditions.

In our hypothetical two-stage scrubbing system we have removed particulates and most of the HCl from the gas stream and have concentrated the product for recovery. However, the gas stream still contains much SO_2 and some residual HCl, which would be highly objectionable if released to the atmosphere. We can add a third-stage packed tower absorber, constructed of polyester fiberglass, which will scrub with a caustic solution to absorb the SO_2 and HCl. The packed bed will contain polypropylene packing and either a polypropylene mesh demister pad or an FRP Chevron-type demister. Although the gas stream originally contained substantial particulate, we are not concerned about plugging the packing at this point, since the particulates have been scrubbed out.

Any of the standard chemical-grade polyester or vinyl ester resins would be suitable construction materials at this point. For outdoor installation, the fiberglass should include an ultraviolet inhibitor. Without such an inhibitor, solar radiation will cause degradation of the plastic. High-energy radiation causes chemical change in the polymers, with the result that tensile strength, percent elongation, and impact strength will be reduced. Discoloration and surface crazing are also indicative of ultraviolet attack. The ultraviolet inhibitor has the ability to absorb ultraviolet rays and dissipate these rays in such a manner that degradation does not occur. The ultraviolet inhibitor need not be pigmented, but may be clear so that it does not obscure the resin for inspection purposes.

Since the packed tower is located downstream of the high-energy scrubber and before the fan, it is under relatively high vacuum. Polyester fiberglass, while being moderately good in tensile strength, is poor in compression. In order to withstand the vacuum conditions of about 60 in.wg negative, the structure will have to be stiffened.

In the actual installation previously mentioned, Hastelloy pumps failed in a short time because of the severe chemical attack of the combination of acids and abrasive conditions. At great expense, the Hastelloy pumps were replaced with rubber-lined pumps. These pumps stood up well in service. This should illustrate a prime lesson to

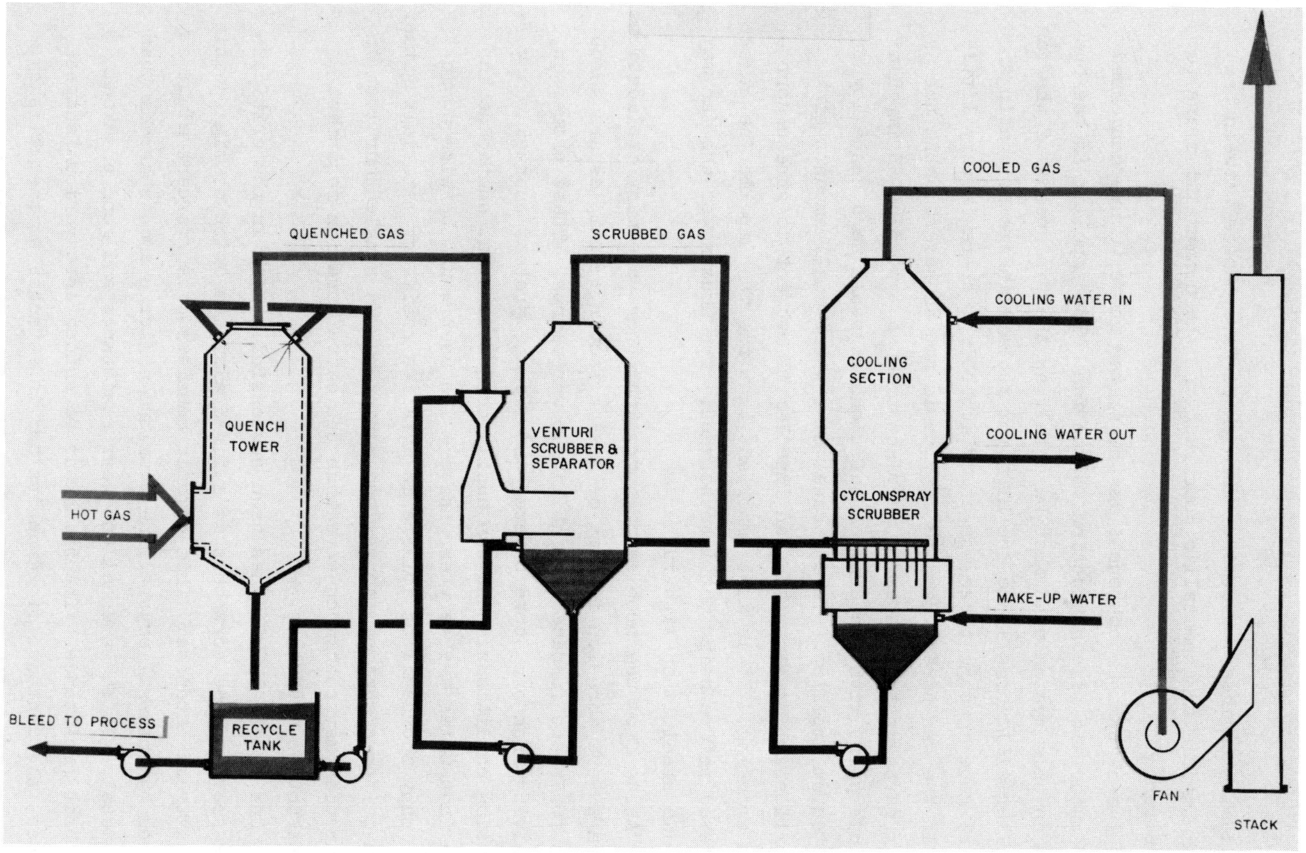

FIG. 40-4. Three-stage scrubbing system. Arrangement of quench tower, venturi scrubber, and gas cooling tower is shown.

the design engineer, that is: when there is a choice between the exotic metals or nonmetallic construction, the nonmetallics are to be preferred.

The next major component in the system where we are concerned with materials of construction is the fan. The impeller is the one item in this system that must be metal, as the tip speeds of the fan impeller will normally be too high for rubber linings or coatings. Having gone through three stages of scrubbing, there is very little HCl in the gas stream, so the service is not so severe as in the other parts of the system. One plant tried a type 316L SS impeller at this point with success. A safer choice would be a Hastelloy or a titanium impeller. In any case, even where small amounts of chlorides are present, the fan impeller is subject to stress corrosion cracking. It is most important that the fan impeller be stress relieved before being put into service. The housing of the fan may be mild steel lined with rubber.

Aside from the problem of corrosion, there are many applications where severe abrasion is a major design factor. Examples are scrubbers for blast furnaces and utility power plants. These processes, involving iron ore, coke breeze, and flyash, present unusually severe abrasion conditions. One can easily conjecture that shutdowns in such industries are very expensive. The user would be wise to invest many thousands of dollars in abrasion-resistant materials rather than suffer a shutdown to replace equipment. A single hour lost in production may cost much more than the extra abrasion-resistant construction materials.

One material which has been successfully used in the critical wear areas of the scrubbers is a fired aluminum brick of high density and extreme hardness. This ceramic consists of 96 percent alumina bonded by internal crystalline aluminum silicate and is harder than any metal. It shows excellent wear resistance and good chemical and heat resistance. The brick can be molded into various shapes to meet the contours of scrubbers and is attached to the walls by epoxy cements. It is usually made in thicknesses of about 1 in.

Other types of abrasion-resistant brick which have been used extensively, particularly in blast furnace scrubbers, are silicon carbide brick. Not as hard as alumina brick, the silicon carbide nevertheless has been used very successfully for abrasion-resistant service. It is installed in thicknesses of 3 in. and is grouted into place over T-clips.

Chapter 41

HOW TO AVOID SCRUBBER CORROSION*

Thomas G. Gleason

Swemco, Inc.
New York, New York

Scrubbers are used for a wide variety of process gas cleaning operations, as well as for the control of air pollution. Scrubbers provide efficient removal of fine particulate matter and also absorb acid gases at high efficiency.

Most high-efficiency scrubbers quench the gas to its saturation temperature. As gas humidification takes place, serious corrosion problems can develop. Acids are formed as the gas is wet. These acids can rapidly attack alloys such as type 304 and 316 stainless steel, as well as the higher alloys such as Alloy-20 and Hastelloy-C. In addition to the scrubber, auxiliary equipment such as fans, pumps, and heat exchangers are subject to attack.

Corrosive attacks are accelerated by high gas inlet temperature, high solids and acid content of recycle liquor streams, as well as by high gas and liquor velocities often necessary to effect good scrubber performance. Scrubber design has an important bearing on corrosion resistance, and the system process flow can also have critical effects on corrosion of materials.

In industrial scrubbing operations, the gases of principal concern are SO_2, SO_3, and HCl. These three acidic components are present in many scrubber applications, and as they enter the liquid phase,

*From Chemical Engineering Progress, 71:3 (1975). Reprinted with permission of American Institute of Chemical Engineers.

difficult corrosion problems begin. The acids formed are sulfurous, sulfuric, and hydrochloric.

When gases are brought directly into the scrubber (Fig. 41-1), an ill-defined junction of wet and dry gases results. This can produce a haphazard interface between the gas stream and the scrubbing liquor. Liquor is readily pulled into the inlet duct by eddy currents, causing high localized acid concentrations. Solids builup can also result, which requires excessive inlet duct maintenance.

To avoid the disadvantages of this type of entry, gases can be prequenched in the scrubber inlet duct (Fig. 41-2). By prequenching the gases, they are already saturated when they enter the scrubber, and the interface and eddy current problems are avoided. Prequenching is achieved by spraying the gases at moderate pressure as they travel vertically downward within the inlet duct. The duct can be protected from acid concentration by a flow of scrubbing liquor along the duct wall.

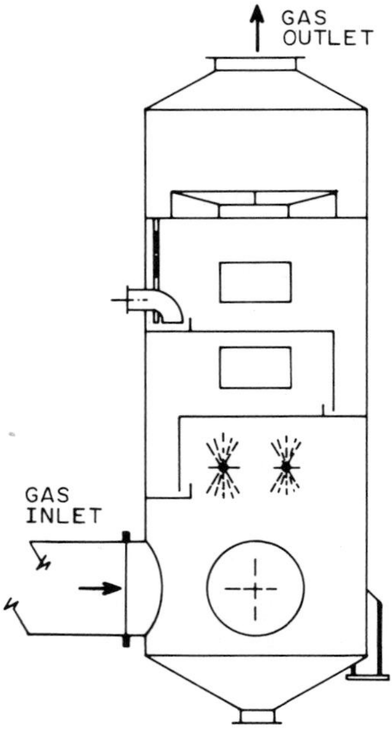

FIG. 41-1. Tray scrubber.

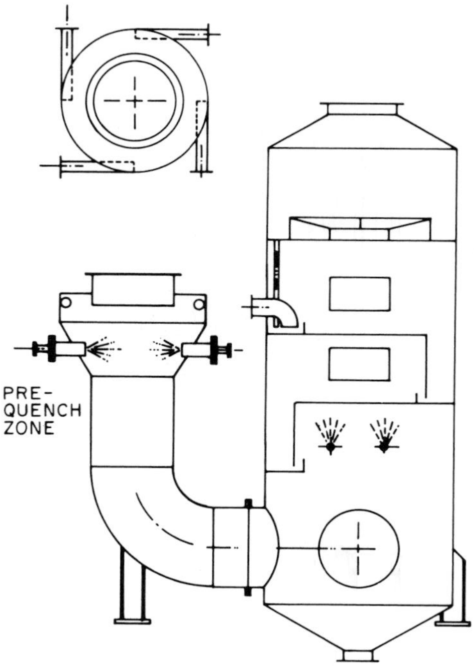

FIG. 41-2.

Tangential feed pipes mounted above the sprays provide a spinning flow of recirculated liquor which protects the duct, particularly opposite the spray nozzles. This prequench arrangement also removes some solids before the gas stream reaches the scrubber. Quench liquor flows into the scrubber and discharges through the scrubber drain.

Figure 41-3 shows a typical venturi scrubber. Gases travel vertically downward and accelerate rapidly as they pass through the converging throat and intimately mix with scrubbing liquor. Scrubbing liquor is injected tangentially above the throat.

Corrosion above the throat is avoided by the flow of liquor along the wall of the duct. The throat itself and the area below the throat can, however, be attacked by acids in the high-velocity gases. Many alloys that are normally corrosion resistant do not

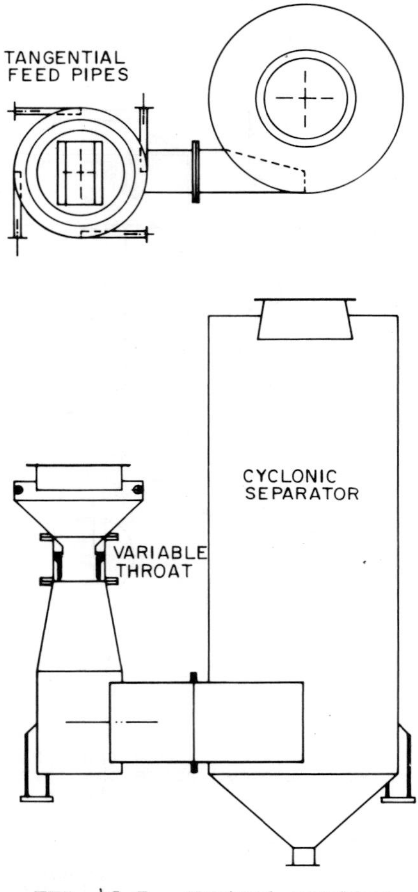

FIG. 41-3. Venturi scrubber.

hold up in this area. The high gas and liquor velocities do not permit a stable protective film to develop on the surface of the metal and it quickly becomes vulnerable to an acid environment. Also particularly vulnerable is the elbow that directs the gas stream into the cyclonic separator following the venturi.

The elbow is protected by tee construction which provides impact surface containing a pocket of liquor which safeguards the elbow from corrosion/erosion attack.

The tangential entry to the cyclonic separator is subject to particularly severe corrosive conditions due to the high entry velocities, which range from 3,500 to 5,000 fpm. In some cases--cleaning gases from sewage sludge incinerators, for example--corrosion can be avoided by using radial entry and relying on alternative entrainment separation equipment higher up in the tower. Radial vane, chevron packing, and other types of separators are available for this purpose.

Whenever reasonable service life can be expected, alloys are preferred for handling corrosive gases. They are lower in cost than brick lining, they can take temperature upset and scrubbing liquor failure, and they permit easier equipment access and maintenance.

In many cases, however, stainless steel alloy construction is an unacceptable risk. In such instances part or all of the scrubber must be lined with brick. Venturi throat and cyclonic separator inlets are scrubber zones which often require brick lining.

A brick lining system employs acid-resistant brick; lead, rubber, or plastic inner lining; and carbon steel supporting structure. The brick provides an acid-resistant, abrasion-resistant barrier and furnishes thermal protection for the waterproof, acid-resistant inner lining. Acid brick is permeable to corrosive acids and the inner lining prevents attack of the carbon steel supporting structure.

Bricks, like stainless steel alloys, have their limitations. Red shale and fireclay acid brick cannot take temperatures above 800 to 1,000°F under gas quenching conditions.

When droplets of scrubbing liquor impinge intermittently on acid brick at inlet gas temperature above about 800°F, spalling can take place. Silicon carbide brick has the best spalling resistance. It will take gas temperatures of 2,500°F or more, but its high installation costs limit its use to critical high-temperature gas-liquor contact zones. Carbon brick is used when fluoride is present in the liquor. However, it has a temperature limitation of 1,000°F. Further, it cannot be placed in direct contact with lead lining because galvanic action will take place.

Acid brick must be carefully applied in combination with the mortar and the backup membrane. When 60- to 80-mil-thick fiberglass-reinforced polyester lining is used, an additional protective layer of material such as asbestos must be employed. Similarly, an additional membrane--Teflon, for example--is required to prevent damage to the lead when carbon brick linings are used.

Brick linings restrict the shape of the scrubber vessel, greatly increase weight, make vessel access difficult, are costly to install, require temperature control and safety backups; and prevent field welding of equipment. Despite these serious limitations, brick linings frequently are vital to the critical zones of the scrubber.

Figure 41-4 shows a metallurgical gas purification system involving a gas stream containing 3.5 to 10 percent by volume of sulfur dioxide at temperatures of 700 to 1,800°F. Gases of high sulfur dioxide content are vented from roasters, copper converters, and acid sludge furnaces. Entrained solids are removed and the gas is cooled in a scrubber-cooler. This is to prepare the gas for the sulfuric acid plant downstream.

In the scrubber-cooler system, an upper tray section is supplied with fresh water. Liquor leaving the upper trays is recycled by a pump to sprays in the lower, humidifying section of the scrubber-cooler. Effluent is stripped of its dissolved sulfur dioxide and recovered. The effluent is further treated to remove solids and neutralize acids before it is discarded. Solids concentration is low and chloride and acid concentrations are usually at a minimum. Therefore, 316 ELC is used throughout for the upper cooler. The lower section is refractory lined due to the high inlet gas temperatures.

Table 41-1 shows materials of construction used on once-through systems involving scrubbing liquors having various chlorine contents. Above 100 to 150 ppm chloride, 316 ELC can show poor service life and Alloy-20 is used with good success. When chloride concentrations are in the 3,000 ppm range and sulfuric acid and solids concentrations are low, Alloy-20 has marginal resistance. Hastelloy-C is used in

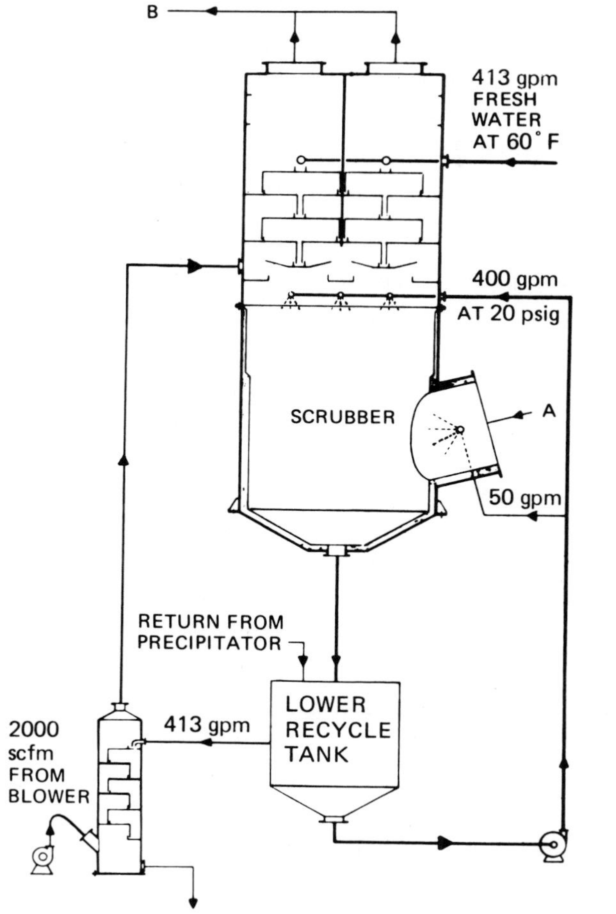

A. INLET GAS CONDITIONS
660°F
9.6 psia
12,000 scfm dry

D.G.	1,036	lb/min
H_2O	126	lb/min
SO_3	2.6	lb/min
SOLIDS	13.6	lb/min

B. OUTLET GAS CONDITIONS
85°F
9.2 psia

D.G.	1,036	lb/min
H_2O	41	lb/min
SO_3	0.78	lb/min
SOLIDS	0.13	lb/min

FIG. 41-4. Metallurgical gas purification system.

TABLE 41-1

Materials of Construction for SO_2 Gas Scrubbers
(Once-Through Scrubbing Liquor)

Chloride concentration in scrubbing liquor (saturated 2 to 10% SO_2, H_2SO_4--0.25%)	Materials of construction
Below 100 to 150 ppm	Humidifier: mild steel, lead, and brick
	Upper tower: 316 ELC stainless steel
150 to 3,000 ppm	Humidifier: mild steel, lead, and brick
	Upper tower: Alloy-20
3,000 ppm and above	Humidifier: mild steel, rubber-lined, and brick
	Upper tower: mild steel, rubber-lined, and brick
	Upper tower: mild steel, rubber-lined with titanium or Hastelloy-C trays and piping

the hot gas quenching zone and may also be used in the upper scrubbing section. When using seawater, once-through mild steel rubber-lined shells are used with titanium scrubbing internals, pipes, and sprays.

It is rare that total effluent can be discarded even after neutralization. Therefore, a scrubber-cooler system is used in which the upper tray cooling water is recycled through external coolers in a closed circuit. The net condensate from the upper circuit is purged to a lower gas cleaning and humidifying section. Two scrubbing circuits may be used to reduce the slurry concentration in the heat exchangers. They also lower acid and chloride levels at the same time so that selection of construction materials is easier. Table 41-2 shows materials of construction normally selected for closed circuit operations.

TABLE 41-2

Materials of Construction for SO_2 Gas Scrubbers (Recycle Systems)

Chloride concentration in scrubbing liquor (as NaCl)	Sulfuric acid concentration (%)		Materials of construction
Below 100 to 150 ppm	<2	Humidifier:	mild steel, lead, brick
		Upper tower:	316 ELC stainless steel
Below 100 to 150 ppm	2 to 20	Humidifier:	mild steel, lead, brick
		Upper tower:	Alloy-20, Hastelloy-C ($>70°C$)
Below 100 to 150 ppm	20 to 30	Humidifier:	mild steel, lead, brick
		Upper tower:	mild steel, lead or rubber, Hastelloy-C trays
Sea water	0.25	Humidifier:	mild steel, rubber, and brick
		Upper tower:	mild steel, rubber, and Hastelloy-C
150 to 1,000 ppm	2 to 5	Humidifier:	mild steel, rubber, brick
		Upper tower:	Alloy-20
1,000 to 5,000 ppm	2 to 5	Humidifier:	mild steel, rubber, brick
		Upper tower:	mild steel, rubber-lined, Hastelloy-C trays and piping

Figure 41-5 shows typical process flows, sulfuric acid concentrations, and solids content of recycled liquor. Alloy-20 is used for shell, internals, piping, and trays when sulfuric acid in the 2 to 20 percent range is handled; Hastelloy-C and Inconel-625 are used for higher acid concentrations. When chloride levels are above the 100 to 150 ppm range, Alloy-20 is required, or possibly Hastelloy-C or Inconel-625 for chloride levels above 2,000 ppm. CD4M and Alloy-20 pumps give good service except when chloride content or solids content becomes too high. Then rubber-lined and reinforced plastic pumps are used.

A schematic process flow is shown in Fig. 41-6 of a scrubber cleaning gases from a refuse incinerator. Hot gas from the incinerator enters a brick-lined venturi. Scrubbing liquor is recycled to the tangential pipes above the throat. Gas accelerates rapidly as it passes through the converging throat and mixes with scrubbing liquor flowing down the venturi wall.

The venturi throat provides the high-speed contact that drives the scrubbing liquor and fine particulates together. The mixture of slurry and captured dust particles is removed in the cyclonic separator. The clean, entrainment-free gas leaving the separator enters a wet exhaust fan which provides draft for the total system and discharges to atmosphere through a stack.

The chloride content of the inlet gas is often high; approximately 1,700 lb/day as sodium chloride in a 200-ton/day refuse incinerator. Makeup water to the system is low. Recycle liquor and the purge from the scrubber system contains chloride concentration in the range of 2,000 to 4,000 ppm, a pH of 2 to 3, and a solids content in the range of 2 to 4 percent by weight.

Because this scrubbing liquor is highly corrosive, serious problems have been encountered using type 316 ELC, 304 ELC, and monel alloys. Preferred construction is a brick-lined venturi and tee-type elbow. The adjustable throat and drive feed pipes and nozzles are fabricated of Alloy-20 or Incoloy-825.

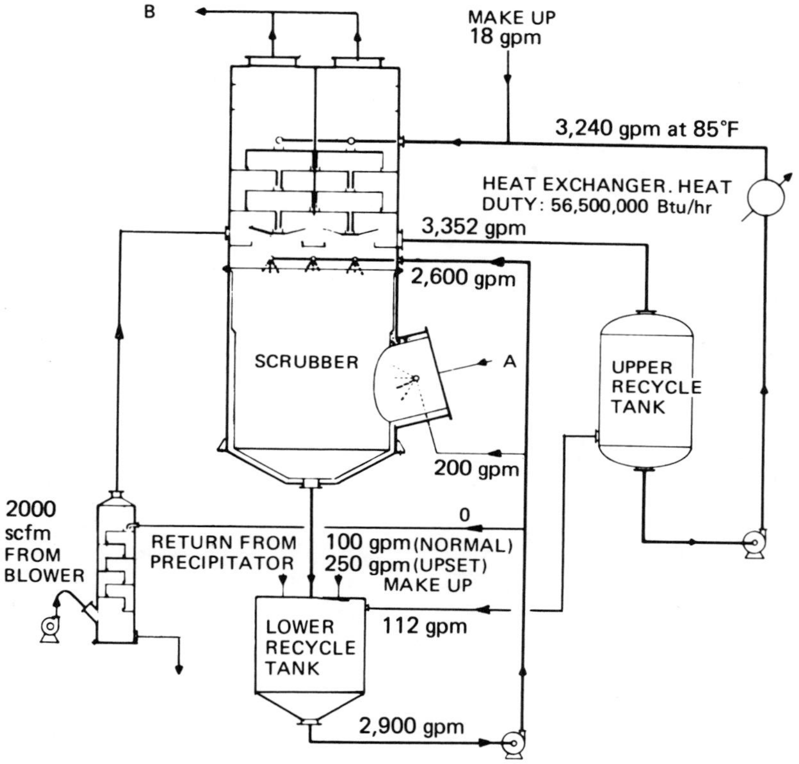

A. INLET GAS CONDITIONS
700°F
11.9 psia
85,150 scfm dry

D.G.	7,500	lb/min	
H_2O	132	lb/min	
SO_3	20.8	lb/min	
SOLIDS	2.5	lb/min	(NORMAL)
	85.15	lb/min	(UPSET)

B. OUTLET GAS CONDITIONS
90°F
11.4 psia
131,000 acfm

D.G.	7,500	lb/min	
H_2O	280	lb/min	
SO_3	6.3	lb/min	
SOLIDS	0.05	lb/min	(NORMAL)
	1.70	lb/min	(UPSET)

	NORMAL TO STRIPPER	UPSET TO DRAIN
WATER gpm	100	250
SOLIDS lb/min	2.45	83.45
H_2SO_4 lb/min	25.5	25.5

FIG. 41-5.

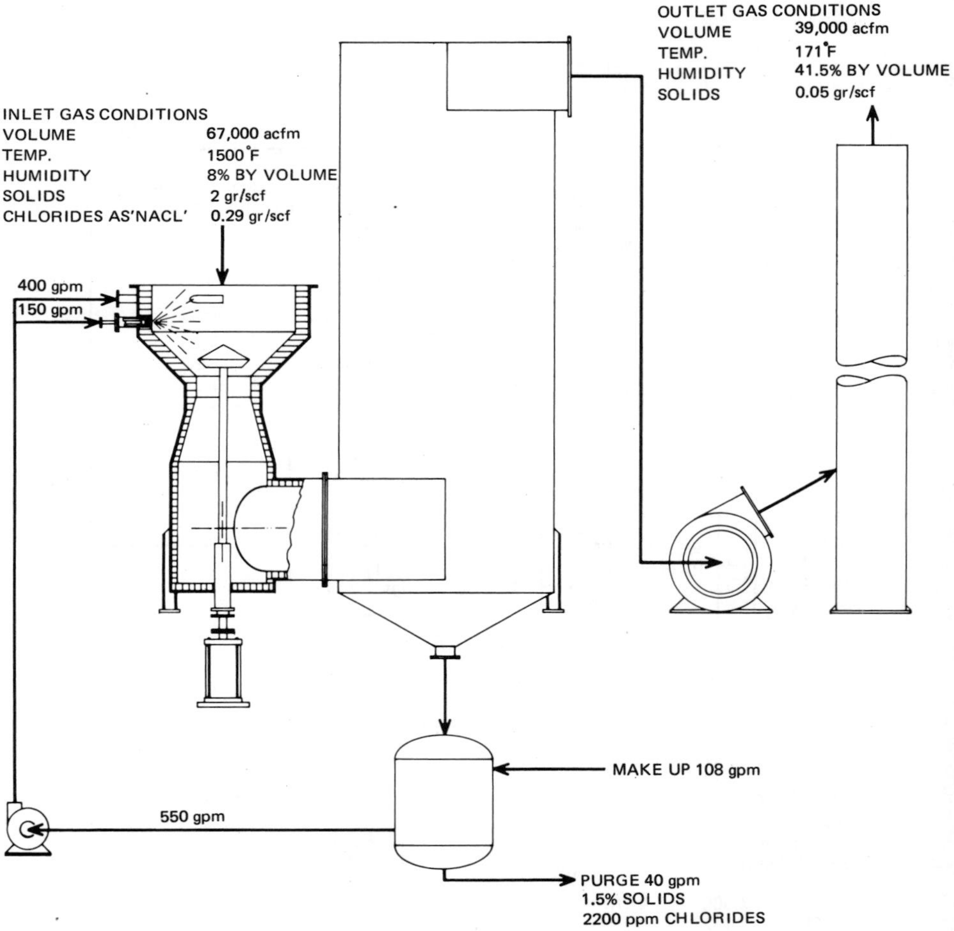

FIG. 41-6.

The cyclonic separator is carbon steel lined with FRP, 60 to 80 mils thick. It may be further lined with acid brick in the tangential gas inlet area. Rubber-lined steel and solid FRP have been used successfully for cyclonic separators. Good service life has been obtained using wet fans with casings of rubber-lined carbon steel and with Incoloy-825 wheel and shaft. Similarly, Hastelloy-C wheels have shown good results. While tests have indicated that service life might be poor with Alloy-20 wheels and shafts, this alloy has in fact worked out quite well on several jobs in service for more than two years.

Rubber-lined pumps are preferred for liquor recycle, and FRP piping provides good corrosion resistance.

Once entrainment has been removed in the final scrubbing stage, there are special problems in selecting corrosion-resistant alloys.

There is no circulation of scrubbing liquor above the entrainment separator, and in this quiescent zone acids and solids can easily build up on the walls of the scrubber and the outlet cone. The acid concentration can easily exceed the corrosion-resistant limits of the alloy, and scrubber failures are often reported above the entrainment separation stage or in the exit duct.

Corrosion in this area is often solved by using higher grade alloy or plastic- or rubber-lined steel. For example, on lime kiln gas scrubbing, carbon steel will hold as well in the main scrubbing zones where the liquor pH is 8 to 9. Type 316 ELC alloy would be required in the outlet duct and fan where the condensate pH might be 2 to 3.

Corrosion problems of this type are more severe in the fan downstream. No scrubbing liquor is available to wash the fan, and the heat of compression generated by the fan causes cyclic, uncontrolled drying of acid on the housing and wheel. This action, combined with mechanical stresses, makes selection of fan materials a difficult one on gas scrubber service.

Consideration must be given to using a higher grade alloy than that used in the main scrubber.

The use of continuous fine misting sprays in the duct upstream of the fan to provide washing and maintain saturation in the fan is of direct benefit. Wall washing and intermittent spraying can also be very helpful in minimizing outlet duct and scrubber outlet corrosion.

Chapter 42

INDUSTRIAL ODOR CONTROL

V. Frega
The Ceilcote Company
Berea, Ohio

Any discussion of odor is complicated from the outset by the fact that odor is a human sensory experience, not a rigorously defined scientific attribute. Attempts to quantify the concept of odor have traditionally taken two tracks: psychophysical measurement of human experience and analysis of odor-bearing agents with scientific instruments. The instruments give a picture which, while satisfactory for the description of a compound, do not reliably relate that picture to the concept of odor. Psychophysical measurements show that the variability of the human sensory experience, both between and within human subjects, is so wide as to make it difficult to use the information as planning data in any program designed to control odors.

Laws and regulations governing the subject of industrial odor emissions reflect lack of a scientific base. They also show a high degree of variability. A recent study [1] shows that the legal principle of "public nuisance" is often applied. Rules for applying this principle are also variable.

Most of the literature skirts the inherent problems of semantics and legalism by concentrating on areas in which there is general agreement that odors do exist and that some control of odor emission is required. The organic waste-processing industries provide most of the documented examples of development of control equipment.

This emphasis also reflects the efforts of these industries to fund research and development projects. Two industry associations, the National Renderers Association, Inc., and the Fats and Proteins Research Institute, are heavily involved in the conduct and funding of research aimed at providing a scientific basis for odor control. The Illinois Institute of Technology Research Institute has conducted much of this sponsored research.

ODOR DETECTION TECHNIQUES

In practice, instrument analysis of odors is a laboratory procedure. When an odor abatement program is considered, the data will usually be derived from measurement of human sensory response. Many state and local governments specify the stimulus presentation equipment for human subjects and indicate response levels which lead to the presumption that an odor requiring abatement is present.

Among the most common devices are the American Society for Testing and Materials (ASTM) dilution syringe and the Barneby-Cheney Scentometer. Illinois Institute of Technology Research Institute (IITRI) has designed the Dynamic Forced-Choice Triangle Olfactometer, which promises to replace many instruments now in field use. These and other instruments are discussed in "Industrial Odor Technology Assessment" by Cheremisinoff and Young [2] to a level necessary to understand the underlying principles of application.

The mechanics of measurement are not as important as the parameters which are measured. Threshold of detectability is one important measurement. This measurement is determined by diluting an emission sample until the odor of the sample is just barely detectable. The specification of this number is in odor units, which are computed by the number of dilutions required to reach the level of detectability. Thus, a dilution of 100 which results in a threshold level is labeled 100 odor units per cubic foot. This is the odor concentration. The cubic foot factor is derived from sample size. It provides a means of comparison between methods when the same sample source is subjected to two different types of measurement. A derived figure of some merit is total odor emission, which is the concentration times total volume of emission in cubic feet.

Another measurement is that of intensity. This measurement often calls for a value judgment by the human observer. Psychophysical measurement analysis techniques exist which can be used to plot the perceived intensity against actual changes in intensity. This function is not linear for any emissions sampled to date. The function has value in determining the degree of abatement required before abatement is perceived.

The third common measurement is of the so-called hedonic response. The human observer makes a value judgment as to the degree of pleasure or displeasure that the odor stimulus arouses in him. Psychophysical analysis techniques yield data which are relatively reliable if sample size is sufficiently large and stimulus presentation is controlled. There are many variations of the measurement technique, including substituting an expert panel for the large random sample and using mathematical weighting and normalizing methods. While careful measurement produces reliable information, that information is strictly limited in its usefulness as an inferential tool. It is highly situationally dependent. It represents data derived in a very specific set of circumstances.

It is the human hedonic response that is responsible for most complaints about offensive odors. This response has been shown to vary from day to day, from hour to hour, and with surrounding circumstances within the same human observer. It varies with distance from the source, with degree of dilution, with intensity, with visual stimuli with which it is associated, and with the general psychological climate of the community and within the individual.

Two examples of the hedonic response will serve to highlight the problem for planners of odor abatement programs. Several years ago, in a midwestern city, a brewery and a chocolate-processing factory were located approximately 1 mile apart. The odors from the cooking process in each plant were exhausted to the open air.

No demonstrably harmful substance appeared in either effluent. Complaints about odor were virtually nonexistent--for either plant. However, on rare occasions, production schedules called for the two plants to cook simultaneously. When that happened, complaints would come in that a disagreeable odor was causing discomfort to residents in the area.

In another midwestern community, a meat-packing plant and a tannery were located about 2 miles apart. Both plants had odors under reasonable control. But complaints came in to each of them whenever hides were transported from the packer to the tanner. The hides were preliminarily processed at the packer's plant and were not considered odorous by the packer. However, the hides were transported in an open truck and were not a pretty sight. Most complaints came from viewers of the open trucks who were psychologically prepared to smell something.

In summary, the measurement of odor level and derivation of data from those measurements for use in odor abatement programs is highly subjective. Attempts to quantify measurement and improve measurement techniques have been concentrated largely on means to control odor stimulus presentation to human subjects. Data derived from psychophysical measurement procedures on human subjects are often reliable considering the wide variability in odor perception in humans, but are limited in their applicability to the specific situation measured. At different times or at different distances from the test situation, human perception will change.

ODOR SOURCES AND SOLUTIONS

Rendering plants provide a good example of an industry in which odor abatement and control have reached a high degree of development. Literature on the subject generally makes the distinction between high-intensity point sources of odor and general background odor in the plant air. At the beginning of odor abatement planning, a choice must be made whether plant air and point source odors will both be controlled or whether only the point sources will be considered.

The choice has to be made on an individual basis. Ries et al. [3] provide an example of a situation in which the control equipment was based on collecting odors at their source. The key in that case was that this degree of control was acceptable to regulating authorities.

In a survey of odor control systems in rendering plants [4], Bethea et al. state: "Odor control systems commonly used at this time are condensation followed by incineration, or scrubbing, or a combination of these. Incineration has been the most effective method to date, especially for low-volume, high-concentration odors. Recent advances in the design and utilization of scrubber systems have significantly increased their odor-destroying ability until they nearly equal the performance of incinerators. Scrubbers are especially useful in controlling high-volume, low-concentration odors. The renderer frequently uses both systems for full plant control of all odors.

"Other odor control methods, used with varying degrees of success by the renderer, include catalytic combustion, adsorption, and ozonation. However, no odor control system can be truly effective unless general cleanliness, housekeeping, raw materials handling and storage, and spill prevention throughout the rendering plant are properly managed by the operator."

This source goes on to discuss incineration and scrubbing in some detail and the other methods in less detail. However, two significant changes have taken place since that writing: the price of fuel for the incineration process has increased greatly and there have been further advances in wet scrubber technology. Wet scrubbers have become more efficient and more economical to install and operate in odor control systems. They will probably be the most widely used devices in the foreseeable future.

WET SCRUBBING SYSTEMS

The various types of wet scrubbers used in odor abatement include air washers, packed towers (both countercurrent and cross-flow), venturis, and combinations of these devices. The composition of the odor-causing agent determines which device or combination of devices is used.

Air Washers

Air washers employ one or more banks of high-pressure spray nozzles which completely cover the gas stream. Generally, the device includes an entrance diffuser that creates a uniform air flow into the first chamber. This chamber contains a densely sprayed chemical solution that is designed to produce high contact efficiency between the gas and the liquid. This chamber is also used to collect and suspend particulate matter in the liquid. The first chamber is followed by a device which provides further intimate contact between the gas and liquid, and separates the liquid droplets from the air stream. A second chamber, containing the same equipment, utilizes a different chemical. It is frequently employed to further enhance odor reduction. Velocities through this equipment are quite high, on the order of 1,000 to 1,500 fpm.

Stainless steel is used extensively in air washer construction. Air washers are generally viewed as low-to-medium-efficiency devices, although efficiencies as high as 98.67 percent have been reported. Due to the high velocity through the system, static pressure drop can range as high as 6 to 7 in.wg. Power requirements are generally 1.5 to 2.0 bhp/1,000 cfm throughput.

Packed Towers

The standard absorption device is the packed tower scrubber. It has been utilized extensively (with a wide variety of chemical solutions) in odor control applications. Two types are utilized--the countercurrent packed tower and the cross-flow packed tower.

The countercurrent packed tower is a device which has a bottom gas entry and a top gas outlet. Scrubbing liquid is introduced above the packed bed and allowed to flow by gravity through the packed bed countercurrent to the gas stream. The packing medium provides intimate contact between the gas and the scrubbing liquid. It also acts as a particulate collector by utilizing the principle of inertial impaction. The selection of the packing medium is extremely critical, since it affects the pressure drop and can provide liquid surface regeneration for maximum absorption.

The odorous contaminants must, of course, be absorbed into the liquid before they can be reacted chemically or be oxidized. Chemicals employed include H_2SO_4, H_2PO_5, NaOCl, NaOH, $KMnO_4$, etc. When using $KMnO_4$, a caustic solution is the common vehicle, and, as a reaction product, MnO_2 is formed as a precipitate. This can cause plugging of the packed bed so that frequent cleaning is generally recommended.

The gas stream in the cross-flow packed tower travels horizontally through the tower, while the scrubbing liquid enters the top and flows by gravity to the bottom across the gas stream. Design parameters in terms of mass flow rates are the same for both countercurrent and cross-flow packed towers. The cross-flow packed towers, however, exhibit a slightly lower pressure drop. Scrubbing efficiencies for either configuration should be the same, provided all design parameters are identical.

Materials of construction used for the shell include stainless steel and fiberglass-reinforced polyester (FRP). It should be noted that in any wet scrubber the presence of halogens can have a drastic effect on stainless steel. In most cases, a finite corrosion rate does exist. Therefore, the stainless steel shell would eventually have to be replaced.

FRP is the most universally accepted construction material for these scrubber shells since FRP is impervious to most chemicals encountered. It has an indefinite life expectancy, is lightweight, and requires no maintenance. Internal components, such as the tower packing, packing support plates, spray headers, and entrainment separators, are generally made of thermoplastic. Thermoplastic materials used include PVC, CPVC, polypropylene, and high-density polyethylene. In some cases, ceramics are used, although they are generally not recommended because of their weight, fragile nature, and limited resistance to caustic compounds.

Gas flow velocities generally range from 300 to 500 fpm. Consequently the pressure drop is relatively low--from 1.5 to 4.0 in.wg., depending on the packing size and type, packing depth, liquid flow, etc.

Since the packed tower is a differential device (i.e., an increase in efficiency can be obtained by adding more packing), efficiencies for this device can be quite high. Tests indicate that 99 percent efficiency is not uncommon.

In addition, since the pressure drop and the amount of power necessary to supply the scrubbing liquid to the tower are low, the overall power consumption is minimal.

Venturis

The venturi wet scrubber employs a constriction in the form of a throat through which the gas is accelerated to extremely high velocities. Scrubbing liquid is introduced ahead of or directly into the venturi throat. The high velocity of the air stream shears the liquid droplets and creates high turbulence in the throat. As a result of turbulence, intimate mixing and effective collisions take place between particulate and scrubbing liquid.

While the venturi has been accepted as one of the best devices for particulate removal, its gas absorption capability is limited. The throat is usually followed by an evase section where some static pressure is regained and contact between the gas and liquid continues. The venturi is then followed by an entrainment separator which removes the liquid droplets entrained by the gas stream.

Pressure drop is a function of throat velocity and entrainment separator design. In general, the higher the pressure drop across the throat, the smaller the particulate which can be effectively removed. But the amount of gas absorption is affected little by pressure drop. Typically, 10-in.wg pressure drop across a venturi throat will result in the removal of 1-μm particles. At the same time it would require 30-in. sp/wg drop to remove 0.6-μm particles, and 100-in. sp/wg to remove 0.1-μm particles.

Overall efficiency in odor removal for a venturi-entrainment separator would be low unless the odors were due primarily to the particulate matter being removed. Although this type of unit has not been tested extensively by itself, it is expected that efficiencies would be in the range of 50 to 80 percent for removal of gas-related odors, but upwards to 99 percent if only particulate-caused odors were present in the gas stream.

Stainless steel and FRP materials have been used for both venturi and entrainment separators. The same comments made concerning packed towers relating to maintenance, longevity, etc., also apply here.

Combination

Several manufacturers have installed systems in the field which use combinations of the wet scrubbing devices already described. Usually these are arranged in series to enhance odor removal efficiencies, permit simultaneous handling of several scrubbing liquids, and/or remove solid particulate prior to a packed column.

Combinations used in the field have included: two countercurrent packed towers; a venturi followed by a countercurrent packed tower; a venturi followed by a cross-flow packed tower; and cross-flow scrubbers with more than one bed. Each of these has proved highly effective when properly applied.

Latest Developments

Recently the Ceilcote Company announced development of a packed tower scrubber employing electrochemical oxidation. This device uses a proprietary scrubbing liquid which readily oxidizes the absorbed gases and is in turn reactivated and recirculated. This system has the distinct advantage of low-cost operation. Chemical costs are minimized, while electrical energy requirements are also maintained at a low level. The use of a venturi or other prewasher is recommended where particulate matter is present since particulate has an adverse effect on the chemical oxidant usage and regeneration.

Another development is a three-step process that utilizes chlorine gas, acid, and alkaline liquids. This three-step process has been applied to obtain the highest levels of odor reduction--99.99+ percent.

The advantage of this system is that chlorine gas provides gas-phase oxidation. When injected ahead of the acid bed, it has a relatively long retention time and is intimately mixed with the entire gas stream. The acid scrubbing liquid removes ammonia- and amine-based compounds. The caustic scrubbing liquid is used to remove the sulfur-bearing compounds and any excess unreacted chlorine.

Several manufacturers offer the three-step process system in either of two configurations. Two countercurrent towers can be arranged in series, or a single cross-flow scrubber (employing two separate packed beds) can be utilized. Operational costs are low for the three-step process, provided that proper precautions are taken to avoid mixing of the scrubbing liquids. The overflow from both recycle tanks is combined to provide an essentially neutral effluent which can be routed directly to a municipal sewage system.

An added advantage of the multiunit approach is that, if it is a modular arrangement, the user can elect to start out with a chlorine-caustic system, then add the acid section at a later date. This approach optimizes the capital investment. It has proved to be a viable method of controlling odors on demand in a number of installations.

Summary

First cost and operational costs should both be considered when selecting equipment. Actual field tests have shown that the most effective means of odor reduction is incineration and multiple-step wet scrubbing. With the high costs and limited supply of fuels, multiple-step wet scrubbing is emerging as the most economical solution. Selection of the type of wet scrubber used should be based on the nature of the odorous constituents and physical nature of the exhaust stream.

FIELD APPLICATIONS, WET SCRUBBING

The following field installations of odor abatement systems, based on the cross-flow packed wet scrubber, illustrate current trends.

Figure 42-1 is an installation that shows the type of system described by Ries et al. [3]. It can be adapted throughout the general rendering industry. This odor control system was installed in the rendering facility of a meat-packing plant. The 10,000-cfm plant effluent is passed through a single scrubber that utilizes two packed beds--one irrigated by caustic, and the other by an acidic solution. The use of front sprays prevents solids buildup. The scrubber handles general exhaust as well as concentrated gases from the "hot well."

Field testing of the inlet and outlet gas streams verified the high efficiency levels possible with this design. The tests also demonstrated that the odor level of the scrubber exhaust gases was well within the allowable limits proposed for this industry by the Environmental Protection Agency (EPA).

Figure 42-2 is a schematic representation of an odor abatement system which was installed in a fish meal-reduction plant. Note the use of available sea water as a scrubbing liquid.

Effective abatement of odors from fish meal-reduction operations is achieved with packed cross-flow scrubbers at several fish-processing plants. These systems have gas-handling capacities of 20,000 to 37,000 acfm at 250° F. The scrubbers use sea water as the scrubbing liquid in the first packed section to cool the air stream and remove particulate matter and effect some absorption. The air stream then passes to a reaction section where raw chlorine gas is sparged in

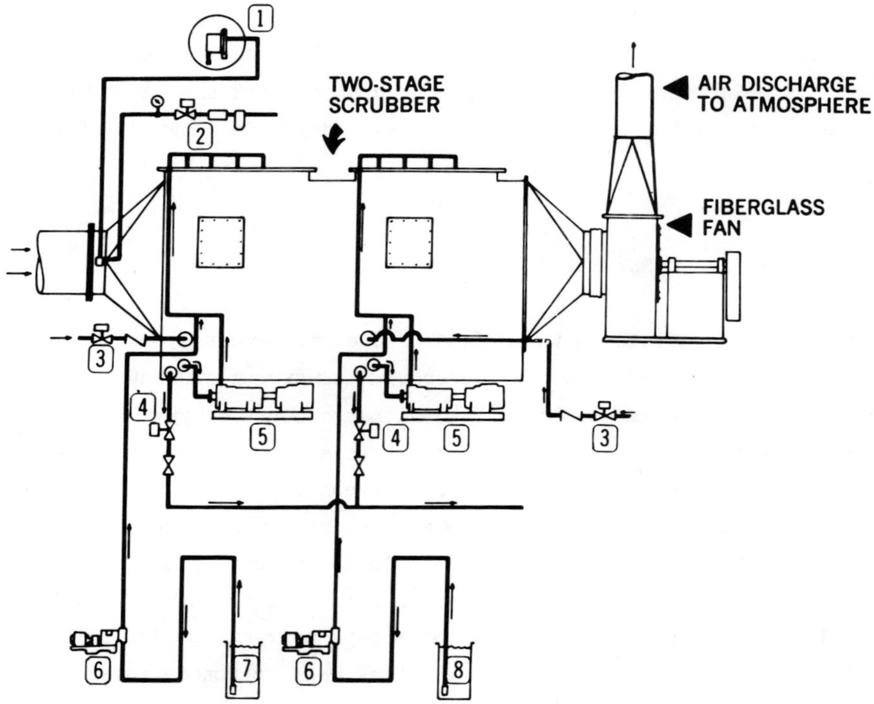

FIG. 42-1. Type of odor abatement system designed for the rendering industry. (1) Chlorine cylinder and regulator, (2) compressed air supply, (3) fresh water makeup solenoid valves, (4) drain solenoid valves, (5) chemical solution recirculation pumps, (6) chemical makeup metering pumps, (7) concentrated sulfuric acid storage, (8) concentrated sodium hydroxide storage.

to oxidize the less soluble mercaptans, amines, etc., to less odorous and/or more soluble compounds. The second packed section is irrigated with a dilute acid solution for absorption of the amine- and ammonia-based compounds. The third packed section uses a recycled dilute caustic solution for absorption of H_2S and mercaptan compounds, as well as any unreacted Cl_2. All scrubbers feature a

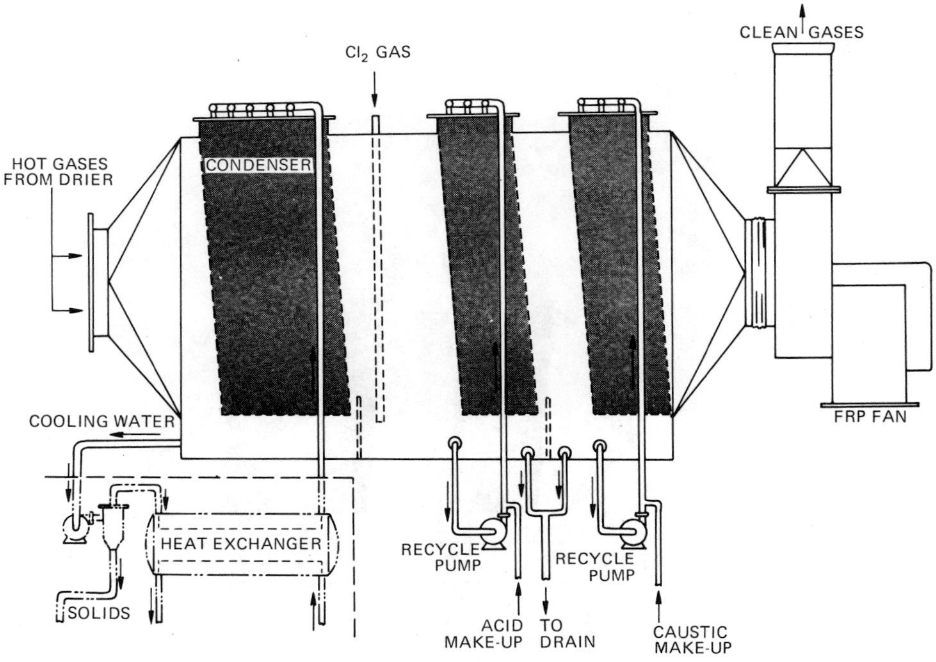

FIG. 42-2. Odor abatement system designed for a fish meal-reduction plant.

prespray arrangement to prevent solids buildup on the front portion of the packaging.

The schematic in Fig. 42-3 illustrates the use of a low-energy venturi as a primary solids collector before a typical three-bed cross-flow odor control scrubber. Application of this arrangement is normally recommended where odoriferous gases are at a high temperature and contain high solids loading. This arrangement removes solids from the recycled solution in the venturi and returns them to the process. The solids loading in the cooling water in the first packed bed stage is also reduced, minimizing the chances of plugging the bed.

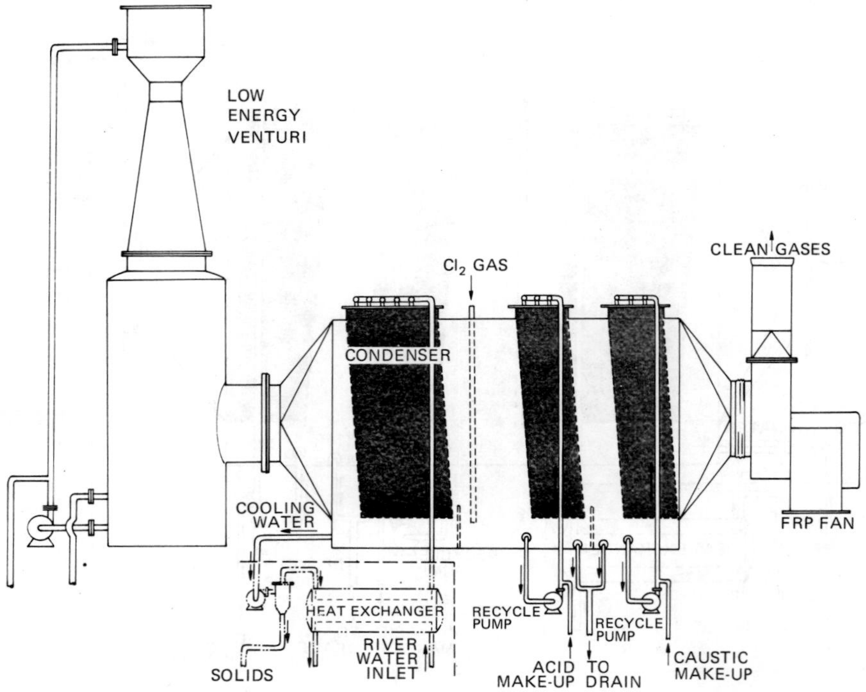

FIG. 42-3. System illustrating the use of a venturi scrubber for the collection of solids before the gas reaches the odor control scrubber.

REFERENCES

1. W. H. Prokop: Status of Regulations for Source Emissions and Ambient Odors. (Paper presented at conference on Odors: Evaluation, Utilization, and Control, New York, Oct. 1973.)

2. P. N. Cheremisinoff and R. A. Young: "Industrial Odor Technology Assessment," Ann Arbor Science Publishers, Ann Arbor, Mich., May 1975.

3. K. M. Ries et al.: Two-Stage, 10,000 cfm Scrubbing System Eliminates Outside Plant Odors, Food Processing Magazine, June 1974.

4. R. M. Bethea et al.: Odor Control for Rendering Plants, Environmental Science & Technology, 7(6):504-510 (June 1973).

Chapter 43

ODOR CONTROL BY ADSORPTION

Thomas M. Hellman

Allied Chemicals Corporation
Morristown, New Jersey

Gas adsorption is a process by which one or more components are removed from a gas stream through adherence to a solid surface. The attractive force holding the gaseous molecule at the surface may be either physical (physical adsorption) or chemical (chemisorption) in nature.

Adsorption finds wide application in the field of odor control. Odors are generally created by the presence of small quantities of contaminants in large volumes of air. The equilibrium relationships associated with adsorption, unlike those involved in absorption, lend themselves to removal of low-concentration contaminants. However, because of the complex mass transfer mechanisms involved in adsorptions, and the great variability in adsorbent physical properties which strongly affect performance, adsorption is much less amenable to generalized design from basic physical data than is absorption. Good design requires specific data for the gas-solid system involved.

TRANSFER MECHANISMS

In industrial operations, adsorption is accomplished primarily on the surfaces of internal passages within small porous particles. Three basic mass transfer processes occur in series: (1) mass transfer from the bulk gas to the particle surface, (2) diffusion through the passages within the particle, and (3) adsorption on the internal particle surfaces. Each of the processes depends on the system operating conditions and the physical and chemical characteristics of the gas stream and the solid adsorbent. Often, one of the transfer processes will be significantly slower than the other two, and will control the overall transfer rate. The other process will operate nearly at equilibrium.

Heat transfer may also play an important role in an adsorption system. The adsorption process is exothermic. Physical adsorption equilibria behave in a manner similar to vapor-liquid equilibria; transfer out of the vapor phase is favored by decreasing temperature. Therefore, rapid dissipation of heat away from the adsorbing surface improves adsorption performance. Chemisorption rates, on the other hand, generally increase with increased temperature.

External mass transfer is the only process of the three involved in adsorption that can be predicted with reasonable accuracy from physical data. Mass transfer from the bulk gas to the particle surface can be considered by the film resistance approach. The rate of mass transfer is proportional to the external surface area of the adsorbent particles and the adsorbate concentration difference between the bulk gas and the particle external surface. The proportionality constant is the mass transfer coefficient, the reciprocal of the resistance to mass transfer of a hypothetical thin film at the particle surface. Correlations of mass transfer coefficients as functions of particle size, gas flows, and properties and operating conditions have been summarized by Smith [15], Levenspeil [6], Treybal [16], and Perry [12].

Transfer of material from the particle surface to internal adsorption sites is accomplished by diffusion through the internal passages of the particle. Diffusion takes place by one or more of several mechanisms. Where passages are sufficiently large that intermolecular collisions are more likely than collisions with the passage walls, bulk diffusion predominates. In smaller passages, and at lower pressures, where collisions with the passage walls are more probable, Knudson diffusion controls. Knudson diffusion rates are often an order of magnitude lower than bulk rates for a given pore length. A third diffusion mechanism, surface diffusion, may predominate in small-passage, high-surface situations. Surface diffusion is the migration of molecules along the passage surface after adsorption. Because of the irregularity and variability of intraparticle passages, and the difficulty of measuring surface concentrations, intraparticle mass transfer is extremely difficult to predict accurately from physical data. Its effects can be determined to a degree by comparison of experimental adsorption rates for different size adsorbent particles, allowing for differences in external mass transfer rates.

The actual adsorption of vapor molecules takes place mainly on the surface of internal passages within the adsorbent particles, since that is where most of the available surface exists. The adsorption process may be either physical or chemical in nature. Physical adsorption is a readily reversible process which occurs as a result of the physical attraction between the gas molecules and the molecules of the solid surface. If the gas-solid intermolecular attraction is greater than the intermolecular attractions in the gas phase, the gas will condense on the solid surface, even though its pressure is lower than its vapor pressure at the prevailing temperature. For example, the equilibrium adsorption pressure of acetone on activated carbon may, under some conditions, be as little as 1/50 to 1/100 of the equilibrium vapor pressure at the same temperature. Therefore, small concentrations of contaminants can often be removed from gas streams by adsorption, but not by absorption. The heat released on adsorption is usually somewhat greater than the latent heat of vaporization and of the order of the heat of sublimation of the gas [16]. Chemical adsorption, or chemisorption, involves formation of chemical bonds between the gas and surface molecules. The strength of the chemical bond may vary considerably, and identifiable chemical compounds in the usual sense may not actually form. However, the adhesive force is much greater than found in physical adsorption. The heat liberated during chemisorption is usually large, of the order of the heat of chemical reaction [16].

EQUILIBRIUM CONSIDERATIONS

Most of the adsorption data available from the literature are equilibrium data. Equilibrium data are useful in determining the maximum adsorbent loading which can be obtained for a specific adsorbate-adsorbent system under given operating conditions. However, equilibrium data by themselves are insufficient for design of an adsorption system. Overall mass transfer rate data are also necessary.

The equilibrium adsorption characteristics of gas or vapor on a solid resemble in many ways the equilibrium solubility of a gas in a liquid. Adsorption equilibrium data are usually portrayed by isotherms--lines of constant temperature on a plot of adsorbate equilibrium partial pressure versus adsorbent loading in mass of adsorbate per mass of adsorbent. Isotherms take many shapes, including concave upward and downward, and S-curves. Equilibrium data for a given adsorbate-adsorbent system cannot generally be extrapolated to other systems with any degree of accuracy.

Several useful methods are available for extrapolating equilibrium data for a given system to various temperatures and pressures. One convenient method is by use of a reference substance plot. Here, the adsorption equilibrium partial pressure of the adsorbate is plotted against a pure substance vapor pressure, preferably that of the adsorbate. If logarithmic coordinates are used on both axes, lines of constant adsorbent loading, isosteres, are linear for most substances. Therefore, only two datum points are required to establish each isostere. Isosteres will also generally be linear on a plot of the logarithm of the ratio of pure substance vapor pressure to adsorption equilibrium vapor pressure against reciprocal absolute temperature. A further condensation of data may be accomplished by plotting absolute temperature times the logarithm of the ratio of pure substance vapor pressure to adsorption equilibrium vapor pressure at that temperature against adsorbent loading. For most single adsorbate systems, a single curve will result for all temperatures, at least over a moderate temperature range. Methods for correlating equilibrium data are discussed by Lewis, Gilliland, Cherton, and Cadogan [7,8], Othmer and Joseforvitz [11], and Treybal [16].

MASS TRANSFER RATE CONSIDERATIONS

As discussed previously, the mass transfer mechanism involved in industrial adsorption processes is complex. Generally, basic physical data on the materials involved are insufficient for design. Experimental mass transfer rate data for the specific adsorbate-adsorbent system are usually required for good design.

GAS-SOLIDS CONTACTING SCHEMES AND EQUIPMENT

Three basic schemes are used in adsorption systems for obtaining effective gas-solid contact: (1) the fluidized bed, (2) continuous moving bed, and (3) unsteady-state fixed-bed techniques. By far the most common, particularly in odor control applications, is the fixed-bed unsteady-state adsorber. In this system, the contaminated gas is passed through a stationary bed of adsorbent. The bed is operated in this manner until the contaminant level in the effluent begins to rise. The adsorbent must then be replaced or regenerated. Generally, gas flow is diverted to a second, parallel bed to allow continued operations during adsorbent change or regeneration. Adsorbent beds range in size and form from small disposable cartridges to dumped beds contained in large vessels.

Unsteady-state fixed-bed adsorbers have the advantage of being relatively simple and economical, particularly at low adsorbate rates. Since the bed is stationary, the adsorbent is handled only during replacement, which should be infrequent in a well-designed system. Continuous solids-handling systems and their inherent high cost and mechanical problems are avoided.

The primary disadvantage of fixed-bed adsorbers arises when contaminant rates are high. Because of the unsteady-state nature of the operation, a large portion of the in-process adsorbent inventory is saturated and, therefore, inactive. Where adsorbate rates are high, unduly large beds are required. In addition, gas flow rates through fixed beds are limited by pressure drop. Extremely high gas rates may require uneconomically large beds.

In fluidized-bed adsorbers, the combination of high gas rate and small adsorbent particle size results in suspension of the adsorbent, giving it many of the characteristics of a fluid. Fluidized-bed adsorbers, therefore, lend themselves to truly continuous, countercurrent, multistage operation. Adsorbent inventory is minimized. As an example, Courtaulds, Ltd., has developed a continuous, fluidized-bed, activated carbon system for recovery of carbon disulfide from the regenerations bath vent in their viscose rayon process [14]. The adsorber and adsorbent regenerator are similar to a liquid-vapor absorption-stripping column in both appearance and operation.

Fluidized-bed adsorbers have several disadvantages. The continuous handling and transport of solids is expensive from an equipment standpoint; fluidized-bed systems must be large to be economical. Solids handling also presents a potential for mechanical problems. Careful control is required to keep the adsorbent fluidized, while minimizing adsorbent loss with the gas-phase attrition of the adsorbent can be high, requiring substantial makeup.

Continuous moving-bed adsorbers, like fluidized-bed systems, lend themselves to true countercurrent, multistage operation. The adsorbent, however, is not fluidized, but is mechanically converged or falls by gravity through the rising stream of gas. Attrition is generally higher than in fluidized-bed systems, but control may be less critical. Other advantages and disadvantages of fluidized-bed adsorbers apply to moving-bed adsorbers.

DESIGN OF ADSORPTION EQUIPMENT

Proper design of an adsorption system must consider both the equilibrium and mass transfer rate aspects of the adsorbate-adsorbent system. A considerable amount of equilibrium data are available from the literature. However, mass transfer data are scarce. Because of the complex mass transfer mechanism involved in industrial adsorption processes, and the strong dependence on highly variable and often irregular physical characteristics of the adsorbent, experimental data for the specific adsorbate-adsorbent system and contacting scheme to be used are usually necessary for good design.

Selection of an adsorption system design method depends on the type of contacting scheme to be used, and whether contaminant removal or actual fractionation of components is sought. Since most odor-abatement adsorbers are unsteady-state, fixed-bed systems for contaminant removal, only the design approach for that type of system will be discussed in detail.

Barry [1] presents a useful approach to the design of unsteady-state, fixed-bed adsorbers--an extension of Michaels' mass transfer zone (MTZ) concept. This concept recognizes that during operation of a fixed bed, the adsorbent near the gas inlet is nearly in equilibrium with the inlet gas, while that near the outlet is relatively adsorbate free. Somewhere within the bed is a zone--the MTZ--where the actual mass transfer takes place. The MTZ gradually moves through the bed from inlet to outlet. When the MTZ reaches the outlet, a sharp rise in effluent adsorbate concentration, or breakthrough, occurs.

The length of the MTZ is independent of total bed length, but is a function of the following parameters:

1. Adsorbent characteristics,
2. Adsorbent particle size,
3. Fluid velocity,
4. Fluid properties,
5. Adsorbate concentration in the entering fluid,

6. Residual adsorbate concentration in the adsorbent (if it has not been fully reactivated),
7. Temperature,
8. Pressure, and
9. Past history of the system.

MTZ lengths can be obtained experimentally, suitable graphical correlations plotted, and the plots used for design purposes. Ordinarily, all the parameters, except fluid velocity and inlet adsorbate concentration, can be held constant in both laboratory and full-scale equipment, so that correlation of MTZ length with those variables can be derived from a few experimental runs.

A more detailed discussion of the MTZ concept and its applications to adsorber design is presented by Campbell, Ashford, Needham, and Reid [2]. The application of the MTZ approach to a commercial adsorption system is described by Mattia [9].

Fair [3] describes a design method for fixed-bed adsorbers based on breakthrough time, derived by integration of the mass balance across a differential adsorber element. The solution presented is strictly applicable only to adsorption of a single component having a linear isotherm, in a plug-flow adiabatic adsorber. The solution is presented in the form of a generalized breakthrough graph first developed by Marshall and Hougen [4]. One of the key independent variables in the solution is the height of transfer unit for the adsorbate-adsorbent system. The height of transfer unit correlation presented applies only to external mass transfer. If external mass transfer is the controlling rate process, and the other conditions limiting the solution are met, the method should provide a reasonably good design based only on basic physical data.

Methods and references for design of steady-state and fractionating adsorption systems are presented by Treybal [16].

Several precautions are necessary in applying experimental adsorption data to full-scale system design:

1. The geometry and flow characteristics of the experimental system should be sufficiently similar to those of the full-scale system that the same mass transfer process controls the rate.

2. Physical characteristics of a given adsorbent may vary somewhat from batch to batch. Normally, this factor can be neglected. However, it might prove worth investigation when a full-scale adsorber departs significantly from design performance.
3. The packed bed will, generally, be compacted during separation. The result will be a shortening of the MTZ, but increased pressure drop.
4. In many systems, heat effects can be ignored to simplify design procedures. However, in such cases, some departure from design performance should be expected.
5. Variations will occur in the rate of adsorbent capacity falloff. Even extensive, well-run adsorbent life studies may yield misleading results. This often occurs when a trace contaminant not present in the experimental system feed appears in the full-scale system feed stream.

ADSORBENT REGENERATION

When adsorbate rates are sufficiently high to make periodic adsorbent replacement uneconomical, regeneration of the adsorbent can usually be justified. In continuous, steady-state systems, regeneration is required for economical operation.

Adsorbent regeneration is normally accomplished by reversing the adsorption process, either by decreasing the system pressure or, more commonly, by increasing the system temperature. In some cases, particularly in chemisorption systems, the adsorbent activity can be restored by reaction with a suitable reagent.

One of the inherent problems in regeneration of adsorbent bods is disposal of the desorbed material. In activated carbon systems, the most common in odor control applications, regeneration is accomplished by heating the bed with a gas or vapor which carries the desorbed contaminant out of the bed. The most commonly used carrier is superheated steam. Normally, the adsorbate is condensed along with the steam. If large quantities of adsorbate are involved, or if the adsorbate is highly water soluble, a secondary liquid waste disposal problem may result. The energy costs for a superheated steam-condensing regeneration system can be appreciale for a large adsorption system. Mattia [9] describes a system in which hot combustion gases are used as the regeneration carrier. The energy required to heat the gas is supplied by combustion of the desorbed material; supplemental fuel is supplied, if necessary. Both the energy cost and secondary disposal problems are solved with this system. However, the cost of the additional equipment required makes this system uneconomical for small adsorption installations.

ADSORBENT SELECTION

Activated Carbon

Activated carbon is by far the most commonly used adsorbent in odor control applications. Because of its relatively uniform distribution of surface electrical charge, activated carbon is not selective toward polar molecules. It can, therefore, be used to remove many organic vapors from gas streams with high water vapor contents. Water molecules, being highly polar, show strong attractions for each other, which compete with their attractions for the nonpolar carbon surface. Consequently, the large, less polar organic molecules are selectively adsorbed [17].

Activated carbon is most effective for adsorbing organic materials which boil at normal ambient temperature or higher. In general, effectiveness increases with increasing molecular weight.

Activated carbons have surface areas on the order of 300 to 700 ft^2/ft^3 [3]. Average pore diameter ranges from 20 to 40 Å [12], typical for most commercially used adsorbents. However, the distribution of pore sizes is substantially broader than found in other adsorbents.

Siliceous Adsorbents

The most commonly used adsorbents of the siliceous class are silica gels and synthetic zeolites, or molecular sieves. These materials are available over a wide range of adsorbent capacities. At best, their capacities are of the same order of magnitude as that of the most highly activated carbons [17]. They exhibit a greater selectivity for polar molecules than does activated carbon.

Silica gel is commonly used to remove water from gas streams. Surface areas average 200 to 700 ft^2/ft^3 [3]. Average pore diameters of various grades range from 20 to 140 Å [12].

Molecular sieves are synthetic zeolites which can be manufactured with extremely close control of pore size. Therefore, they can be tailored to suit specific applications. In addition to gas drying applications, molecular sieves are used for the separation of gases and vapors on the basis of molecular size and shape. Surface areas range from 350 to 1,000 ft^2/ft^3 [3].

Metal Oxides

Since metals are less electrophilic than silicon, metal oxide adsorbents show even stronger selectivity for polar molecules than do siliceous materials [17]. The most commonly used metal oxide adsorbent is activated alumina, used primarily for gas drying. Occasionally, metal oxides find applications in specific chemisorption systems. For example, several processes are under development utilizing lime or limestone for removal of sulfur oxides from flue gases.

Activated aluminas have surface areas in the 200- to 1,000-ft^2/ft^3 range [3]. Average pore diameters range from 30 to 80 Å [12].

Impregnated Adsorbents

For some applications, an adsorbent may be impregnated with a material which enhances its contaminant-removal ability. The improved effectiveness may be related to any of several mechanisms. The impregnating material may react with the vapor contaminant to form a compound or complex which remains on the adsorbent surface. Some impregnants react with the contaminant, or catalyze reactions of the contaminant with other gas constituents to form less noxious vapor-phase substances. In some instances, the impregnant acts as a catalyst intermittently, for example, under regeneration conditions.

In this case, the contaminant is adsorbed by physical adsorption and destroyed by a catalytic reaction during regeneration.

REFERENCES

1. H. N. Barry: Fixed-Bed Adsorption, Chemical Engineering, p. 105 (Feb. 8, 1960).

2. J. M. Campbell, F. E. Ashford, R. B. Needham, and L. S. Reid: More Insight into Adsorption Design, Hydrocarbon Processing and Petroleum Refiner, 42(12):89-96 (Dec. 1963).

3. J. R. Fair: Sorption Processes for Gas Separation, Chemical Engineering, pp. 90-110 (July 14, 1969).

4. O. A. Hougen and K. M. Marshall: Adsorption from a Fluid Stream Flowing Through a Stationary Granular Bed, Chemical Engineering Progress, 43:197 (1967).

5. C. L. Humphries: Now Predict Recovery from Adsorbents, Hydrocarbon Processing, 45(12):88 (Dec. 1966).

6. Octave Levenspiel: "Chemical Reaction Engineering," John Wiley & Sons, New York, 1962.

7. W. K. Lewis, E. R. Gilliland, B. Cherton, and W. P. Cadogan: Adsorption Equilibria--Hydrocarbon Gas Mixtures, Industrial and Engineering Chemistry, 42:1319 (1950).

8. W. K. Lewis, E. R. Gilliland, B. Cherton, and W. P. Cadogan: Adsorption Equilibria--Pure Gas Isotherms, Industrial and Engineering Chemistry, 42:1326 (1950).

9. M. M. Mattia: Process for Solvent Pollution Control, Chemical Engineering Progress, 66(12):74-79 (Dec. 1970).

10. A. S. Michaels: Simplified Method of Interpreting Kinetic Data in Fixed-Bed Ion Exchange, Industrial and Engineering Chemistry, 44(8):1922 (1952).

11. Donald F. Othmer and Samuel Josefowitz: Correlating Adsorption Data, Industrial and Engineering Chemistry, 40:723 (1948).

12. Robert H. Perry, Cecil H. Chilton, and Sidney D. Kirkpatrick: "Chemical Engineer's Handbook," 4th ed., McGraw-Hill Book Co., New York, 1963.

13. G. C. Ray and E. O. Box, Jr.: Adsorption of Gases on Activated Charcoal, Industrial and Engineering Chemistry, 42:1314 (1950).

14. H. M. Rowson: Fluid Bed Adsorption of Carbon Disulfide, British Chemical Engineering, 8(3):180 (March 1963).

15. J. M. Smith: "Chemical Engineering Kinetics," 2nd ed., McGraw-Hill Book Co., New York, 1970.

16. Robert E. Treybal: "Mass Transfer Operations," 2nd ed., McGraw-Hill Book Co., New York, 1968.

17. Amos Turk: Source Control by Gas-Solid Adsorption and Related Processes, in A. C. Stern (ed.), "Air Pollution," Vol. 3, Academic Press, New York, 1948.

Chapter 44

ODOR CONTROL BY ODOR MODIFICATION

Thomas M. Hellman

Allied Chemicals Corp.
Morristown, New Jersey

Odor counteraction is the phenomenon by which odor intensity, however measured, is reduced by adding a nonchemically reactive controlling agent to a malodor. The concept of odor counteraction pairs odoriferous gases in proportions that render the mixture odorless or nearly odorless [1,2]. Some of the more interesting pairs of odors compensating each other are ethyl mercaptan and eucalyptol, skatole and coumarin, butyric acid and oil of juniper.

Odor cancellation implies complete counteraction, i.e., reduction to an odorless condition. The action of odor cancellation can be divided into three basic categories: (1) temporary inactivation of the smelling nerves, thus lessening the organoleptic response to odors; (2) odor compensation, e.g., odor pairs; and (3) chemical combination.

Odor masking is the method by which the quality of a malodor is changed by mixing with a control agent. Usually the control agent has a stronger pleasant odor quality which when mixed with the malodor results in a pleasant odor uncharacteristic of the initial malodor.

The term masking is less fallacious than counteraction in that it implies deception in the sense that a mask conceals but does not eliminate. Control by masking and counteraction have similar operational requirements, and are frequently indistinguishable from each other, even though their objectives are presumably different. To avoid confusion between the two terms, it has been suggested that odor modification be used for the process and odor modifier be used to denote the control agent.

ODOR MODIFICATION

Odor modification (especially in the sense of counteraction) has also been rationalized as a means by which a malodor is incorporated into an overall odor mixture that is pleasant and/or mild. For example, sewage odor is considered to be caused to a major degree by indole and skatole. The same compounds are components of oil of jasmine. By adding an odorant that has the floral attributes of jasmine but is lacking the indole component to an appropriate sample of sewage odor, the malodor is said to disappear into the floral note to a considerable degree [4]. The resulting odor is said to be no longer unpleasant and less intense. However, these claims have not been checked by independent investigators, as the compositions of odor modifiers remain proprietary information with the companies that sell them.

ODOR MODIFICATION SYSTEMS AND THEIR APPLICATIONS

Direct Addition to the Odor Source

In certain situations, such as cooking operations, in a rendering plant or digesting operations in kraft sulfate plants, odor modifiers may be added directly to the product in process, or metered automatically by means of proportioning pumps. Concentrations may range from 1 to 10 ppm based on the weight of the process charge under treatment. This mode suffers from the disadvantage that the odor modifier may be altered by the process, especially when high temperatures are involved.

Discharge to Atmosphere with the Malodor

Odor modifiers may be atomized by air pressure through properly designed spray nozzles into the duct in which malodors are discharged. The odor modifier is first diluted to a practical concentration, usually about 1 percent, either dissolved in oil or emulsified in ether. Injection is made well below the top of the stack to ensure good mixing, but beyond any other control devices such as cyclones, scrubbers, etc.

An odor modifier may be sprayed or sprinkled along the perimeter of a settling tank or lagoon, or poured on the surface where it can spread rapidly. Normal evaporation disperses the modifier continuously over the entire area.

Discharge to Atmosphere at a Distance from the Source

Odor modifiers should be discharged into the atmosphere in a manner designed to protect or screen residents from the effects of a malodor. In practice, this involves placement of the discharge along the perimeter of the property line of the facility from which the malodor originates. Usually, a line of stovepipe 6 to 12 in. long is laid along the periphery of the offending units as close as possible to the offending receptors. Air containing the volatilized modifying agent is injected into the line. The volatilization of odorant may be controlled by passage of air over a pan containing the modifier held at the appropriate temperature. The stovepipe contains a series of properly spaced holes through which the modifier is emitted into the atmosphere.

The most generally accepted application of odor modification is the unconfined odor source. Manufacturers of odor modification systems claim successful odor control in chemical plants, waste treatment plants, lagoons, refineries, etc. However, it should be recognized that odor modification does not represent an acceptable control technique in the eyes of some state pollution enforcement agencies. Therefore, the circumstances under which odor modification is employed as a control method should be carefully considered.

Odor modification agents should not be used when malodors may be toxic or mixed with toxic compounds. The most important example of a toxic odorant is hydrogen sulfide (TLV 10 ppm, odor recognition threshold 0.0005 ppm).

ODOR MODIFICATION ECONOMICS

The economics of odor modification is specific to the application. However, it can be generalized that odor modification is usually relatively less expensive as compared with other control methods. One case for which the economics can be estimated is the discharge into the atmosphere at some distance from the source. In a typical installation, 400 ft of 6-in. pipe was laid around the periphery of a unit. Holes 1/2 in. in diameter were drilled at 3-ft intervals along the pipe. An air flow of 300 was supplied at a power consumption of 1/3 hp. Typical cost of such piping installed varies from $1.50 to $3.50/lineal foot. Cost of the chemical modifiers was approximately $5.00/day (24-hr operation).

REFERENCES

1. P. N. Cheremisinoff and R. A. Young: "Industrial Odor Technology Assessment," Ann Arbor Science Publishers, Ann Arbor, Mich., 1975.
2. H. Zwaardemaker: "Die Physiologie des Gerüchs," Engelmann, Leipzig, 1895.
3. F. N. Jones and M. H. Waskow: On the Intensity of Odor Mixtures, Annals of the New York Academy of Sciences, 116(2):484 (1964).
4. S. Elinsky, Rhodia Inc., West Chester, Pa., private communication.

INDEX

Part 1: Pages 1 to 607
Part 2: Pages 609 to 1008

Part 1: Pages 1 to 607
Part 2: Pages 609 to 1008

Part 1: Pages 1 to 607
Part 2: Pages 609 to 1008